Student Workbook

Beginning Algebra

FIRST EDITION

Mark Clark
Palomar College

Cynthia Anfinson
Palomar College

Written and Prepared by

Maria H. Andersen
Muskegon Community College

Australia • Brazil • Japan • Korea • Mexico • Singapore • Spain • United Kingdom • United States

ISBN-13: 978-1-111-56890-0
ISBN-10: 1-111-56890-1

Brooks/Cole
20 Davis Drive
Belmont, CA 94002-3098
USA

Cengage Learning is a leading provider of customized learning solutions with office locations around the globe, including Singapore, the United Kingdom, Australia, Mexico, Brazil, and Japan. Locate your local office at: **www.cengage.com/global**

Cengage Learning products are represented in Canada by Nelson Education, Ltd.

To learn more about Brooks/Cole, visit **www.cengage.com/brookscole**

Purchase any of our products at your local college store or at our preferred online store **www.cengagebrain.com**

Printed in the United States of America
1 2 3 4 5 17 16 15 14 13

Algebra is weightlifting for the Brain!

Table of Contents: Algebra Activities

RNUM: Real Numbers

Student Activity

Building Blocks

Fill In The Blanks: Fill in the missing numerators with whole numbers to build equivalent fractions to the fraction in the "Goal" box. If there is not a whole number numerator that will work, then cross out the fraction.

Example:

Goal: $\frac{1}{2}$	$\frac{\boxed{15}}{30}$	~~$\frac{\_\_}{7}$~~	$\frac{\boxed{5}}{10}$

Goal: $\frac{1}{4}$	$\frac{\_\_}{24}$	$\frac{\_\_}{36}$	$\frac{\_\_}{16}$

Goal: $\frac{3}{16}$	$\frac{\_\_}{32}$	$\frac{\_\_}{4}$	$\frac{\_\_}{80}$

Goal: $\frac{3}{2}$	$\frac{\_\_}{50}$	$\frac{\_\_}{8}$	$\frac{\_\_}{1000}$

Goal: $\frac{5}{8}$	$\frac{\_\_}{48}$	$\frac{\_\_}{32}$	$\frac{\_\_}{60}$

Goal: $\frac{3}{3}$	$\frac{\_\_}{9}$	$\frac{\_\_}{15}$	$\frac{\_\_}{99}$

Goal: $\frac{7}{9}$	$\frac{\_\_}{99}$	$\frac{\_\_}{81}$	$\frac{\_\_}{54}$

Goal: $\frac{2}{3}$	$\frac{\_\_}{18}$	$\frac{\_\_}{15}$	$\frac{\_\_}{1}$

Goal: $\frac{7}{5}$	$\frac{\_\_}{25}$	$\frac{\_\_}{35}$	$\frac{\_\_}{14}$

Student Activity

Factor Pairings

Directions: In each diagram, there is a number in the top box and exactly enough spaces beneath it to write all the possible factor-pairs involving whole numbers. See if you can find all the missing factor-pairs. The number 30 has been done for you.

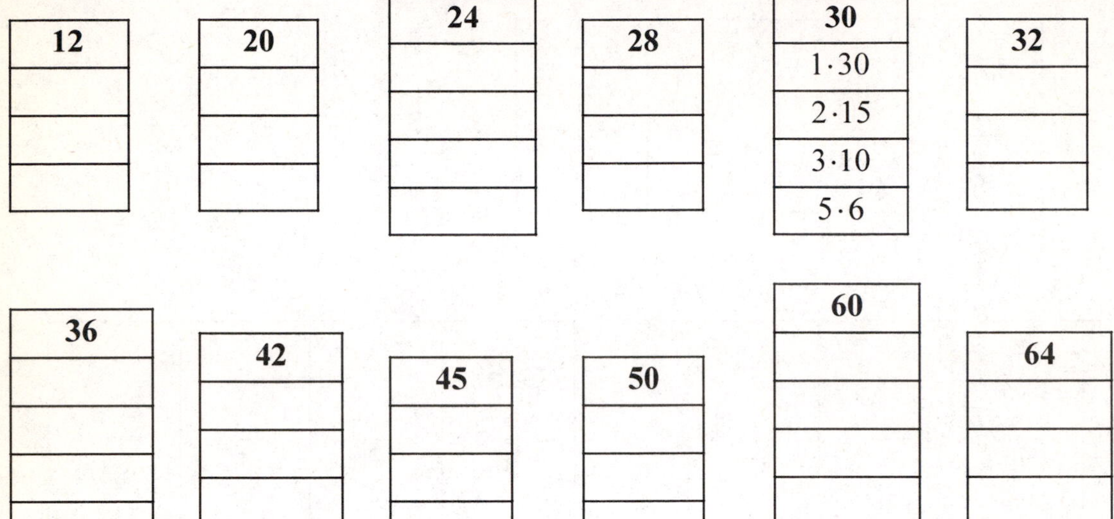

1. What is the largest number that is a factor of 20 and 30? _____

Simplify: $\frac{20}{30}$

2. What is the largest number that is a factor of 28 and 36? _____

Simplify: $\frac{28}{36}$

3. What is the largest number that is a factor of 36 and 60? _____

Simplify: $\frac{36}{60}$

4. What is the largest number that is a factor of 24 and 42? _____

Simplify: $\frac{24}{42}$

Student Activity

Match Up on Fractions

Match-up: Match each of the expressions in the squares of the table below with its simplified value at the top. If the solution is not found among the choices A through D, then choose E (none of these).

A 1 **B** $\frac{3}{4}$ **C** $\frac{7}{8}$ **D** 0 **E** None of these

$\left(\frac{3}{2}\right)\left(\frac{1}{2}\right)$	$\frac{7}{6} \div \frac{4}{3}$	$\frac{7}{8} \div 0$	$2\frac{1}{8} - \frac{5}{4}$
$\frac{15}{8} - 1$	$6\left(\frac{1}{6}\right)$	$\frac{2}{3} \div \frac{8}{9}$	$0 \div \frac{3}{4}$
$\frac{1}{2} + \frac{3}{8}$	$\frac{19}{12} - \frac{5}{6}$		
$\frac{1}{5} + \frac{8}{10}$	$\frac{3}{4}(0)$		
$\frac{1}{2} + \frac{2}{2}$	$\frac{1}{3} \div \frac{1}{3}$		

I thought we shared a common denominator, but he was only a fraction of the person I thought he was.

Student Activity

Fractions Using a Calculator

When you input fractions into a calculator, you must be careful to tell the calculator which parts are fractions. Each calculator has a set of algorithms that tell it what to do first (later on, we will learn the mathematical order of operations, which is similar). In order to ensure that fractions are treated as fractions, for now, you need to tell your calculator which parts ARE fractions.

1. For example, first show that $\frac{3}{4} \div \frac{2}{5}$ is $\frac{15}{8}$ by hand:

2. To get the decimal value of $\frac{15}{8}$ on the calculator, we type $15/8$ or $15 \div 8$ (depending on the calculator). Practice by finding the decimal values for:

$\frac{15}{8}$ $\frac{3}{20}$ $\frac{1}{4}$ $\frac{7}{8}$ $\frac{2}{3}$

3. Now try using your calculator to evaluate $\frac{3}{4} \div \frac{2}{5}$, but do it without using any parentheses. Do you get the decimal value equal to 15/8?

4. Find the button(s) on your calculator that allow you to input parentheses and write down how to use them on your calculator.

5. Try it on your calculator like this now: $\left(\frac{3}{4}\right) \div \left(\frac{2}{5}\right)$

On my calculators, I type $(3/4)/(2/5)$ or $(3 \div 4) \div (2 \div 5)$ to enter this expression. But each calculator is a little different. When you have done it correctly, you should get 1.875.

Write down how to do it on your calculator:

6. Now try these fraction problems *using parentheses* to tell your calculator which numbers represent fractions:

$\frac{1}{2} \cdot \frac{4}{9}$ $\frac{3}{4} \div \frac{2}{15}$ $\frac{3}{8} + \frac{4}{5}$ $\frac{7}{12} - \frac{1}{5}$

7. The operation in mixed numbers is **addition**, so when you input $2\frac{3}{5}$ into your calculator, you must treat it like $2 + (3/5)$. What is $2\frac{3}{5}$ as a decimal? ____

Student Activity

Linking Rational Numbers with Decimals

Let's investigate why we say that decimals that terminate and repeat are really rational numbers. You will need a calculator and some colored pencils for this activity.

Rational numbers consist of all numbers that can be expressed as a fraction (or *ratio*) of *integers* (except when zero is in the denominator).

In the grid below are a bunch of fractions of integers.

1. Work out the decimal equivalents using your calculator. If the decimals are repeating decimals, use an overbar to indicate the repeating sequence (like in the example that has been done for you).
2. Shade the grid squares in which fractions were equivalent to repeating decimals in one color and indicate the color here: ___________.
3. Shade the grid squares in which fractions were equivalent to terminating decimals in another color and indicate the color here: ___________.
4. In the last row of the grid, write some of your own fractions built using integers and repeat the steps above.

$\frac{2}{3} = 0.\overline{6}$	$\frac{1}{4} =$	$\frac{7}{8} =$	$\frac{5}{9} =$
$\frac{1}{2} =$	$\frac{4}{9} =$	$\frac{8}{5} =$	$\frac{17}{25} =$
$\frac{1}{1000} =$	$\frac{5}{12} =$	$\frac{3}{4} =$	$\frac{7}{27} =$
$\frac{1}{5} =$	$\frac{12}{5} =$	$\frac{4}{3} =$	$\frac{1}{6} =$

5. Are there any fractions in the grid that were not shaded as either terminating or repeating?

6. If you write one of these fractions as its decimal equivalent, what kind of decimal do you get?

Guided Learning Activity

Using Addition Models

Part I: The first model for addition of real numbers that we look at is called the "colored counters" method. Traditionally, this is done with black and red counters, but we make a slight modification here to print in black and white.

 Solid counters (black) represent positive integers, $+1$ for each counter.

 Dashed counters (red), represent negative integers, -1 for each counter.

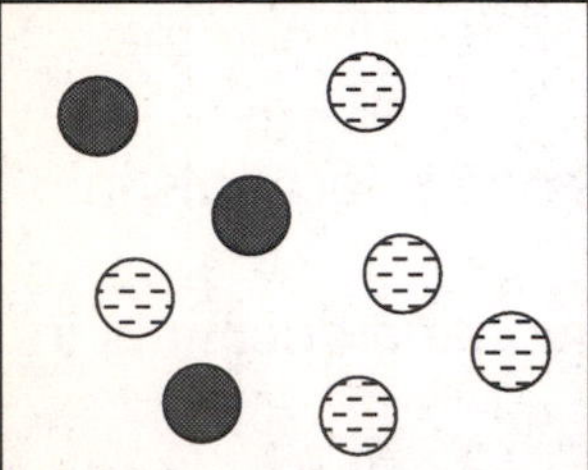

When we look at a collection of counters (inside each rectangle) we can write an addition problem to represent what we see. We do this by counting the number of solid counters (in this case 3) and counting the number of dashed counters (in this case 5). So the addition problem becomes $3+(-5)=\_\_\_\_$.

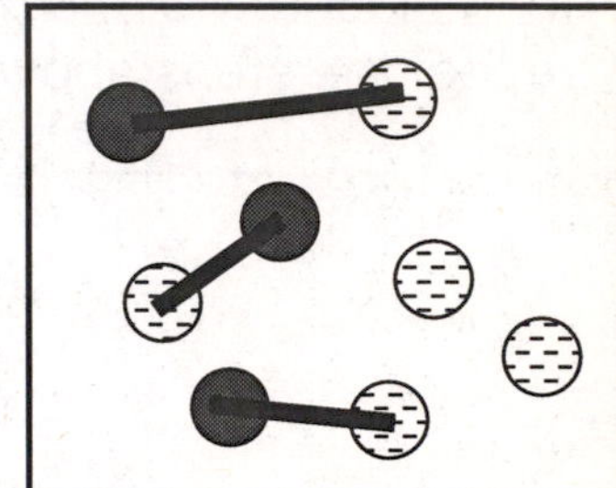

To perform the addition, we use the Additive Inverse Property, specifically, that $1+(-1)=0$. By matching up pairs of positive and negative counters until we run out of matched pairs, we can see the value of the remaining result. In this example, we are left with two dashed counters, representing the number -2. So the collection of counters represents the problem $3+(-5)=-2$.

Now try to write the problems that represent the collections below.

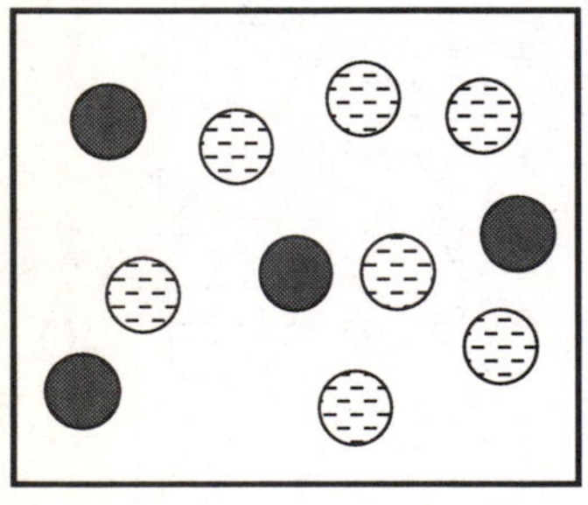

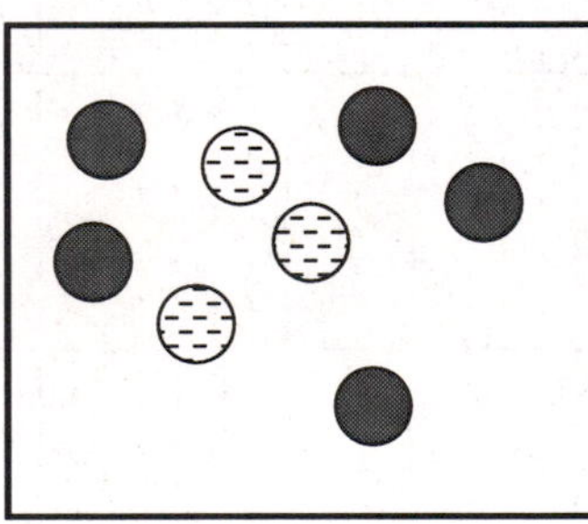

b. ____+____=____

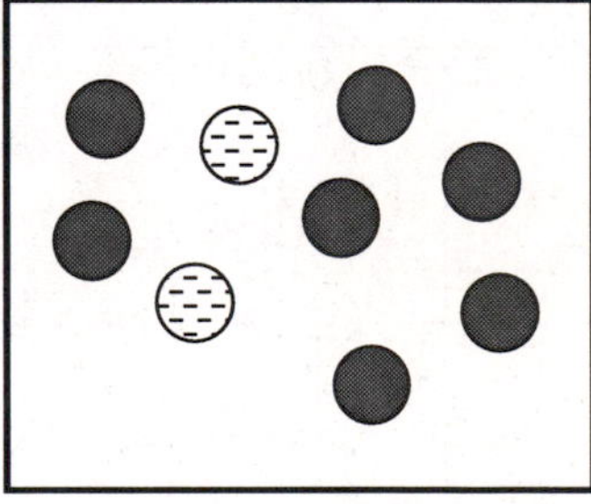

c. ____+____=____

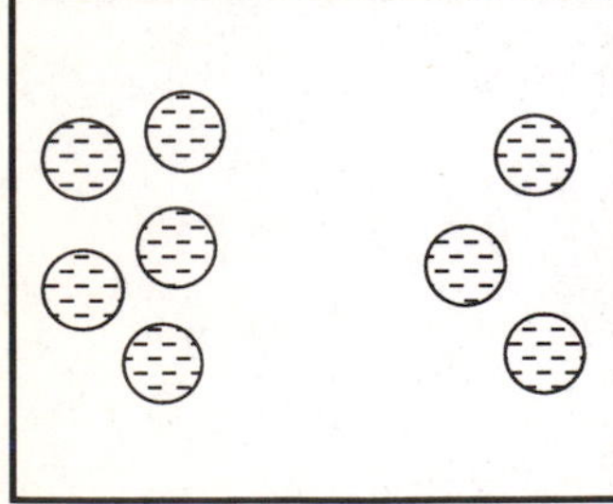

d. ____+____=____

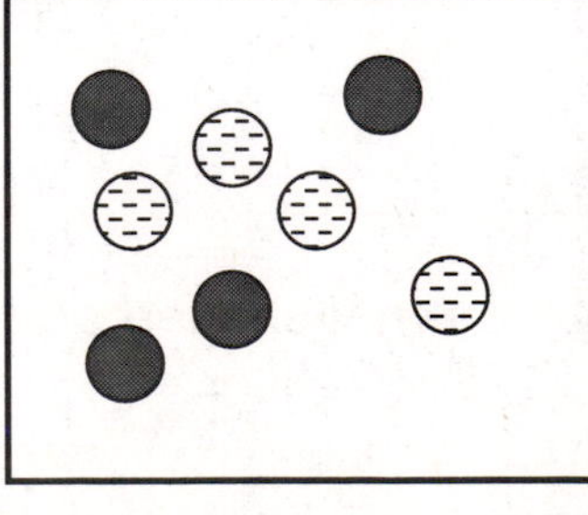

e. ____+____=____

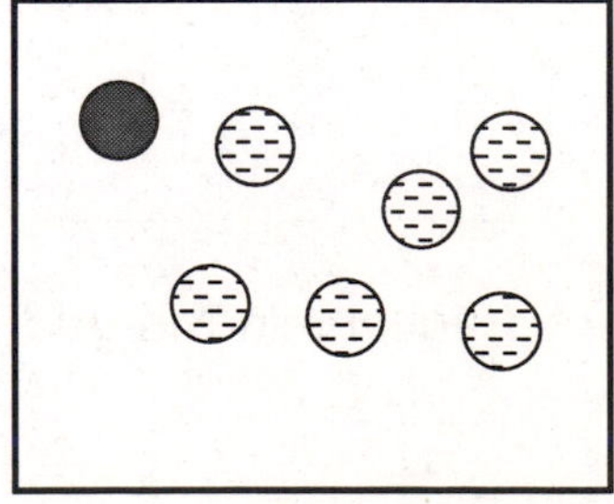

f. ____+____=____

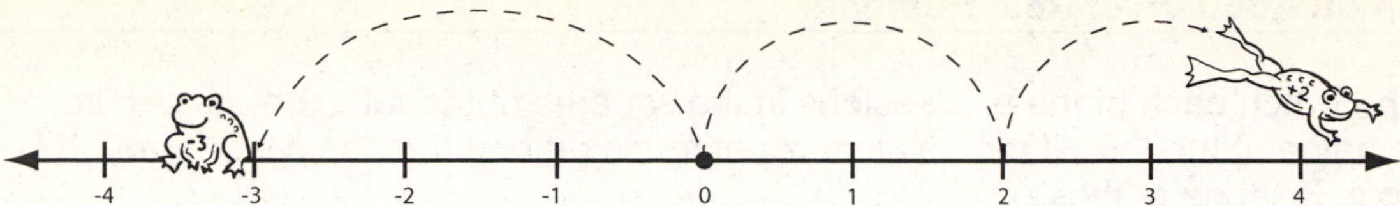

Part II: The second model for addition of real numbers that we look at is called the "number line" method. We use directional arcs to represent numbers that are positive and negative. The length that the arc represents corresponds to the magnitude of the number.

When a directional arc indicates a positive direction (to the right), it represents a positive number. In the diagram below, each arc represents the number 2, because each arc represents a length of two and each arrow points to the right.

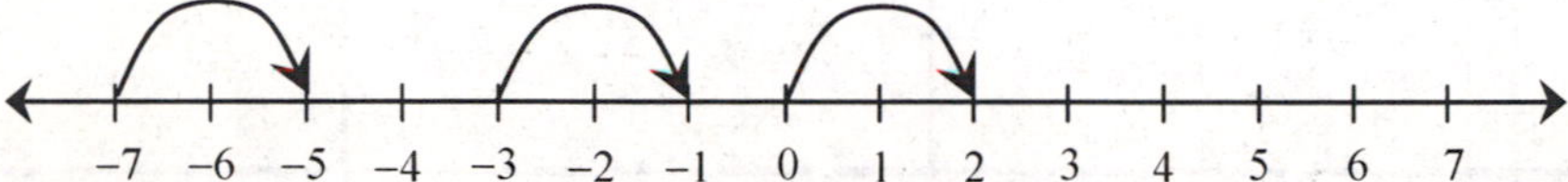

In the next diagram, each directional arc represents the number -5, since each arc represents a length of five and each arrow points to the left.

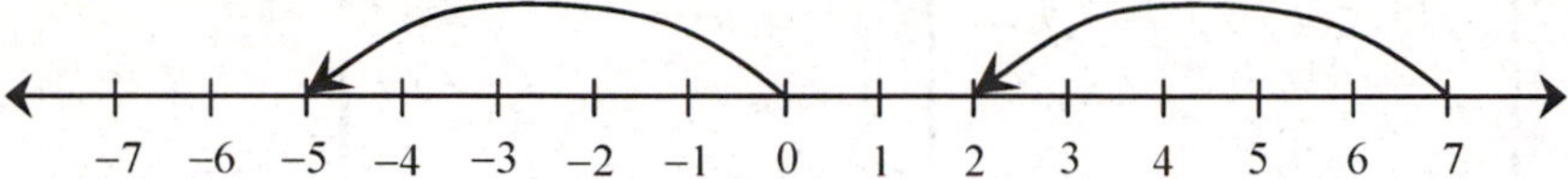

When we want to represent an addition problem, **we start at zero**, and travel from each number to the next using a new directional arc. Thus, the following number line diagram represents $2+(-5)=-3$. The final landing point is the answer to the addition problem.

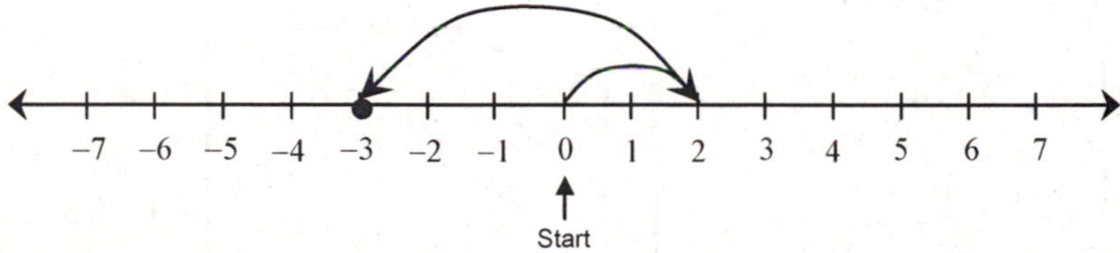

Now try to solve these addition problems on a number line using directional arcs.

a. $-3+7=$ _____

−7 −6 −5 −4 −3 −2 −1 0 1 2 3 4 5 6 7

d. $-6+2=$ _____

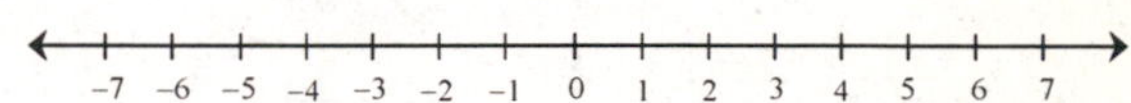

b. $-2+(-3)=$ _____

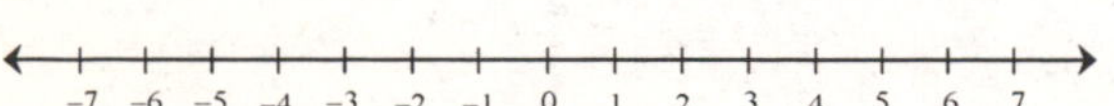

e. $-2+(-2)+(-1)=$ _____

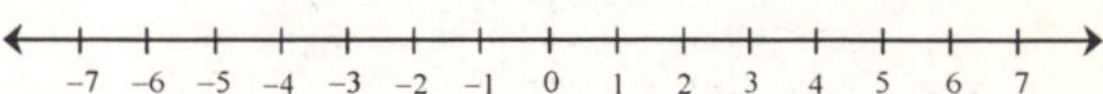

c. $5+(-5)=$ _____

−7 −6 −5 −4 −3 −2 −1 0 1 2 3 4 5 6 7

f. $4+(-8)+3=$ _____

−7 −6 −5 −4 −3 −2 −1 0 1 2 3 4 5 6 7

Student Activity

Match Up on Addition of Real Numbers

Match-up: Match each of the expressions in the squares of the table below with its simplified value at the top. If the solution is not found among the choices A through D, then choose E (none of these).

A 3 **B** -6 **C** -1 **D** 1 **E** None of these

$-8+2$	$-9+8$	$-7+10$	$-3+(-3)$
$-2+(-1)+6$	$3+(-3)$	$-10+(-4)+8$	$-\frac{1}{2}+\left(-\frac{1}{2}\right)$
$6+(-7)$	$5+(-6)$	$-1+(-1)+(-1)+4$	$-11+12$
$-4+\left(-\frac{1}{2}\right)+\frac{1}{2}$	$8+(-5)+(-9)$	$10+(-6)+(-1)$	$-6+12$
$0+(-6)$	$-1+7$	$30+(-27)$	$1+\frac{1}{4}+\frac{1}{2}+\left(-\frac{3}{4}\right)$

Student Activity
Scrambled Addition Tables

Here is a simple addition tables with natural number inputs.

Addition:

+	1	2	3	4
1	2	3	4	5
2	3	4	5	6
3	4	5	6	7
4	5	6	7	8

Directions: The first table that follows is an addition table involving integer inputs. The second table is a *scrambled* addition table with integer inputs (this means that the numbers in the first row and column do not increase nicely like 1, 2, 3, 4). Fill in the missing squares with the appropriate numbers.

+	−3	−2	−1	0	1	2	3
−3							
−2							
−1							
0							
1							
2							
3							

+	15	−5	20	−10	0	5	10
10							
5							
−5							
−15							
20							
0							
15							

Directions: The following tables are ***scrambled*** addition tables with the additional challenge of missing numbers in the shaded rows and columns. Fill in the missing squares with the appropriate numbers.

+	2			3	
−5					−5
−2			−3		
				7	
	8				
		−4		3	

+		−10	15		−5
−20	−15			−20	
			5		
				0	
10					
	10				

Student Activity

Signed Numbers Magic Puzzles

Directions: In these "magic" puzzles, each row and column adds to be the same "magic" number. Fill in the missing squares in each puzzle so that the rows and columns each add up to be the given magic number.

Magic Puzzle #1

−2	8	
	−9	4

Magic Number = 5

Magic Puzzle #2

		4
−5		
8		−6

Magic Number = 0

Magic Puzzle #3

		$2\frac{1}{4}$
	$2\frac{1}{2}$	$-2\frac{1}{2}$
$\frac{1}{4}$		

Magic Number = $\frac{1}{2}$

Magic Puzzle #4

−2		9	−7
	−9		4
8	−4		−5
		−6	

Magic Number = 1

Magic Puzzle #5

−8	−3		7
	−11	2	6
8	−2		−7

Magic Number = −2

Guided Learning Activity

Language of Subtraction

How do you interpret the – sign? It is a minus sign if it is **between** two numbers as a mathematical operation. Otherwise, it is a negative.

Other ways to signify minus: difference, less than, subtract … from …
Other ways to signify negative: opposite

How do you tell if *less* means < or –? Look for the distinction between "**is** less than" and "less than." See the two examples in the table below.

	Expression? Equation? Or Inequality?	Equivalent statement or phrase in words
$-(-4)$	Expression	the opposite of the opposite of 4 the opposite of negative 4
$8-3$	Expression	the difference of 8 and 3 subtract 3 from 8 8 minus 3 3 less than 8
$8-3=5$	Equation	The difference of 8 and 3 is 5. Subtract 3 from 8 to get 5. 8 minus 3 is 5. 3 less than 8 is 5.
$-2-4$	Expression	the difference of negative 2 and 4 subtract 4 from negative 2 negative 2 minus 4 4 less than negative 2
$-5<-2$	Inequality	Negative 5 is less than negative 2.
$9-(-2)$	Expression	9 minus negative 2 subtract negative 2 from 9 the difference of 9 and negative 2

Note that expressions are represented in words by phrases (no verb) and equations and inequalities are represented by sentences (with verbs).

Now try these! For any problem with subtraction, find at least two ways to write it in words.

		Expression? Equation? Or Inequality?	Equivalent statement or phrase in words
a.			Zero is less than 8.
b.	$10-2=8$		
c.			the difference of 2 and negative 5
d.	$-(-10)$		
e.	$5-(-10)=15$		
f.			Negative 6 is less than negative 3.
g.			The opposite of negative 5 is 5.
h.	$3<-(-6)$		
i.	$\frac{1}{2}-\frac{3}{4}$		
j.	$-19<-18$		

Student Activity

Match Up on Subtraction of Real Numbers

Match-up: Match each of the expressions in the squares of the table below with its simplified value at the top. If the solution is not found among the choices A through D, then choose E (none of these).

A 3 **B** -6 **C** -1 **D** 1 **E** None of these

$-1-(-4)$	$0-(-1)$	$-30-(-25)+(-1)$	$-9-(-8)$
$2-(-1)$	$-3+(-3)$	$3-(-3)$	$-6-(-3)$
$0-(-3)$	$12-(-6)$	$0-(-1)$	$3-6$
$-\frac{9}{4}-\left(-\frac{1}{2}\right)+\frac{3}{4}$	$\frac{6}{4}-\frac{1}{2}$	$\frac{5}{2}-4-2\frac{1}{2}+3$	$5\frac{2}{3}-\left(-\frac{1}{3}\right)+(-3)$
$\frac{1}{2}-\left(-\frac{1}{2}\right)$			

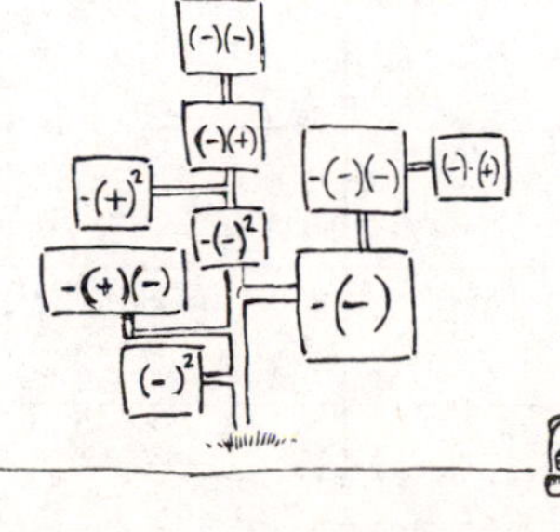

Watch the SIGNS!

Student Activity

Signed Numbers Using a Calculator

When you input expressions with signed numbers into a calculator, you must be careful to tell the calculator which " – " signs represent a minus, and which represent a negative.

1. The minus button on your calculator looks like this: [–]. It is found with the addition, multiplication, and division functions. The button on your calculator that is used to denote a negative may look like [+/–] or [(–)], or it may be above one of the keys, accessed with a 2nd function, [2nd]. Locate where your calculator input for a negative is, and draw it here:

2. On some calculators, the negative is typed before the number, and on some it is typed after the number. We need to figure out which type you have. We'll calculate $-2+5$ (which should be ___). Try it both ways. Write down exactly how to do $-2+5$ on your calculator here:

3. Let's try something more complicated now. How would we write $-8-3$ in words using the word minus? __________________________________
What should the answer be? ___ Now write down the keystrokes for inputting this expression into your calculator here:

4. Work out each of these expressions by hand, then write down how to express them in words, and finally, write down how to input the keystrokes properly into your calculator.

Expression	Answer	In words (using negative and/or minus)	Keystrokes
$-6+3$			
$5-(-4)$			
$-19-6$			
$-1.25-0.25$			
$10-(-8)$			
$-3-(-3)$			

Student Activity

Match Up on Multiplying and Dividing Real Numbers

Match-up: Match each of the expressions in the squares of the table below with its simplified value at the top. If the solution is not found among the choices A through D, then choose E (none of these). Note that some of the expressions involve other operations besides multiplication and division, so be careful!

A 2 **B** 0 **C** –8 **D** 12 **E** None of these

$\frac{-6}{-3}$	$-\frac{4}{3}\left(-\frac{3}{2}\right)$	$(-4)(-2)(-1)$	$-2 \div 0$
$(-0.5)(-24)$	$(-0.25)(-8)$	$(0)(-2)(-1)$	$32 \div (-4)$
$36 \div (-3)$	$-3\left(\frac{4}{6}\right)$	$2 \div \left(-\frac{1}{2}\right)$	$-36\left(-\frac{2}{6}\right)$
$4-(-3)+5$	$5 \div \left(-\frac{5}{8}\right)$	$\left(-\frac{4}{3}\right)(6)$	$0(-8)-(-2)$
$-2(-2)(-3)(-1)$	$-\frac{1}{3}(-6)$	$-\frac{3}{2} \div \left(-\frac{1}{8}\right)$	$\frac{0}{-5}$

Student Activity

The Exponent Trio

Directions: In each of the "trios" below, place three equivalent expressions of the following format:

Expanded Expression using multiplication	
Compact exponential expression	Simplified expression

The first one has been done for you. Sometimes there are two possible trios for a simplified exponential expression, so you will see some of these listed twice.

$(-3)(-3)$	
$(-3)^2$	9

$3\cdot 3$	

$2\cdot 2\cdot 2\cdot 2$	

-3^2	

$(-2)^4$	

$-(2\cdot 2\cdot 2\cdot 2)$	

	8

$(-2)(-2)(-2)$	

-2^3	

$-(4\cdot 4)$	

$(-4)^2$	

-10^3	

$\left(-\frac{1}{2}\right)^2$	

	$\frac{1}{4}$

$2\cdot 2\cdot 2\cdot 2\cdot 2\cdot 2\cdot 2\cdot 2$	

$(-9)^2$	

	125

5·5

25

5²

The Exponent Trio

Student Activity

Order Operation

Directions: With a highlighter, shade the numbers and operation that comes first in the order of operations in each problem. For example, for $7-3\cdot 2$, you would highlight $3\cdot 2$. If more than one operation could be done first (at the same time), shade both sets. Once you are certain that you have chosen the **first** steps correctly, then simplify each expression.

1. $5+3\cdot 5-2$

2. $12\div 2\cdot 6+1$

3. $4-3^2+6$

4. $6\cdot 2-8\div 2+4$

5. $-12-(4+3)$

6. $5-2(4\cdot 3)-5$

7. $9\cdot 2\div 6-5(2)^2$

8. $\dfrac{2-4}{(-3)^2+1}$

9. $\left[4+2(2-5)^2\right]-3$

10. $\dfrac{3}{4}-\left(\dfrac{2}{3}\right)\left(\dfrac{1}{2}\right)^2$

Assess Your Understanding

Real Numbers

For each of the following, describe the strategies or key steps that will help you **start** the problem. You do **not** have to complete the problems.

		What will help you to start this problem?
1.	Add: $-3+(-7)$	
2.	Use < or > to make this statement true: $-3\ \square\ -4$	
3.	Divide: $\frac{3}{4}\div\frac{1}{6}$	
4.	Simplify: $7-2\cdot 3+4$	
5.	Subtract: $-2-(-4)$	
6.	Is 1.5 rational or irrational?	
7.	Add: $\frac{1}{3}+\frac{4}{5}$	
8.	Fill in the missing value to make the statement true: $\frac{2}{5}=\frac{\square}{20}$	
9.	Is -3 a natural number? A whole number? An Integer?	
10.	Simplify: $(-4)^4$	

Metacognitive Skills

RNUM-23

Real Numbers

Metacognitive skills refer to the ability to judge how well you have learned something and to effectively direct your own learning and studying. This is a self-evaluation tool designed to help you focus your studying and to improve your metacognitive skills with regards to this math class.

Fill the 1st column out **before** you begin studying. Fill the 2nd column out after you study for your test.

Go back to this assessment after your test and circle any of the ratings that you would change – this identifies the "disconnects" between what you **thought** you knew well and what you **actually** knew well.

Use the scale below to assign a number to each topic.

5 *I am confident I can do any problems in this category correctly.*
4 *I am confident I can do most of the problems in this category correctly.*
3 *I understand how to do the problems in this category, but I still make a lot of mistakes.*
2 *I feel unsure about how to do these problems.*
1 *I know I don't understand how to do these problems.*

Topic or Skill	Before Studying	After Studying
Finding the prime-factored form of a number; finding a factor-pair for a number.		
Simplifying a fraction to write it in lowest terms.		
Multiplying or dividing (signed) fractions and simplifying the result.		
Building equivalent fractions.		
Adding and subtracting (signed) fractions and simplifying the result.		
Working with mixed numbers in mathematical expressions.		
Knowing what opposite, inverse and reciprocal mean in terms of real numbers.		
Categorizing numbers in different number sets (Real, rational, natural, etc.)		
Evaluating expressions involving absolute value.		
Using an inequality symbol (< or >) to determine the order of real numbers (like they would be found on a number line).		
Adding and subtracting real numbers (signed numbers).		
Identifying which addition or multiplication property has been used in a statement (commutative, associative, identity, inverse, etc.).		
Solving application problems that involve addition or subtraction of signed numbers.		
Multiplying or dividing real numbers (signed numbers).		
Identifying division by zero and understanding why division by zero is undefined.		
Evaluating or rewriting exponential expressions.		
Distinguishing between exponential expressions that involve parentheses, for example: $(-3)^2$ and -3^2.		
Knowing the rules of the order of operations.		
Applying the order of operations to a numerical expression.		
Applying the order of operations when it involves absolute value or a fraction bar.		
Finding the average of a set of data.		

ALG: Algebraic Expressions

Student Activity

Translating Mathematical Operations

Directions: For each line, fill in the missing boxes with the proper words or notation.

ADDITION: Written in words.	**Using the word *sum*.**	**Using the + sign.**
5 plus 9 is 14.		
		$3+8=11$
	The sum of 2 and x is 9.	

SUBTRACTION: Written in words.	**Using the word *difference*.**	**Using the – sign.**
12 minus 4 is 8.		
		$11-7=4$
	The difference of 9 and 3 is 6.	

MULTIPLICATION: Written in words.	**Using the word *product*.**	**Using a raised dot.**	**Using parentheses.**
3 multiplied by 5 is 15.			
		$8\cdot5=40$	
	The product of 2 and n is 14.		
			$(6)(7)=42$ or $6(7)=42$

DIVISION: Written in words.	**Using the division symbol.**	**Using long division notation.**	**Using the fraction bar.**
The quotient of 16 and 2 is 8.			
		$5\overline{)35}$ with quotient 7	
	$27\div3=9$		
			$\frac{32}{4}=8$

Student Activity

Translating Expressions and Equations

Directions: For each line, fill in the missing boxes with the proper words or notation. The first one has been done for you.

	Phrase or sentence	Expression or Equation?	Write the expression or equation.
a.	the number of feet, n, times 12	Expression	$12n$
b.			$17-x=3$
c.	The sum of the measures of angles x and y is 180°.		
d.	the quotient of m and 100		
e.			$z+100$
f.	The product of y and 10 is 1000.		
g.			$20\div 4$
h.			$u+v=90°$
i.	The number of inches, n, divided by 12 is the number of feet, f.		
j.	100 less than x		
k.	the difference of 100 and x		

Student Activity

Count Consecutive Integers

1. Fill in the table below:

Let $x = ...$	$x+1$	$x+2$	$x+3$
5			
10			
22			
a			
$2n$			

a. How would you represent the two integers **after** x: ______ and ______

b. How would you represent the two integers **before** x: ______ and ______

n, ah, ah ah ah, ah ...
n+1, ah ah ah ah, ah ...
n+2 ...

Billy didn't care! He loved algebra more than candy!

2. Fill in the table below:

Let $x = ...$	$2x$	$2x+1$	$2x+2$	$2x+3$	$2x+4$
1					
2					
3					
18					
25					
a					
Odd or even?					

a. How would you represent an unknown even integer? ________

b. How would you represent an unknown odd integer? ________

c. How would you represent the two integers that follow $2x$? ______ and ______

d. How would you represent the two **even** integers after $2x$? ______ and ______

e. How would you represent the two **odd** integers after $2x+1$? ______ and ______

Student Activity

Follow the Multiplication Road

Match-up: Find your way from start to finish along the multiplication road by shading in matching pairs of algebraic expressions. The first pair has been shaded for you!

START					
$3(4a)$	$7a$	$\frac{4}{3}(4x)$	$\frac{x}{3}$	$\frac{4x}{3}(4)$	$\frac{z}{5}$
$12a$	$\frac{2}{3}(3x)$	$2x$	$-x(-9)$	$-9x$	$5\left(\frac{2z}{5}\right)$
$\frac{1}{12}(2a)$	$6x$	$\frac{4x^2}{5x^2}$	$9x$	$9-x$	$5\left(\frac{5z}{2}\right)$
$(0.8x)(-2)$	$1.6x$	$\frac{4x^2}{5}$	$\frac{4}{5}(x^2)$	$8x$	$5z$
$(5t)(-4)$	$12a$	$-3(-4a)$	$-7a$	$4(4x)$	$4x^2$
$-20t$	$(-5t)(-4)$	$-9t$	$\frac{2}{3}(12t^2)$	$9t^2$	$\frac{x^2}{7x^2}$
$5(-2x)$	$-10x$	$\frac{3}{4}(8t^2)$	$6t^2$	$\frac{1}{7}(x^2)$	$\frac{x^2}{7}$
					FINISH

Student Activity

Match Up on Like Terms and Distribution

Match-up: Match each of the expressions in the squares of the lower table with an equivalent expression from the top. If the solution is not found among the choices A through D, then choose E (none of these).

A $4x+16-12y$ **B** $5+4x+6y$ **C** $x+y-z$ **D** $3x+3y+3z$ **E** None of these

$3(x+y)+z$	$5+2x+6y+2x$	$3y+3(x+z)$	$2(8+2x-6y)$
$(x-z)+y$	$4(x-3y+4)$	$4(x+3y)+16$	$6y+5+4x$
$4(x+4)-12y$	$2(2x+8-6y)$	$(5+6y)+4x$	$\frac{1}{3}(-3z+3y+3x)$
$5+2(3y+2x)$	$2y-(z+y)+x$	$4x+12y-16$	$(4x+16)-12y$
$2(2x+8)+6y$	$3x+3(z+y)$		
$5x+3y+3z-2x$	$6y+5+4x$		
$6x+6y-3(x+y-z)$	$x-(z-y)$		
$(z+y)-z$	$4x-(12y-16)$		

Student Activity

Language of Parentheses

For each problem below either write out the mathematical expression in words, or write the missing expression. Then simplify the expressions. Remember to use words like "the quantity of" or "the sum of" or "the difference of" when you describe more than one term in parentheses.

	Expression	Write the expression in words
1.		8 times the difference of $5y$ and 3
2.	$8-5(y-3)$	
3.		8 minus the quantity of $5y$ minus 3
4.	$8-5y-3$	
5.		the difference of 8 and $5y$, times -3
6.	$-6-2(x+7)$	
7.		negative 6 times the sum of $-2x$ and 7
8.	$-6-(2x+7)$	
9.		negative 6 minus $2x$ plus 7
10.	$(-6-2x)\cdot 7$	

Assess Your Understanding

Algebraic Expressions

For each of the following, describe the strategies or key steps that will help you **start** the problem. You do **not** have to complete the problems.

		What will help you to start this problem?
1.	Write an expression to represent: y increased by 10	
2.	Simplify: $9y+3z-3y-z$	
3.	How many terms are in the expression $7t^2-8t+5$?	
4.	Evaluate $4a^2-2$ when $a=-5$.	
5.	Simplify: $-2(4c+5)-3c$	
6.	Write an algebraic expression to represent the value in cents of q quarters.	
7.	Write $4(x-3)$ in words.	
8.	Simplify: $9-(3x+6)$	
9.	Write using algebra: *Five more than twice a number is 10.*	
10.	What is the coefficient of the x-term in $4x^3-2x+5$?	

Metacognitive Skills

Algebraic Expressions

Metacognitive skills refer to the ability to judge how well you have learned something and to effectively direct your own learning and studying. This is a self-evaluation tool designed to help you focus your studying and to improve your metacognitive skills with regards to this math class.

Fill the 1st column out **before** you begin studying. Fill the 2nd column out after you study for your test.

Go back to this assessment after your test and circle any of the ratings that you would change – this identifies the "disconnects" between what you **thought** you knew well and what you **actually** knew well.

Use the scale below to assign a number to each topic.

5 *I am confident I can do any problems in this category correctly.*
4 *I am confident I can do most of the problems in this category correctly.*
3 *I understand how to do the problems in this category, but I still make a lot of mistakes.*
2 *I feel unsure about how to do these problems.*
1 *I know I don't understand how to do these problems.*

Topic or Skill	Before Studying	After Studying
Identifying the number of terms in an expression, its coefficients, or its factors.		
Identifying words that translate into mathematical operations.		
Translating a phrase or sentence into a mathematical expression or equation and vice versa.		
Evaluating an algebraic expression for a given number or numbers.		
Distinguish between equations and expressions.		
Using algebra to represent a varying quantity in an application problem.		
Simplifying expressions involving like and unlike terms.		
Multiplying an algebraic term by a constant.		
Applying the distributive property.		
Applying a negative that is in front of a set of parentheses.		

UNIT: Unit Conversions and Significant Digits

Guided Learning Activity

Understanding Significant Digits

Rules for Counting Significant Digits

1. We do **not** apply the rules of significant digits in definitions of relationships.

   Example: The relationship of 12 inches in 1 foot is a defined relationship; it is in no way an estimation, so we would not apply the rules of significant digits to the numbers 12 or 1.

   Example: There are 4 tires on my car. Not 4.1 tires, not 3.9 tires, but exactly 4 tires. So again, we would not apply the rules of significant digits to the number 4.

   Example: I calculated that my car gets 24.6 miles per gallon of gas. This is not a definition, most likely it is 24.6 miles ± a tenth of a mile or two. Thus, we would apply the rules of counting significant digits to the number 24.6.

2. As long as there is a decimal point in the number, start counting significant digits at the first non-zero number on the left; finish counting when you run out of digits to count.

   Try these examples:

0.8904 g	3.04 g	4.20 mL
240. mL	620.4 L	5.2 gal
0.00304 mol	0.00040 mol	8.004 mg
1200. miles	30.00 m	0.05000 in.

3. If there is not a decimal point in the number, start counting significant digits at the first non-zero number on the left; finish when you run out of non-zero digits.

   Try these examples:

1200 ft	200 mm	624,000 in.
120.0 ft	4.00 m	400 m

Significant Digits in Addition and Subtraction Calculations

1. Make sure that the numbers have the same units. If they don't, convert one of the numbers to the correct units, keeping the same number of significant digits. *It is usually easier to convert both numbers to the smaller unit of measure.*

2. In the calculation, it is the number with the least precision, that is the least number of decimal places, that "dominates" the calculation. The result of the calculation should be rounded to the same number of decimal places as the dominating number. A nice way to see this is to line up the decimal places.

Try these examples:

a) $\begin{array}{r} 4.28\text{ m} \\ +\,5.6\ \ \text{m} \\ \hline \end{array}$

c) $\begin{array}{r} 0.0342\text{ g} \\ +\,0.02\ \ \ \ \text{g} \\ \hline \end{array}$

b) 4,200 ft + 6,258 ft =

d) 78 in + 3.0 ft =

Significant Digits in Multiplication and Division Calculations

1. It is the number in the calculation with the least number of significant digits that "dominates" the calculation. The result of the calculation should be rounded (or extended) to the same number of significant digits as the dominating number.

Try these examples:

a) $92.8\text{ mi} \div 3.5\text{ gal} =$

b) $1.04\text{ g} \div 0.90\text{ mL} =$

c) $0.00636\text{ mol} \times 1.008\ \dfrac{\text{g}}{\text{mol}} =$

d) $0.00004\text{ g} \times 68{,}000{,}000\ \dfrac{\text{molecules}}{\text{g}} =$

Student Activity

Match Up on Significant Digits

Directions: Match each of the numbers in the squares of the table below with its number of significant digits in choices **A**-**E**. If the solution is not found among the choices A through E, then choose F (none of these).

A 1 **B** 2 **C** 3 **D** 4 **E** 5 **F** None of these

1.0	1.25	0.3	25
0.00003	0.300000	30000	1200.0
1001	0.030	30	1,200,000
7	5.20	520	52.0
0.070	0.70	7000	25.5
5,400.	5,400	0.0540	5.40000
0.0000009	0.0009	0.9	90.

Student Activity

Paint by SDs

Paint by SDs: Write each number with the desired number of significant digits (SDs). Then find and shade the correct answer below.

1. Write 7,252 with 3 SDs.	**8.** Write 0.031 with 2 SDs.
2. Write 10,001 with 1 SD.	**9.** Write 2000 with 4 SDs.
3. Write 0.00500 with 2 SDs.	**10.** Write 0.00500 with 1 SD.
4. Write 40.4 with 2 SDs.	**11.** Write 88.88 with 3 SDs.
5. Write 123.456 with 4 SDs.	**12.** Write 424.9 with 2 SDs.
6. Write 399.99 with 4 SDs.	**13.** Write 0.00622 with 1 SD.
7. Write 30.4 with 1 SD.	**14.** Write 96.00069 with 5 SDs.

4.04	40	40.	30	30.	3.04
88.8	88.9	90.0	10,000.	10,000	1.0001
420.	420	425	0.05	0.005	0.0005
7,250.	7,250	725	0.050	0.0050	0.00500
0.0031	0.031	0.03100	123.4560	123.5	123.4
96.00069	96.001	96.0007	0.0062	0.006	0.00622
399.9	400.00	400.0	2000.	2000	2000.0

Student Activity

Significant Decisions

Directions: Karen needs help with precision. Decide how many decimal places the solution to each addition or subtraction problem should have. Then decide if Karen's solution is correct, too precise, or not precise enough.

Problem #1	Karen's Solution	Precision
Add 3.01 cm and 2.5 cm.	3.01 cm + 2.5 cm 5.51 cm	**a.** Too precise **b.** Correct **c.** Not precise enough

Problem #2	Karen's Solution	Precision
Add 4.0 g and 4.012 g.	4.0 g + 4.012 g 8 g	**a.** Too precise **b.** Correct **c.** Not precise enough

Problem #3	Karen's Solution	Precision
Subtract 0.062 cm from 3.4 cm.	3.4 cm - 0.062 cm 3.3 cm	**a.** Too precise **b.** Correct **c.** Not precise enough

Problem #4	Karen's Solution	Precision
Add 1001 pounds and 0.1200 pounds.	0.1200 lb + 1001. lb 1001.12 lb	**a.** Too precise **b.** Correct **c.** Not precise enough

Problem #5	Karen's Solution	Precision
Find the sum of 722.0 mL and 6510 mL.	722.0 mL + 6510 mL 7230 mL	**a.** Too precise **b.** Correct **c.** Not precise enough

Problem #6	Karen's Solution	Precision
Subtract 100.09 kg from 102.1 kg.	102.1 kg - 100.09 kg 2 kg	**a.** Too precise **b.** Correct **c.** Not precise enough

Student Activity

Measuring the Significance

Application: Determine the correct number of significant digits for each calculation result. Then carry out the calculation and round appropriately.

1. To estimate the amount of fertilizer to purchase, Steve measures the dimensions of his rectangular garden. The dimensions are 1.32 meters by 1.4 meters. What is the area of the garden?

2. A fish tank measures 36.0 inches by 12.5 inches by 24.25 inches. How many cubic inches of water will the tank hold?

3. The diameter of a round water balloon is 22.2 centimeters. If the volume of the water balloon is estimated by using $\frac{4}{3}\pi r^3$, what is the volume of the balloon?

4. At the store, Mark purchases a jug of liquid hummingbird food that contains 4.5 quarts of pre-mixed nectar. After he fills the bird feeder, which holds 3.25 quarts, to the top, how much nectar will be left in the jug?

5. Bobbie has a large tractor with tires that are 3.4 feet in diameter. Find the circumference of these tires using the formula $C = \pi \cdot d$.

Guided Learning Activity

Riding the Train

One way to perform conversion calculations is to use a factor table (or "train track")

1. ***Start with what you know.*** Start with the number you are given to convert and create an empty factor table.
2. ***Know where you want to go.*** Write the units that you are trying to get for the answer on the answer-side of the = sign so that you know where your calculation is going.
3. ***Insert factor units.*** In the next empty factor space, write in the units that you want to cancel (in the appropriate location), then, in the same factor, write in the units you can easily convert to.
4. ***Insert factor numbers.*** Insert the correct numbers for this conversion.
5. ***Cancel units.*** *Repeat?* Cancel the units that are exactly the same (units only). Repeat steps 3 and 4 until you have reached the appropriate units.
6. ***Calculate.*** To perform the calculation, simply multiply by all numbers in the numerator, and divide by all numbers in the denominator (see example below).

$$\begin{array}{c|c|c} a & b & c \\ \hline & d & e \end{array}$$ could be found by calculating $a \cdot b \cdot c \div d \div e$ OR $(a \cdot b \cdot c) \div (d \cdot e)$

NOTE: The answer to a unit conversion problem can have no more significant digits than the *measured number* in the problem with the *least* number of significant digits.

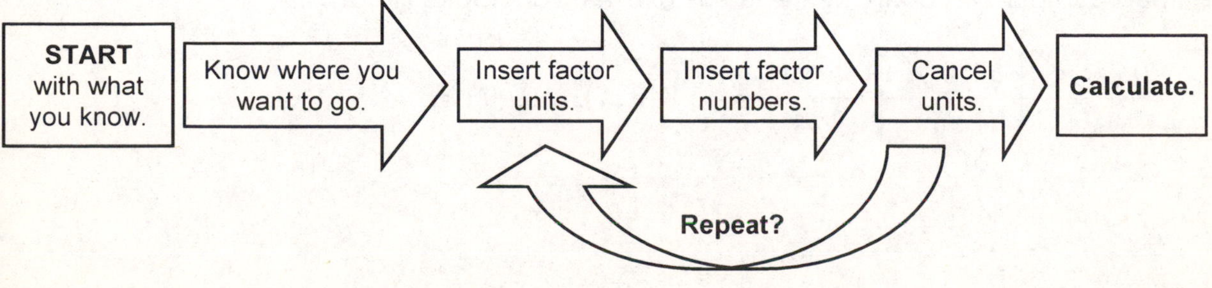

Detailed Example: Convert 0.25 miles to inches.

Start with what you know:

$$\begin{array}{l|l} 0.25\text{ mi} & \\ \hline & \end{array} =$$

Know where you want to go:

$$\begin{array}{l|l} 0.25\text{ mi} & \\ \hline & \end{array} = \quad \text{in}$$

Insert factor units:

$$\begin{array}{l|c|l} 0.25\text{ mi} & \text{ft} & \\ \hline & \text{mi} & \end{array} = \quad \text{in}$$

Insert factor numbers:

$$\begin{array}{l|c|l} 0.25\text{ mi} & 5280\text{ ft} & \\ \hline & 1\text{ mi} & \end{array} = \quad \text{in}$$

Cancel units. Repeat?

$$\begin{array}{l|c|l} 0.25\ \cancel{\text{mi}} & 5280\text{ ft} & \\ \hline & 1\ \cancel{\text{mi}} & \end{array} = \quad \text{in}$$

We need to *repeat* because if we stop right now, the final units would be feet.

Insert more factor units:

$$\begin{array}{l|c|c} 0.25\ \cancel{\text{mi}} & 5280\text{ ft} & \text{in} \\ \hline & 1\ \cancel{\text{mi}} & \text{ft} \end{array} = \quad \text{in}$$

Insert factor numbers:

$$\begin{array}{l|c|c} 0.25\ \cancel{\text{mi}} & 5280\text{ ft} & 12\text{ in} \\ \hline & 1\ \cancel{\text{mi}} & 1\text{ ft} \end{array} = \quad \text{in}$$

Cancel units. Repeat?

$$\begin{array}{l|c|c} 0.25\ \cancel{\text{mi}} & 5280\ \cancel{\text{ft}} & 12\text{ in} \\ \hline & 1\ \cancel{\text{mi}} & 1\ \cancel{\text{ft}} \end{array} = \quad \text{in}$$

Our final units now would be inches, so we're done writing the conversion steps.

On a calculator, we would find $0.25 \boxed{\times} 5280 \boxed{\times} 12 \boxed{\div} 1 \boxed{\div} 1$ (of course, it is not necessary to divide by the 1's as it does not alter the calculation).

Remember the answer can only have 2 SD, because the measured number, 0.25 mi, had only 2 SD.

The final result is 16,000 inches and the overall work looks like this:

$$\begin{array}{l|c|c} 0.25\ \cancel{\text{mi}} & 5280\ \cancel{\text{ft}} & 12\text{ in} \\ \hline & 1\ \cancel{\text{mi}} & 1\ \cancel{\text{ft}} \end{array} = (15{,}840\text{ in}) = 16{,}000\text{ in}$$

Common Length Conversions

English	Metric	Bridges
12 in = 1 ft 3 ft = 1 yd 5280 ft = 1 mi	1 km = 1000 m 1 m = 1000 mm 1 m = 100 cm 1 cm = 10 mm	1 in = 2.54 cm 1 mi = 1.609 km

Now try these!

1. Convert 6600 feet to miles.

=

2. Convert 3.0 miles to inches.

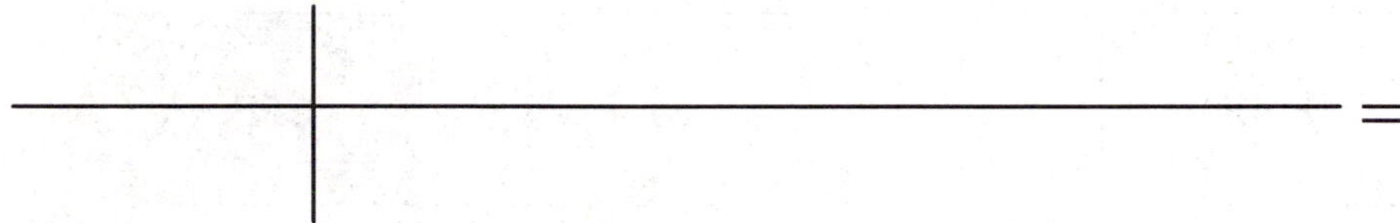

=

3. Convert 2.6 meters to millimeters.

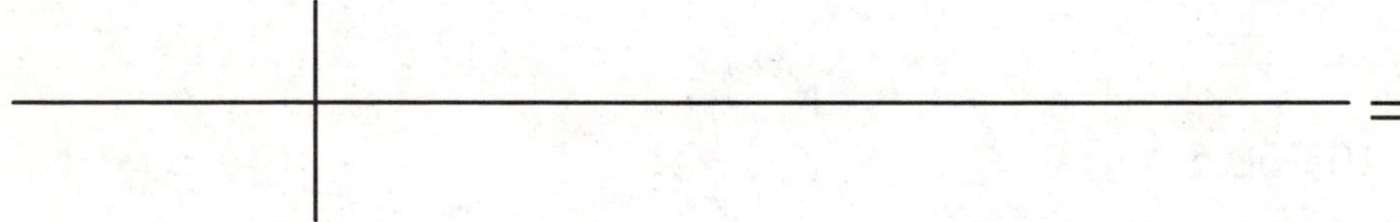

=

4. Convert 3.28 cm to kilometers.

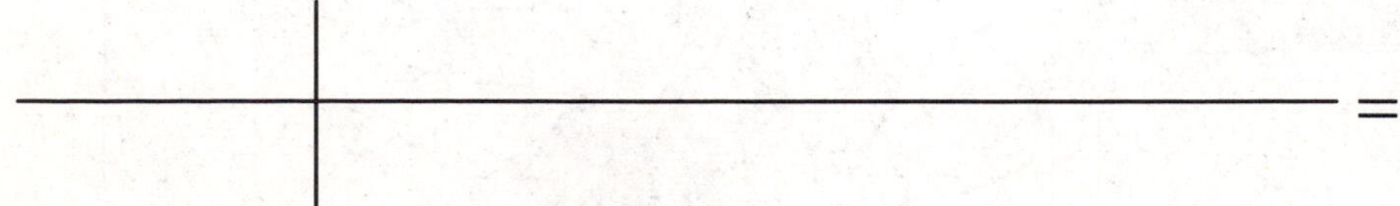

=

5. Convert 124,000 mm into inches.

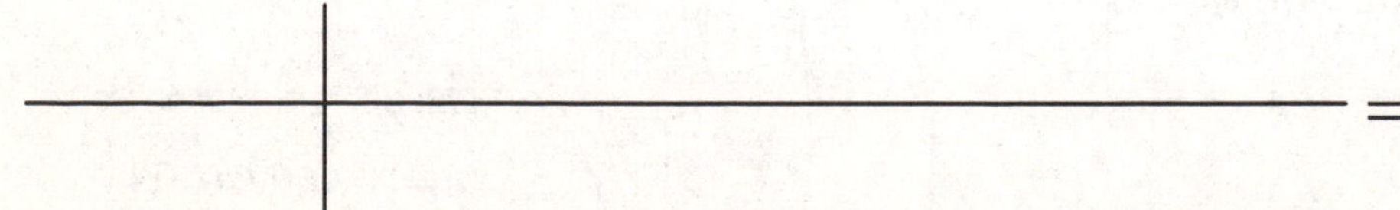

=

Student Activity

Four Square on Length

Directions: In each set of four, find and circle the measurements that are equivalent to the measurement in the **shaded square**. Make sure you watch out for significant digits!

1.25 in	0.104 ft
0.035 yd	0.0000197 mi

163,000 in	13,554 ft
4518 yd	**2.567 mi**

170.0 in	**14.25 ft**
4.722 yd	0.002683 mi

7200 in	600 ft
200 yd	0.1 mi

1,500,000 in	130,000 ft
41000 yd	24 mi

63360 in	**5280 ft**
1760 yd	1 mi

203,400 in	16,950 ft
50850 yd	**3.210 mi**

25,900 in	2160 ft
719 yd	0.500 mi

Student Activity

Race to the Finish

Directions: Fill in the missing unit conversion in each of the calculations below using the choices **A**-**F**. Then finish the calculation, including the correct number of significant digits.

A $\dfrac{1\text{ m}}{100\text{ cm}}$ **B** $\dfrac{1\text{ km}}{1000\text{ m}}$ **C** $\dfrac{1000\text{ m}}{1\text{ km}}$ **D** $\dfrac{100\text{ cm}}{1\text{ m}}$ **E** $\dfrac{1\text{ cm}}{10\text{ mm}}$ **F** $\dfrac{10\text{ mm}}{1\text{ cm}}$

1. ☐ $$\begin{array}{l|l|l} 2.5\text{ m} & & 10\text{ mm} \\ \hline & & 1\text{ cm} \end{array} = \underline{\qquad\qquad} \text{ mm}$$

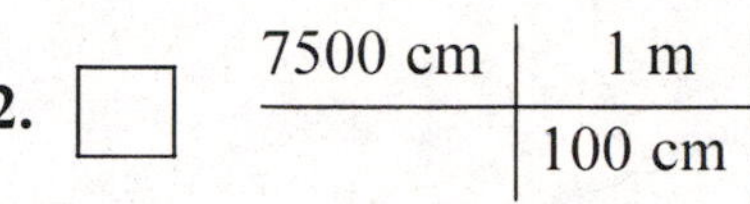

2. ☐ $$\begin{array}{l|l|l} 7500\text{ cm} & 1\text{ m} & \\ \hline & 100\text{ cm} & \end{array} = \underline{\qquad\qquad} \text{ km}$$

3. ☐ $$\begin{array}{l|l|l} 0.05\text{ km} & & 100\text{ cm} \\ \hline & & 1\text{ m} \end{array} = \underline{\qquad\qquad} \text{ cm}$$

4. ☐ $$\begin{array}{l|l} 1.25\text{ km} & \\ \hline & \end{array} = \underline{\qquad\qquad} \text{ m}$$

5. ☐ $$\begin{array}{l|l} 2.012\text{ km} & \\ \hline & \end{array} = \underline{\qquad\qquad} \text{ m}$$

6. ☐ $$\begin{array}{l|l} 27.2\text{ cm} & \\ \hline & \end{array} = \underline{\qquad\qquad} \text{ mm}$$

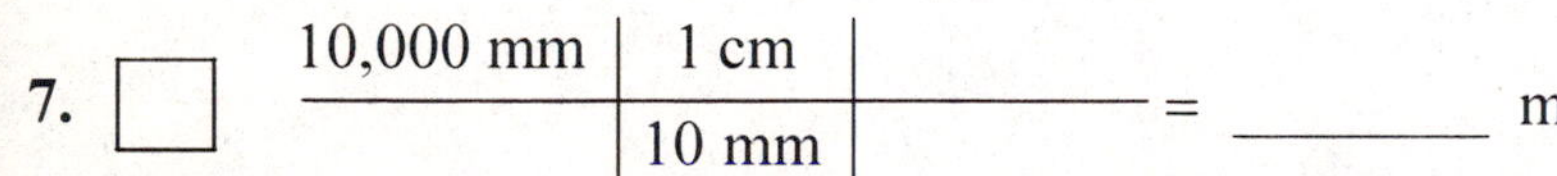

7. ☐ $$\begin{array}{l|l|l} 10{,}000\text{ mm} & 1\text{ cm} & \\ \hline & 10\text{ mm} & \end{array} = \underline{\qquad\qquad} \text{ m}$$

8. ☐ $$\begin{array}{l|l|l|l} 52{,}000\text{ mm} & 1\text{ cm} & 1\text{ m} & \\ \hline & 10\text{ mm} & 100\text{ cm} & \end{array} = \underline{\qquad\qquad} \text{ km}$$

9. ☐ $$\begin{array}{l|l|l|l} 10{,}005\text{ mm} & 1\text{ cm} & & 1\text{ km} \\ \hline & 10\text{ mm} & & 1000\text{ m} \end{array} = \underline{\qquad\qquad} \text{ km}$$

10. ☐ $$\begin{array}{l|l|l} 2.012\text{ km} & 1000\text{ m} & \\ \hline & 1\text{ km} & \end{array} = \underline{\qquad\qquad} \text{ cm}$$

Student Activity

For the Birds

Application: Margaret works designing pet houses and furniture. When she sends plans to clients in Europe, she must convert the lengths from English to Metric units. Help her convert these plans, making sure to use the correct number of significant digits

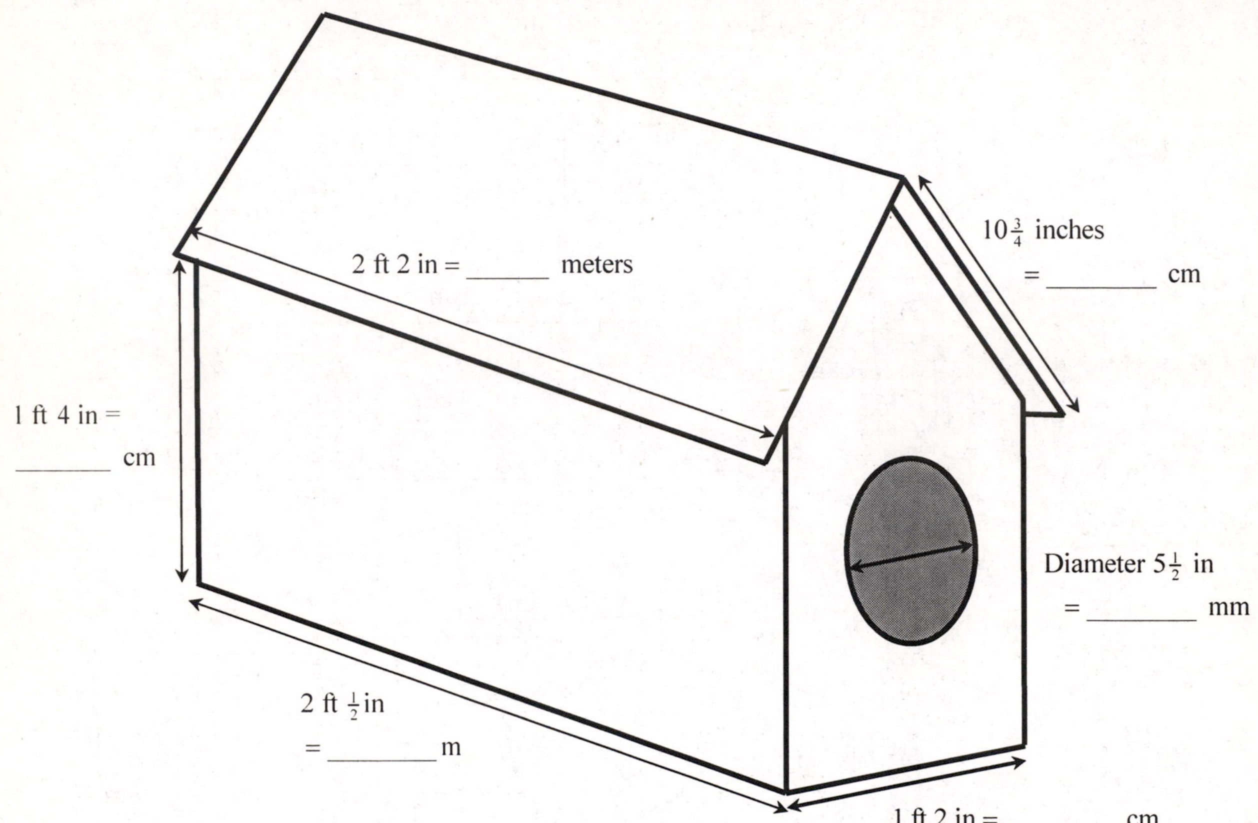

Student Activity

UNIT-13

MASSachusetts

Directions: Perform each conversion and then find the answer in the grid at the bottom of the page. When you find the answer, fill in the missing units.

1. Convert 2.0 tons into pounds.

2. Convert 32,000 ounces into tons.

3. Convert 3.25 pounds into ounces.

4. Convert 2.500 kilograms to milligrams

5. Convert 0.62 tons to ounces.

6. Convert 100 grams to kilograms

7. Convert 100 grams to milligrams.

8. Convert 2.5 kilograms to grams.

9. Convert 1 gram to milligrams.

10. Convert 250 milligrams to kilograms.

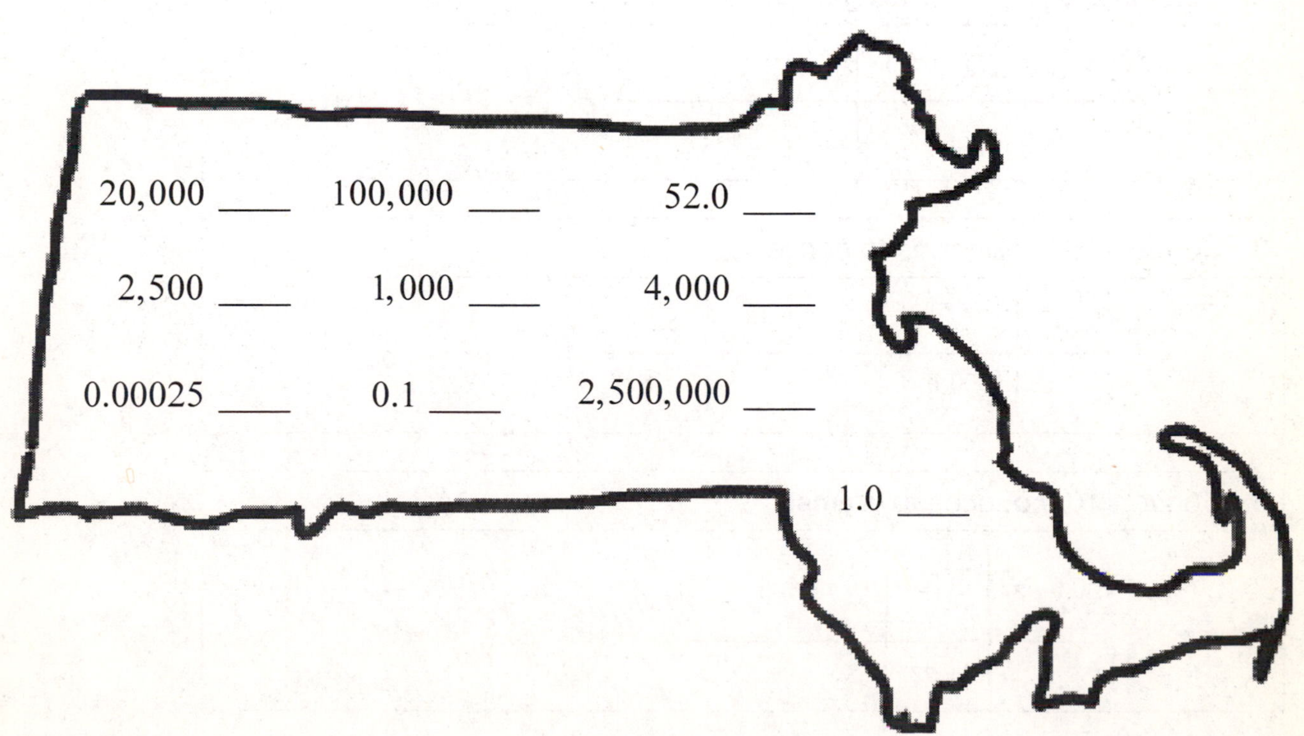

Common Mass Conversions

English	Metric	Bridges
16 oz = 1 lb 2,000 lb = 1 ton	1000 mg = 1 g 1000 g = 1 kg	454 g = 1 lb * 1 kg = 2.2 lb *

* These conversions are not exact. Use proper significant digits.

Directions: Fill in the correct English-metric bridge to complete each conversion. Then calculate the solution using the correct number of significant digits.

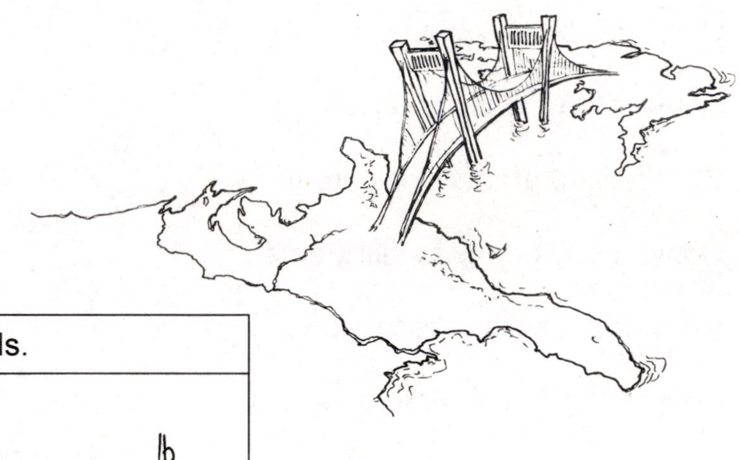

1. Convert 3.2 kilograms to pounds.

3.2 kg	

= lb

2. Convert 0.050 tons to kilograms.

0.050 tons	2,000 lb	
	1 ton	

= kg

3. Convert 100 milligrams to ounces.

100 mg	1 g		16 oz
	1000 mg		1 lb

= oz

4. Convert 10.0 ounces to grams.

10.0 oz	1 lb	
	16 oz	

= g

Student Activity

Time to Change

Directions: At each position on the clock, find the desired time conversion. Assume all clock values are exact.

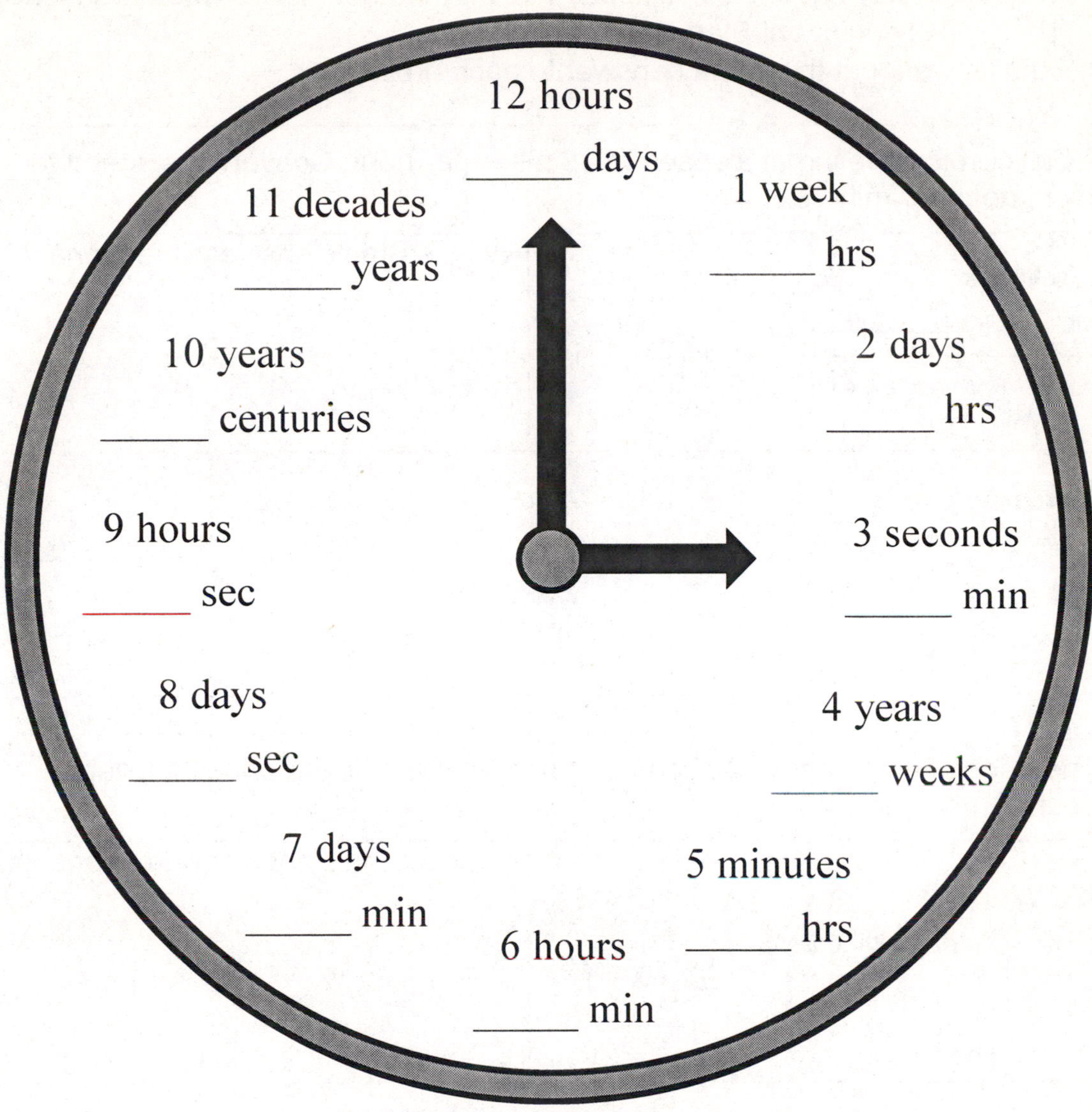

Common Time Conversions

60 sec = 1 min 60 min = 1 hr 24 hr = 1 day 7 days = 1 wk	52 wk = 1 yr * 365 days = 1 yr* 10 yr = 1 decade 100 yr = 1 century

* These conversions are not exact. Use proper significant digits.

Student Activity

Saving Stan

Which way: Stan is not having much luck with his conversion homework. His instructor told him that every single problem is wrong!

- First tell Stan what should've tipped him off that the answer was unreasonable.
- Then, nicely point out Stan's error.
- Finally, work out the correct answer to each problem.

1. A bicyclist is traveling at a speed of 32 miles per hour. Convert this speed to kilometers per minute.

Stan's Work:

$$\frac{32\text{ mi}}{1\text{ hr}} \cdot \frac{60\text{ min}}{1\text{ hr}} \cdot \frac{1.609\text{ km}}{1\text{ mi}} = 30{,}000\text{ km / min}$$

Why is Stan's answer unreasonable?

Your turn:

2. The price of cinnamon is 39 cents per ounce. What is the price in dollars per pound?

Stan's Work:

$$\frac{39\text{ cents}}{1\text{ oz}} \cdot \frac{16\text{ oz}}{1\text{ lb}} \cdot \frac{100\text{ dollars}}{1\text{ cent}} = \$62{,}000\text{ / lb}$$

Why is Stan's answer unreasonable?

Your turn:

3. On July 24, 2009, the minimum wage in the United States went up to $7.25 per hour. Convert this wage into cents per minute.

Stan's Work:

$$\frac{\$7.25}{1\ \text{hour}} \cdot \frac{60\ \text{hours}}{1\ \text{min}} \cdot \frac{100\ \text{cents}}{\$1} = 43,500\ \text{cents / min}$$

Why is Stan's answer unreasonable?

Your turn:

4. In 2009, the IRS mileage rate was 55 cents per mile. Convert this to dollars per kilometer.

Stan's Work:

$$\frac{55\ \text{cents}}{1\ \text{mile}} \cdot \frac{1\ \text{mi}}{1.609\ \text{km}} \cdot \frac{100\ \text{cents}}{\$1} = \$3418\ /\ \text{km}$$

Why is Stan's answer unreasonable?

Your turn:

5. Bamboo can grow as much as 60 cm per day. Convert this rate to inches per hour.

Stan's Work:

$$\frac{60\ \text{cm}}{1\ \text{day}} \cdot \frac{24\ \text{hours}}{1\ \text{day}} \cdot \frac{2.540\ \text{in}}{1\ \text{cm}} = 3700\ \text{in / hr}$$

Why is Stan's answer unreasonable?

Your turn:

Guided Learning Activity

Evolution of Cubed Units

Squared units measure area and cubed units measure volume.

Units like ft, ft^2, and ft^3 do not measure the same thing.

Conversions are a little different too:

$$1 \text{ ft} = 12 \text{ in}$$
$$(1 \text{ ft})^2 = (12 \text{ in})^2$$
$$(1 \text{ ft})^3 = (12 \text{ in})^3$$

Let's see how this plays out.

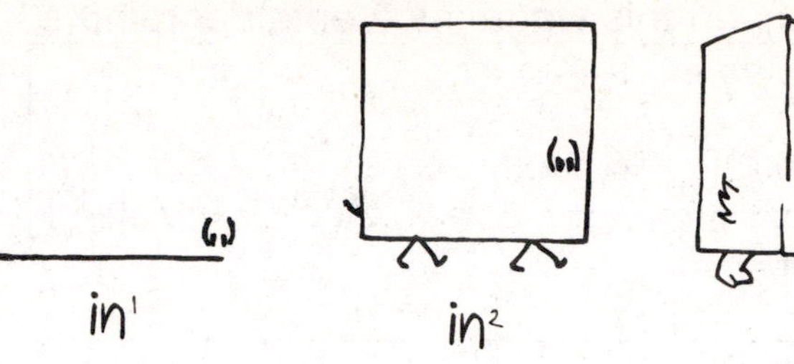

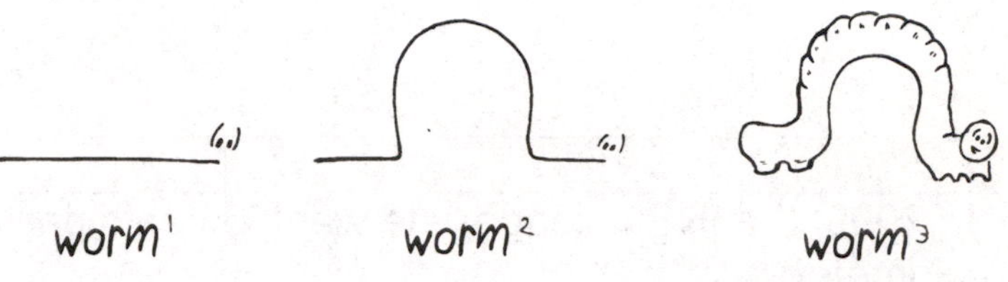

$$(1 \text{ ft})^2 = (12 \text{ in})^2$$
$$(1 \text{ ft})(1 \text{ ft}) = (12 \text{ in})(12 \text{ in})$$
$$(1)(1)(\text{ft})(\text{ft}) = (12)(12)(\text{in})(\text{in})$$
$$1 \text{ ft}^2 = 144 \text{ in}^2$$

$$(1 \text{ ft})^3 = (12 \text{ in})^3$$
$$(1 \text{ ft})(1 \text{ ft})(1 \text{ ft}) = (12 \text{ in})(12 \text{ in})(12 \text{ in})$$
$$(1)(1)(1)(\text{ft})(\text{ft})(\text{ft}) = (12)(12)(12)(\text{in})(\text{in})(\text{in})$$
$$1 \text{ ft}^3 = 1{,}728 \text{ in}^3$$

To see these conversions visually, we can easily look at a comparison of 1 in^2 and 1 ft^2.

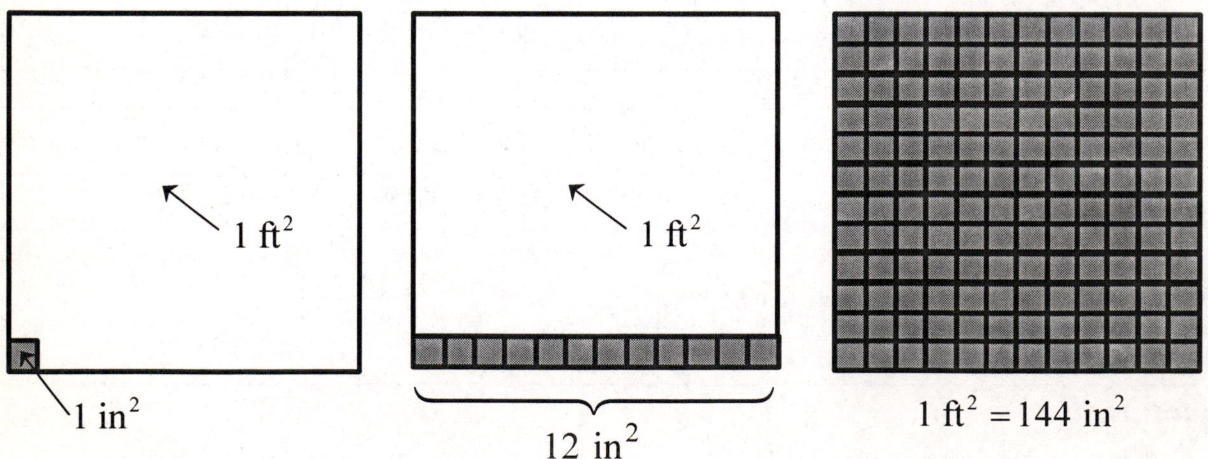

Notice that the area of the square foot can be found by taking the length (12 in) times the width (12 inches), which is also 144 in^2.

Let's say the cube to the right represents 1 cubic foot ($1\ \text{ft}^3$). Draw 1 cubic inch in one of the corners of the cubic foot.

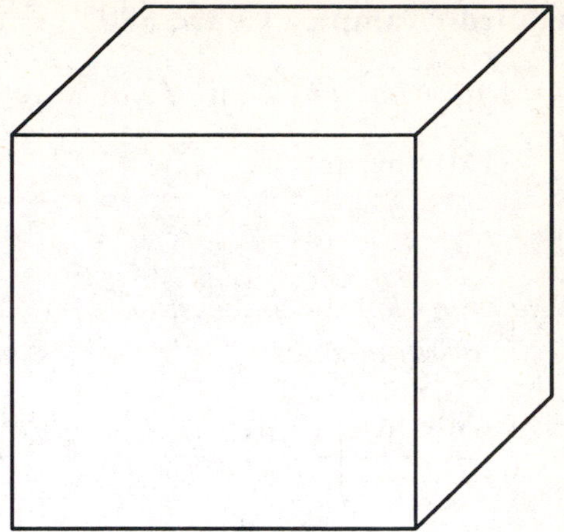

Write the length of each edge of the cubic foot in <u>inches</u>.

For a rectangular solid, recall that $V = \ell wh$.

What is the volume of this cube in inches?

This leads us again to the conversion: $1\ \text{ft}^3 = 1728\ \text{in}^3$

Notice that we can arrive at the same conversion by taking our normal length conversions and either squaring both sides, or cubing both sides.

The conversions for yards and feet are also shown below.

$1\ \text{ft} = 12\ \text{in}$	$1\ \text{ft} = 12\ \text{in}$	$1\ \text{yd} = 3\ \text{ft}$	$1\ \text{yd} = 3\ \text{ft}$
$(1\ \text{ft})^2 = (12\ \text{in})^2$	$(1\ \text{ft})^3 = (12\ \text{in})^3$	$(1\ \text{yd})^2 = (3\ \text{ft})^2$	$(1\ \text{yd})^3 = (3\ \text{ft})^3$
$1^2\ \text{ft}^2 = 12^2\ \text{in}^2$	$1^3\ \text{ft}^3 = 12^3\ \text{in}^3$	$1^2\ \text{yd}^2 = 3^2\ \text{ft}^2$	$1^3\ \text{yd}^3 = 3^3\ \text{ft}^3$
$1\ \text{ft}^2 = 144\ \text{in}^2$	$1\ \text{ft}^3 = 1728\ \text{in}^3$	$1\ \text{yd}^2 = 9\ \text{ft}^2$	$1\ \text{yd}^3 = 27\ \text{ft}^3$

You should **not** try to memorize the conversions for squared or cubed units, simply remember that the length conversions have to be squared or cubed.

When you want to use a squared- or cubed-unit conversion:

1. Write in the length conversion factor.
2. Square or cube both the number <u>and</u> the units (like in the 3rd line of the conversions above).
3. Make sure to use the square or cube in the final calculation.

Detailed Example: Convert 802 in^3 to cubic feet.

Start with what you know. Know where you're going.

$$\frac{802\ \text{in}^3 \;\Big|\;}{\;\Big|\;} = \ \text{ft}^3$$

Write in the length conversion. You can see that the units would not cancel properly if we left the factor alone as is.

$$\frac{802\ \text{in}^3}{} \;\Big|\; \frac{1\ \text{ft}}{12\ \text{in}} = \ \text{ft}^3$$

Cube both the numbers and units of this factor to get the appropriate units for cancellation.

$$\frac{802\ \text{in}^3}{} \;\Big|\; \frac{1^3\ \text{ft}^3}{12^3\ \text{in}^3} = \ \text{ft}^3$$

Checking unit cancellation, we see that we would be left with the appropriate units. The calculation would be $802 \cdot 1^3 \div 12^3$ *which gives us* ≈ 0.46412.

$$\frac{802\ \cancel{\text{in}^3}}{} \;\Big|\; \frac{1^3\ \text{ft}^3}{12^3\ \cancel{\text{in}^3}} = \ \text{ft}^3$$

Rounding to the nearest thousandth: (3 significant digits)

$$\frac{802\ \cancel{\text{in}^3}}{} \;\Big|\; \frac{1^3\ \text{ft}^3}{12^3\ \cancel{\text{in}^3}} \approx 0.464\ \text{ft}^3$$

Now try these!

1. Convert 1650 cubic inches to cubic feet.

2. Convert 864 square inches to square feet.

3. Convert 2.0 cubic yards to cubic inches.

 Clark/Anfinson Beginning Algebra 1e, Student Workbook, M. Andersen,

Student Activity

Out of this World

Common English Volume and Area Conversions

Volume (non-cubic units)	Volume (cubic units)	Area
1 pint = 2 cups 1 quart = 2 pints 1 gallon = 4 quarts 1 pint = 16 fl.oz.	$1^3\ ft^3 = 12^3\ in^3$ $1^3\ yd^3 = 3^3\ ft^3$ 1 gallon = 231 in^3	$1^2\ ft^2 = 12^2\ in^2$ $1^2\ yd^2 = 3^2\ ft^2$ 1 mi^2 = 640 acres

Directions: Perform each unit conversion calculation and then round the answer to the appropriate number of significant digits.

1. Area 51 occupies approximately 150 square miles of land (a dried-up lakebed in the Great Basin Desert of Nevada). How many acres of land is this?

2. The Space Shuttle Endeavor can hold 534,900 gallons of propellant. How much propellant is that in cubic feet?

3. The Apollo 11 command module (the crew compartment) measured 210 cubic feet. How many cubic yards of space was this?

4. If it's legal to sell land on the moon (and if his 1980 claim to the moon is legal), then Dennis Hope, head of the Lunar Embassy Corporation, has sold 2,500,000 acres of land on the moon (in small plots of land) to 3.7 million people. How many square miles of land has he sold?

   On a side note, the surface of the moon is approximately 26,000 square miles. So what percentage of the moon's surface has Dennis Hope sold?

   Source: http://news.nationalgeographic.com/news/2009/07/090720-apollo-11-who-owns-moon.html

5. The Qwiggle "Alien Autopsy" Gelatin Mold is approximately 15" x 4" x 4" in size. Estimate the maximum volume of this gelatin mold in cups (if it were a rectangular solid, $V = \ell \cdot w \cdot h$).

Student Activity

Spaces of Sports

Common Metric Volume and Area Conversions

Volume	Area
$1\text{ L} = 1000\text{ mL}$ $1\text{ mL} = 1\text{ cc} = 1\text{ cm}^3$ $1^3\text{ m}^3 = 100^3\text{ cm}^3$ $1^3\text{ cm}^3 = 10^3\text{ mm}^3$	$1^2\text{ m}^2 = 100^2\text{ cm}^2$ $1^2\text{ km}^2 = 1000^2\text{ m}^2$ $1^2\text{ cm}^2 = 10^2\text{ mm}^2$ $1\text{ hectare} = 10{,}000\text{ m}^2$

Directions: Perform each unit conversion calculation and then round the answer to the appropriate number of significant digits.

1. The volume of an Olympic swimming pool is 2,500 cubic meters. How many liters is this?

2. A regulation international soccer field measures 105 meters long and 68 meters wide. How many hectares is this?

3. Doubles tennis is played on a rectangular court measuring 260.8 square meters. How many square centimeters is this?

4. Athletes should consume approximately 8.62 liters of water per day. How many cubic centimeters is this?

5. The smallest regulation-size soccer ball has a volume of 5310 cubic centimeters. What is this volume in cubic meters?

Student Activity

Powering Through It

English-Metric Bridges for Volume and Area

Volume	Area
1 gallon = 3.785 L $1\text{ in}^3 = 16.39\text{ cm}^3$	1 acre = 4047 m^2 $1\text{ m}^2 = 1550\text{ in}^2$

Directions: Perform each unit conversion calculation and then round the answer to the appropriate number of significant digits.

1. In 2005, the per capita gasoline consumption of Americans was 1618.6 liters. How many gallons is this?

2. The average American uses 0.069 barrels of oil per day. If there are 42 gallons in a barrel, how many liters does the average American consumer use a day?

3. The Horse Hollow Wind Energy Center in Texas consists of 421 wind turbines spread over 47,000 acres. How many square kilometers does the wind farm occupy?

4. The wall of the Three Gorges Hydroelectric Dam in China measures 2,335 meters long by 101 meters tall. How many square miles is the area of the dam wall?

5. When the Hanford Site for Nuclear Production was decommissioned at the end of the Cold War, approximately 53 million gallons of nuclear waste was left behind at the site. How much waste is this in cubic meters?

Assess Your Understanding

Unit Conversions and Significant Digits

For each of the following, describe the strategies or key steps that will help you **start** the problem. You do **not** have to complete the problems.

		What will help you to start this problem?
1.	Round 0.024263 to three significant digits.	
2.	Add 12.3 cm + 1.27 cm and round to the appropriate number of significant digits.	
3.	Convert 4200 mg to grams.	
4.	Convert 9.2 L to gallons.	
5.	Convert 4.2 yards to meters.	
6.	Convert 9.25 gallons to cubic meters.	
7.	Convert 120 cc to liters.	
8.	Convert 8.2 acres to square feet.	
9.	Convert 8,520 seconds to hours.	
10.	Convert 12.8 m/s to mi/hr.	

Metacognitive Skills

Unit Conversions and Significant Digits

Metacognitive skills refer to the ability to judge how well you have learned something and to effectively direct your own learning and studying. This is a self-evaluation tool designed to help you focus your studying and to improve your metacognitive skills with regards to this math class.

Fill the 1st column out **before** you begin studying. Fill the 2nd column out after you study for your test.

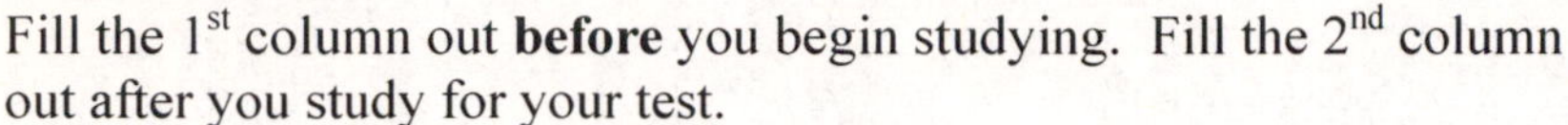

Go back to this assessment after your test and circle any of the ratings that you would change – this identifies the "disconnects" between what you **thought** you knew well and what you **actually** knew well.

Use the scale below to assign a number to each topic.

5 *I am confident I can do any problems in this category correctly.*
4 *I am confident I can do most of the problems in this category correctly.*
3 *I understand how to do the problems in this category, but I still make a lot of mistakes.*
2 *I feel unsure about how to do these problems.*
1 *I know I don't understand how to do these problems.*

Topic or Skill	Before Studying	After Studying
Counting the significant digits in a given number.		
Rounding a number to a given number of significant digits.		
Rounding to the correct number of significant digits after addition or subtraction.		
Rounding to the correct number of significant digits after multiplication or division.		
Converting length measurements involving English and/or Metric units.		
Converting mass measurements involving English and/or Metric units.		
Converting between various units of time.		
Converting measurements made up of two or more units (like mi/hr).		
Performing calculations with squared or cubed units.		
Converting area measurements involving English and/or Metric units.		
Converting volume measurements involving English and/or Metric units.		

EQN: Solving Linear Equations and Inequalities

Student Activity

Is it a Solution?

Tic-tac-toe Directions: If the number in the square **IS** a solution of the equation, then put an **O** on the square. If it **IS NOT** a solution, then put an **X** on the square.

$x+5=9$ 4	$\lvert y-3 \rvert=5$ 2	$4.3+x=7.7$ 3.4
$10-\frac{x}{2}=4$ 6	$3z+7=-1$ -2	$x^2-3x-4=0$ 4
$0.2x=3$ 0.6	$\lvert 6-a \rvert=9$ 15	$x^2-3x-4=0$ -1

Student Activity

Checking Solutions with a Calculator

When you check the solution to an equation, you hope to see an equality (a true statement) when you're finished. However, when the solution is a rounded decimal value, you may only see an approximate equality. To denote this, use the symbol $\approx$ instead of $=$.

For example, we check two solutions in the equation $7x-2=4$.

Check $x=\frac{6}{7}$:	Check $x=0.86$:
$7x-2=4$ $7\left(\frac{6}{7}\right)-2\stackrel{?}{=}4$ $6-2\stackrel{?}{=}4$ $4=4$ True.	$7x-2=4$ $7(0.86)-2\stackrel{?}{=}4$ $6.02-2\stackrel{?}{=}4$ $4.02\approx 4$ True.

Note that $\frac{6}{7}=0.8571428...\approx 0.86$, so both answers should work, but in the second check, we get only an approximate equality. These numbers are close enough for us to believe that $6/7$ is a solution of the equation.

Directions: In each box, determine if the given number creates an equality, an approximate equality, or is not a solution. Use the symbols $=$, $\approx$, and $\neq$ where appropriate.

$8x=4-x$ 0.44	$2y+1=4$ 1.5	$3u^2+5u-2=0$ 0.33	$\lvert 2s+3\rvert=3.5$ -0.25
$w^2=\frac{9}{16}$ 0.75	$x=\frac{x+2}{4}$ 0.67	$3z+3=2-z$ -0.12	$a^3=7$ 1.91

Student Activity

Match Up on One-step Equations

Match-up: Match each of the equations in the squares in the lower table with its solution from the top. If the solution is not found among the choices A through D, then choose E (none of these).

A -2

B 8

C -1

D 0

E None of these

There's going to be some give and take in this process, and some things may need to be divided up, but I'm here to make sure each side gets treated the same. Hopefully we can all arrive at a solution.

$2 = x + 4$	$-5a = -40$	$\frac{m}{4} = 0$	$-3 + t = 5$
$\frac{5r}{2} = 5$	$\ell - 5 = -7$	$-\frac{2}{3}u = \frac{2}{3}$	$1 = h + 3$
$8 = -8 + g$	$-x = -8$	$b - 4 = -4$	$32 = 4d$
$-7 = n - 6$	$\frac{3}{4}f = -\frac{3}{2}$	$-2 = \frac{w}{-4}$	$z - \frac{1}{6} = -\frac{7}{6}$

Student Activity
Match Up on Solving Equations

Match-up: Match each of the equations in the squares of the table below with its solution from the top. If the solution is not found among the choices A through E, then choose F (none of these).

A $x=-2$

B $x=3$

C $x=2$

D $x=0$

E $\varnothing$

F None of these

$5x-4=3x+2$	$-2(1-4x)=3x+8$	$5x-3x=0$	$5x+4x=x-4$
$5-4x=-(1+x)$	$5-(3x+2)=-6$	$5=5-3x$	$2(x+3)=-x$
$2(x-1)-x=x+5$	$3x=4-3x$	$2x-4=-8$	$3x-1=2x+x$

Student Activity

Pie Charts

Hints: Remember that all the pieces of a pie chart must add up to 100%. It may also help to first write a percent-sentence before solving each problem.

On a recent exam, Sarah's algebra class received the grade distribution shown in the pie chart to the right.

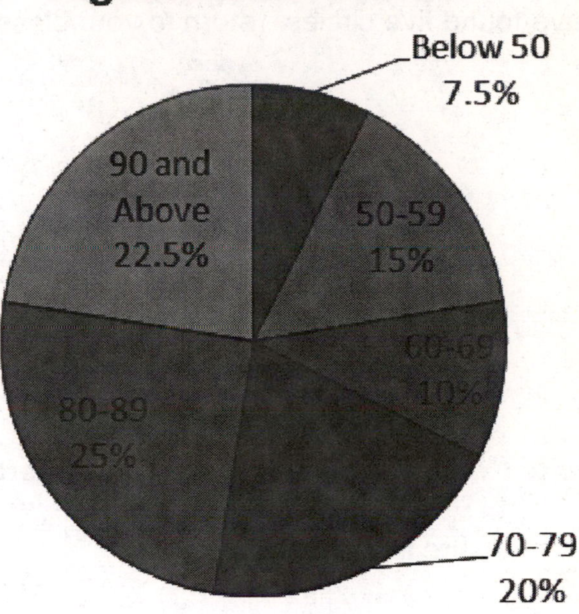

1. If 10 students scored in the 80-89 range on the test, how many students took the exam?

2. What percentage of the class scored a 70 or better?

3. How many students scored at least a 70?

The pie chart shown on the left represents Richard's monthly household budget.

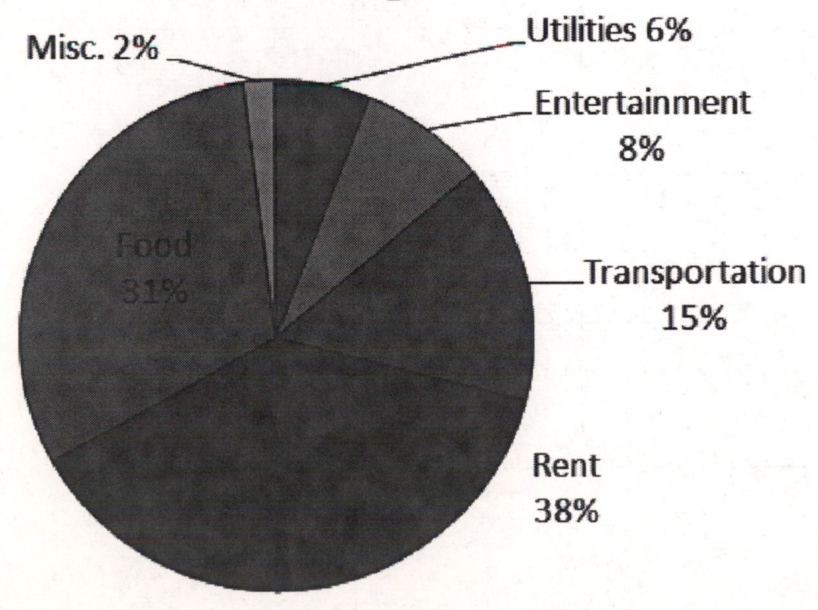

4. If Richard's total take home pay is $1500 per month, how much does he spend on food and rent together?

5. How much does he spend on entertainment?

6. What percentage of his income is not spent on utilities?

Student Activity

Circle Scavenger Hunt

Materials: Your instructor will provide you with a long piece of string and a ruler.

Directions: You have 10 minutes to find five circular objects to measure on your campus. Use the string and ruler to measure the circumference and diameter of each circle. Measure the first two objects in inches and the rest in centimeters. Once you have found five circles, return to your classroom to do the rest of the calculations.

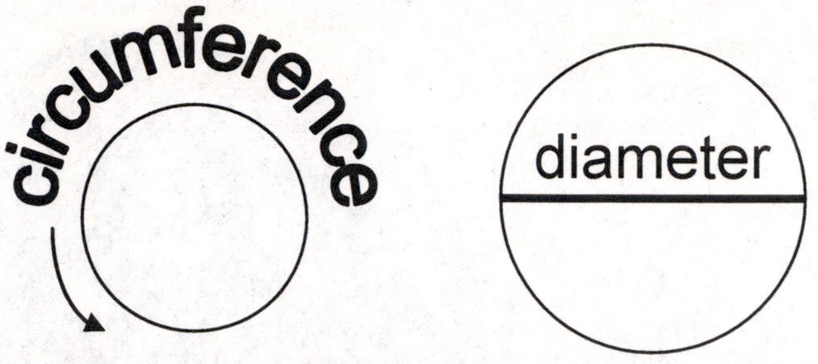

Object	Circumference	Diameter	**Calculate:** $\frac{\text{circumference}}{\text{diameter}}$
1. (measure in inches)			
2. (measure in inches)			
3. (measure in centimeters)			
4. (measure in centimeters)			
5. (measure in centimeters)			
			Average:

Conclusion:

Student Activity

Is it Length, Area, or Volume?

Measuring Length

$P = 2w + 2\ell$ (rectangle)

$P = 4s$ (square)

$C = 2\pi r = \pi d$ (circle)

$P = a + b + c$ (triangle)

English units: in, ft, yd, mi
Metric units: mm, cm, m, km

Measuring Area

$A = \ell w$ (rectangle)

$A = s^2$ (square)

$A = \pi r^2$ (circle)

$A = \frac{1}{2}bh$ (triangle)

English units: in^2, ft^2, yd^2, mi^2
Metric units: mm^2, cm^2, m^2, km^2

Measuring Volume

$V = \ell wh$ (rectangular solid)

$V = s^3$ (cube)

$V = \frac{4}{3}\pi r^3$ (sphere)

$V = \pi r^2 h$ (cylinder)

English units: in^3, ft^3, yd^3, mi^3
Metric units: mm^3, cm^3, m^3, km^3

Directions: Read the problems and decide whether you will have to find a length, area, or volume. Then identify the appropriate formula to use and decide what the resulting units will be. You do not have to solve the problems.

Problem	Length, Area, or Volume?	Formula	Units of answer
1. Mark is going to paint a large rectangular wall in a museum that measures 20 feet by 9 feet. How many square feet will Mark be painting?			
2. Alice wants to build a fence around her circular garden. If the diameter of the garden is 6 meters, how long does the fence need to be?			

Problem	Length, Area, or Volume?	Formula	Units of answer
3. The cylindrical mixing barrel on a cement truck measures 4 yards long and has a radius of 0.8 yards. How much cement can the truck hold in its mixing barrel?			
4. A rectangular fish tank measures 24 inches by 10 inches by 10 inches. How much water can the tank hold?			
5. A real-estate agent is pricing a New York studio apartment for someone. If the apartment is square, with 15 feet on each side, and apartments cost approximately $6.00 per square foot per month, how much will this one rent for?			
6. Stephen wants to buy fabric for a triangular sailboat sail that will measure 20 feet tall by 10 feet wide. How much fabric does he need?			
7. A glass globe hummingbird feeder has a diameter of 12 centimeters. How much sugar-water will it hold?			
8. Sarah is buying a circular cage for her 3 foot long python. She needs one that has a diameter at least twice as long as the snake. Will a cage with a 36 square foot floor be big enough?			
9. Brian is installing an invisible dog fence at his house (a wire that is buried underground). How much wire will it take to fence in his rectangular yard, which measures 20 meters by 30 meters?			
10. How many 8-oz (approximately 236.6 cubic centimeters) servings of iced tea can fit in a jar with diameter 25.4 centimeters and height 38.1 centimeters?			

Problem 2: The terminal bus at an airport drives a 4.2 mile loop between the three terminals (A, B, and C). If Terminal A is 1.6 miles from Terminal B, and Terminal B is 0.9 miles from Terminal C, then how far is the drive from Terminal C back to Terminal A?

Analyze the problem:

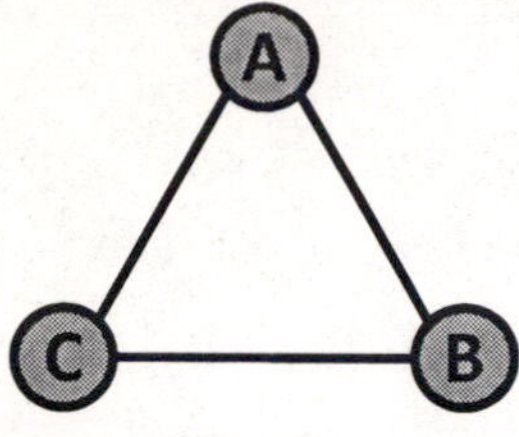

Total:

Write an equation:

Solve the equation:

State the conclusion in a complete sentence:

Explain why your solution makes sense:

Problem 3: Three consecutive even integers sum to 78. What are the three integers?

Analyze the problem:

First even integer	
Second	
Third	
Sum	

Write an equation:

Solve the equation:

State the conclusion in a complete sentence:

Explain why your solution makes sense:

Problem 4: In April, the water level in a mountain reservoir increased 0.6 m as the snow began to melt. In May, the water level increased another 0.3 m. In June the water level remained constant, but in July the water level decreased by 0.2 m because of a dry spell. At the end of July, the water level was 14.2 m; what was the water level at the beginning of April?

Analyze the problem:

Beg. of April	
Change in April	
Change in May	
Change in June	
Change in July	
End of July	

Write an equation:

Solve the equation:

State the conclusion in a complete sentence:

Explain why your solution makes sense:

Problem 5: The campsite fee at a state park is $27 per day plus $8 to make your campsite reservation online. Based on a budget of $170 to make the reservations and pay for camping, how many days can a family camp at the state park?

Analyze the problem:

Cost for 1 day	
Cost for 2 days	
Cost for 3 days	
Cost for 4 days	
Cost for x days	

Write an equation:

Solve the equation:

State the conclusion in a complete sentence:

Explain why your solution makes sense:

Student Activity
What was the Problem?

For each of the solutions below, **write a problem** that could go with the solution. Luckily for you, each student has done a particularly good job of showing their work!

Problem 1:

Student work:

Let $x =$ the measure of one of the isosceles angles.
Then $2x+20$ is the measure of the large angle.

$$\begin{aligned} x+x+(2x+20)&=180 \\ 4x+20&=180 \\ 4x&=160 \\ x&=40 \end{aligned}$$

The angles measure 40°, 40°, and 100°.

Check: $40+40+(2\cdot 40+20)\stackrel{?}{=}180$

$180=180$

Problem 2:

Student work:

Let $C =$ commission and $P =$ selling price of house., then $C=0.07P$.
The owner would receive $P-C$ or $P-0.07P$.

$$\begin{aligned} P-0.07P&=250{,}000 \\ 0.93P&=250{,}000 \\ P&\approx \$268{,}817 \end{aligned}$$

The selling price of the house should be \$268,817.

Check: $268{,}817-0.07(268{,}817)\stackrel{?}{=}250{,}000$

$250{,}000=250{,}000$

Problem 3:

Student work:

Susan has a budget of \$65. For 30 pages, the scrapbook price is \$30.
Let $p =$ the number of extra pages in the scrapbook.
Then $1.75p$ is the cost of the extra pages.

$$\begin{aligned} 30+1.75p&=65 \\ 1.75p&=35 \\ p&=20 \end{aligned}$$

Susan can get 20 extra pages, for a total of 50 pages in the scrapbook.

Check: $30+1.75(20)\stackrel{?}{=}65$

$65=65$

Student Activity
Working with the Smaller Pieces

Practice with Distance-Rate-Time Problems

Distance, rate and time can be related using the formula $d = rt$.

1. If you travel 8 miles from your house to visit a friend, what is your round-trip distance? ___________

2. Will you run faster when you run *with* the wind or *against* it? ___________________

3. How far can you drive in 90 minutes at a rate of 70 miles per hour? ______________

4. If you work 30 miles away from your home, how long will it take you to get there if your average rate is 45 miles per hour? ______________

5. If you drive for 3 hours and travel 210 miles, what was your average rate? _________________

6. If you are in a boat that can travel 4 miles per hour in still water, will you be traveling *faster* or *slower* than 4 miles per hour when traveling upstream? _____________

Practice with Value Mixture and Percent Mixture Problems

7. If a store owner mixes peanuts that cost \$3 per pound with cashews that cost \$8 per pound, a pound of mixed nuts should cost
   a. less than \$3 **b.** between \$3 and \$8 **c.** more than \$8

8. Simon has to mix two solutions in his chemistry class. One contains 3% sulfuric acid, and the other contains 10% sulfuric acid. How much sulfuric acid will the resulting mixture contain?
   a. less than 3%
   b. between 3% and 10%
   c. between 10% and 13%
   d. more than 13%

9. Lucy has to combine plain water with a solution that contains 15% ammonia. The resulting solution will be

a. less than 15% ammonia **b.** more than 15% ammonia

10. If you wanted to obtain 5 liters of a solution that is 4% nitric acid, which solution could you **not** add?

a. 3 liters of a 3% solution
b. 2 liters of a 5% solution
c. 6 liters of a 2% solution

11. At Meg's coffee shop, frozen coffees cost $3.50 each. If Meg took in a total of $42 on Tuesday, what is the maximum number of frozen coffees she could have sold?________

Practice with Investment Problems

Interest earned, principle invested, interest rate per year, and time invested (years) can be related using the formula $I = Prt$.

12. If you invest $2,500 for one year at an annual rate of 6%, how much interest does the investment earn? _______

13. If you invested a total of $15,000 in two different accounts, and you put $6,000 in the first account, how much did you put in the second account?________

14. If you invest $1,000 in an account that pays 5% annual interest, how much money will you have after one year?_________

15. Martin invests $2,000 in a bank CD. If the interest after one year is $160, what was the annual interest rate? __________

Guided Learning Activity

Organizing Information into Tables

For each problem:

a) Fill in each table with numbers in the appropriate spaces (there will be a few unfilled spaces).
b) One of your unfilled spaces should be filled with the unknown variable or expressions involving the unknown variable (like x, $2x$, $x+5$, etc.).
c) The last column of each table is calculated using the previous columns. Go ahead and fill this in.

1. A man decided to invest the $15,000 inheritance he received so that he could use the annual interest earned to pay the annual taxes on his home, which are $1,200. The highest bank rate that he could find was 6% annual simple interest, but this is not high enough to make the $1,200. So, instead of investing all the money at the bank, he invested some of the money in a riskier, but more profitable, investment offering a 10% return. How much should the man invest in each account to make exactly $1,200 in one year?

Investment	P	r	t	Interest
Bank				
Riskier investment				

2. A chemistry experiment calls for a 40% hydrochloric acid solution. If the lab supply room has only 60% and 10% hydrochloric acid solutions, how much of each should be mixed to obtain 12 liters of the 40% acid solution?

Solution	Solution Amount	Strength	Acid Amount
Strong Acid			
Weak Acid			
Mixed Acid			

3. A bulk-foods department purchaser wants to mix up 20 pounds of trail mix consisting of peanuts, chocolate candies, and raisins. The mixture should contain twice as many peanuts as raisins (by weight) and the amount of raisins and chocolate candies should be equal. If peanuts cost $2.40 per pound, chocolate candies cost $1.80 per pound, and raisins cost $3.20 per pound, how many pounds of each ingredient should be used in order to create the trail mix and how much will it cost per pound?

Ingredient	Amount	Price per pound	Total Value
Peanuts			
Raisins			
Chocolate			
Trail Mix			

4. A feed store sells sunflower seeds for $2.50 per pound and feed corn for $0.60 per pound. The manager wants to create 100 pounds of small animal feed that costs $1.50 per pound by mixing these two ingredients. How much of each should be used?

Ingredient	Amount	Price per pound	Value
Sunflower seed			
Feed corn			
Small Animal Feed			

5. Two runners leave traveling in opposite directions on a 22-mile loop. The first runner sets a pace of 5 mph and the second runner sets a pace of 6 mph. How long will it take them to meet back up on the loop?

	Rate	Time	Distance
First Runner			
Second Runner			

6. Two tractor trailer trucks are 540 miles apart and their speeds differ by 5 mph. Find the speed of each truck if they are traveling toward each other and will meet in 4 hours.

	Rate	Time	Distance
First truck			
Second truck			

7. At an electronics store, the number of 37-inch HDTVs that is expected to be sold in a month is double that of 58-inch HDTVs and 50-inch HDTVs combined. Sales of 58-inch and 50-inch HDTVs are expected to be equal. 58-inch HD-TVs sell for $4,800, 50-inch HD-TVs sell for $2,000, and 37-inch HDTVs sell for $1800. If total sales of $168,000 are expected this month, how many of each HDTV should be stocked?

	Number sold	Price	Sales $
37-inch HDTV			
50-inch HDTV			
58-inch HDTV			

Student Activity

Tic-Tac-Toe with Inequalities

Tic-tac-toe #1: If the inequality in the square is true, then put an **O** on the square. If it is false, then put an **X** on the square.

$-5 < -1$	$8 > 8$	$-6 > -2$
$2 - 3 < -1$	$\frac{1}{2} < \frac{1}{3}$	$7 \leq 7$
$a \geq a$	$0 < -4$	$\lvert -3 \rvert > 1$

Tic-tac-toe #2: If the number in the square is a solution to the inequality, then put an **O** on the square. If it is false, then put an **X** on the square.

$x - 2 < 7$ 9	$x + 4 \geq -2$ -7	$x - 3 \leq -1$ 1
$-2x < 4$ -1	$\frac{1}{2}x \geq 2$ 6	$-1 \leq 2 - x$ 3
$8 > 4 - x$ -5	$-x < 4$ -3	$2x + 3 < 3x$ 2

Student Activity

Match Up on Equations and Inequalities

Match-up: Match each of the equations and inequalities in the squares of the table below with an equivalent equation or inequality. If one is not found among the choices A through D, then choose E (none of these).

A $x>-1$ **B** $x=-1$ **C** $x<4$ **D** $x=4$ **E** None of these

$2x+4(2x-1)=-14$	$-8(x-1)=4x+5$	$-4x+1>-15$	$\frac{x}{3}-\frac{x}{4}>-\frac{1}{12}$
$3x+7(2x+5)=103$	$\frac{1}{3}x-2<\frac{1}{4}x-\frac{5}{3}$	$19x-21=-(4x-71)$	$\frac{x+1}{3}>\frac{-3x-3}{2}$
$\frac{5(x+2)}{2}+1=4x$	$-\frac{5(2-x)}{3}<2x-3$	$0.03x+0.08>0.2x-0.6$	$\frac{3}{2}(x+5)=\frac{17+5x}{2}$
$\frac{11+3(2x-1)}{2}=20-x$	$6>\frac{5}{4}x+1$	$-2(x-3)<-(x-2)$	$\frac{4x+1}{3}=\frac{2x-3}{5}$
$-2x-1>5x+6$	$-\frac{3}{4}x+1=-2+\frac{9}{4}x$	$-2x>7-(13x+18)$	$1.2x+7.3=-5.1x+1$

Assess Your Understanding

Solving Linear Equations

For each of the following, describe the strategies or key steps that will help you **start** the problem. You do **not** have to complete the problems.

		What will help you to start this problem?
1.	Solve: $3x+5=11$	
2.	Solve: $-4x=12$	
3.	Solve: $\frac{3x}{4}-2=\frac{x}{6}$	
4.	Is -3 a solution to $4x+6=-2$?	
5.	Solve: $3(-2n+4)=-30$	
6.	Write $-2<x\le 4$ in interval notation	
7.	Solve: $\frac{b}{2}-\frac{1}{3}=\frac{3}{4}$	
8.	Solve: $-4x\le 24$	
9.	Solve $A=bh+Bh$ for B.	
10.	Is $x=-2$ a solution of $2x+1<-3$?	

		What will help you to start this problem?
11.	Change 0.25% to a decimal.	
12.	Find 2.5% annual simple interest on $500.	
13.	Find the volume of a sphere if the radius is 4 mm.	
14.	If the circumference of a circle is 14π cm, find the radius.	
15.	25 is what percent of 60?	
16.	If the sales tax on an item is 6%, and the price is $129.95, find the total paid by the customer.	
17.	How many liters of a 2% acid solution must be added to 30 liters of a 10% acid solution to dilute it to an 8% acid solution?	
18.	A car averaged 60 mph for part of a trip and 70 mph for the remainder. If the 7 ½ -hour drive covered 500 miles, for how long did the car average 60 mph?	
19.	If the vertex angle of an isosceles triangle is $24°$, find the measure of each base angle.	
20.	The cost of a stereo to an electronics store is $268. If the markup is 20%, what is the sales price of the stereo?	

Metacognitive Skills

Solving Linear Equations and Inequalities

Metacognitive skills refer to the ability to judge how well you have learned something and to effectively direct your own learning and studying. This is a self-evaluation tool designed to help you focus your studying and to improve your metacognitive skills with regards to this math class.

Fill the 1st column out **before** you begin studying. Fill the 2nd column out after you study for your test.

Go back to this assessment after your test and circle any of the ratings that you would change – this identifies the "disconnects" between what you **thought** you knew well and what you **actually** knew well.

Use the scale below to assign a number to each topic.

5 *I am confident I can do any problems in this category correctly.*
4 *I am confident I can do most of the problems in this category correctly.*
3 *I understand how to do the problems in this category, but I still make a lot of mistakes.*
2 *I feel unsure about how to do these problems.*
1 *I know I don't understand how to do these problems.*

Topic or Skill	**Before Studying**	**After Studying**
Checking the solution to an equation.		
Applying the Addition Property of Equality to solve an equation.		
Applying the Multiplication Property of Equality to solve an equation.		
Setting up an equation to represent an application problem.		
Setting up a percent-sentence and then translating it into an equation.		
Setting up and solving application problems that involve percents, including those with a percent increase or decrease.		
Solving linear equations like $ax+b=c$.		
Solving linear equations that involve parentheses and like terms.		
Clearing the fractions from equations involving fractions.		
Understanding the difference between solving an equation and simplifying an expression.		
Solving a formula for a selected variable.		
Choosing a formula and solving a problem involving business or science formulas.		
Choosing the formula and solving a problem involving length, area, or volume.		
Using problem solving skills to solve application problems.		
Placing the correct units on the results to application problems.		
Setting up a table to organize the information in an application problem.		
Deciding whether an inequality is true or false.		
Solving an inequality.		
Graphing an inequality.		
Writing the answer to an inequality in interval notation.		
Solving a compound inequality (like $-3<x+1<5$).		

ABS: Absolute Value Equations and Inequalities

Student Activity

Working with Intersection and Union

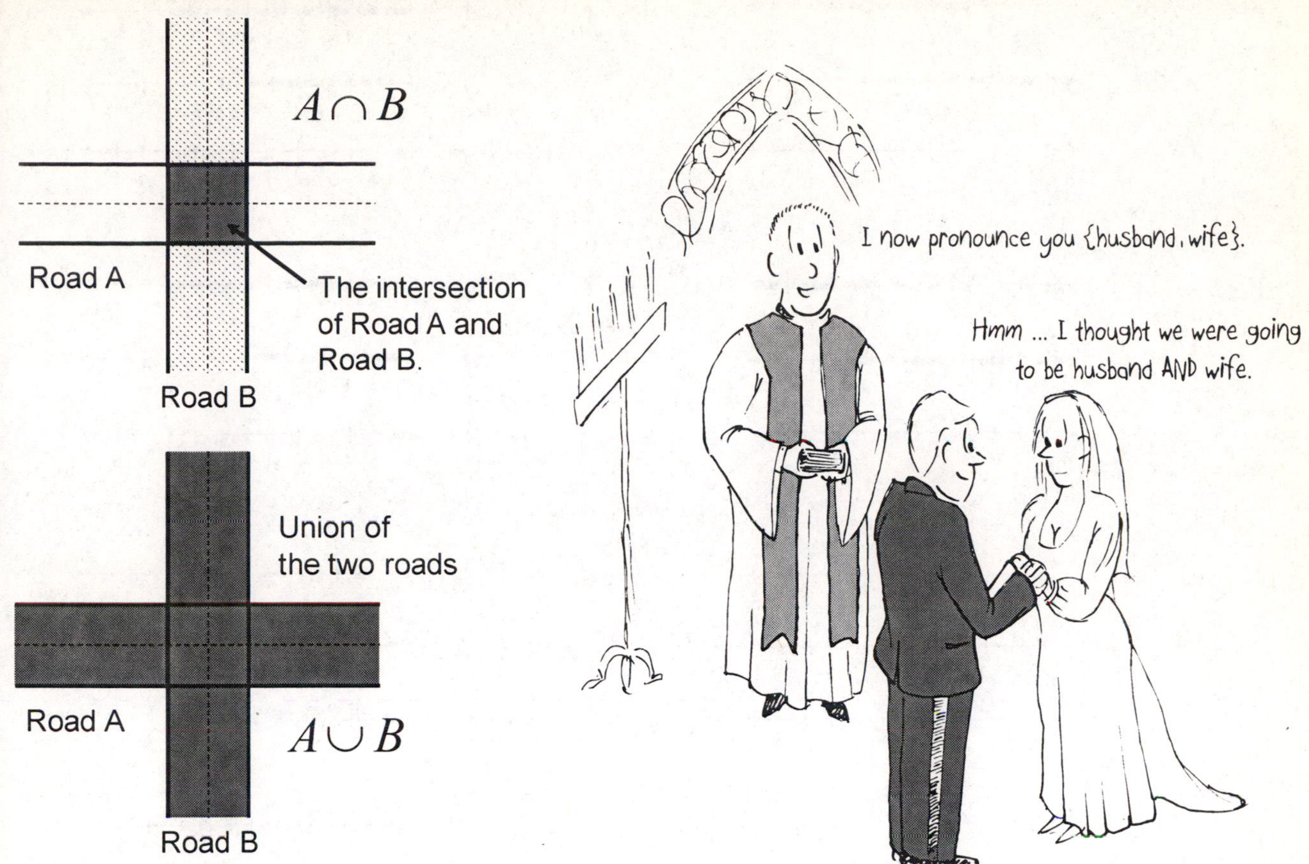

Section I: Working with Sets of Numbers

	Set A	**Set** B	**Intersection** $A \cap B$ (A and B)	**Union** $A \cup B$ (A or B)
a.	$\{-2,0,2,4\}$	$\{0,1,2,3\}$		
b.	$\{-3,-2,-1\}$	$\{1,2,3\}$		
c.	$\{-3,1,2,4\}$	$\{-2,-1,2\}$		
d.	$\{1,2,3\}$	$\{1,2,3\}$		
e.	$\{0\}$	$\{-1,0,1\}$		
f.	$\mathbb{Z}$	$\mathbb{N}$		
g.	$\{\ldots,1,2,3\}$	$\{2,3,4,\ldots\}$		

Section II: Working with Number-line Graphs

a. A: , B: , $A \cap B$:

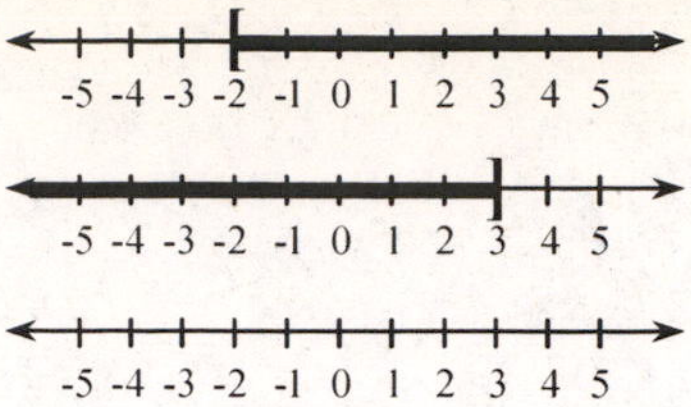

b. A: , B: , A or B :

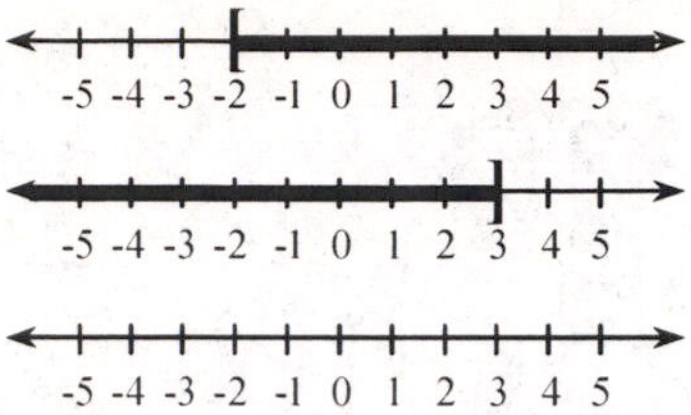

c. A: $x < 0$, B: $x \le -2$, $A \cap B$:

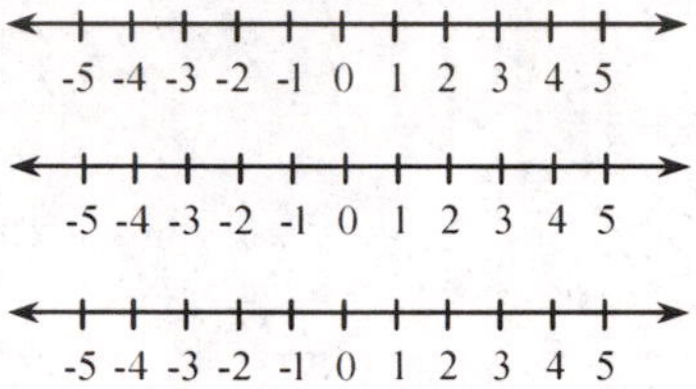

d. A: $x < 0$, B: $x \le -2$, $A \cup B$:

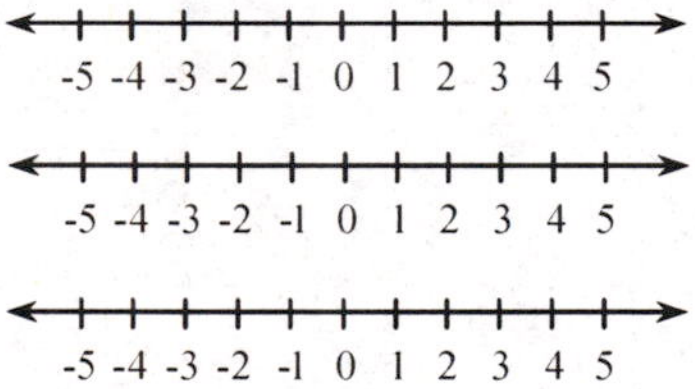

e. A: , B: , A and B :

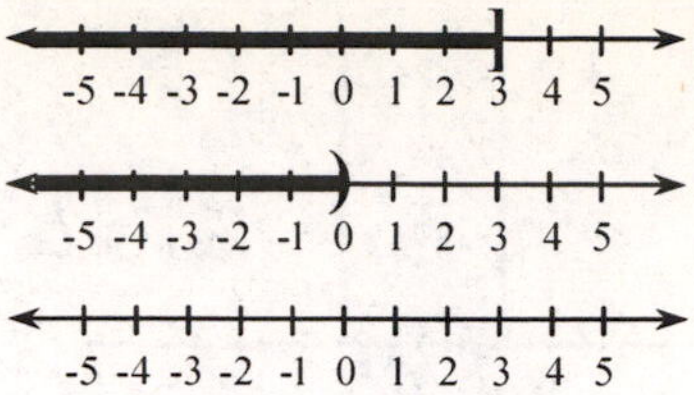

f. A: , B: , $A \cup B$:

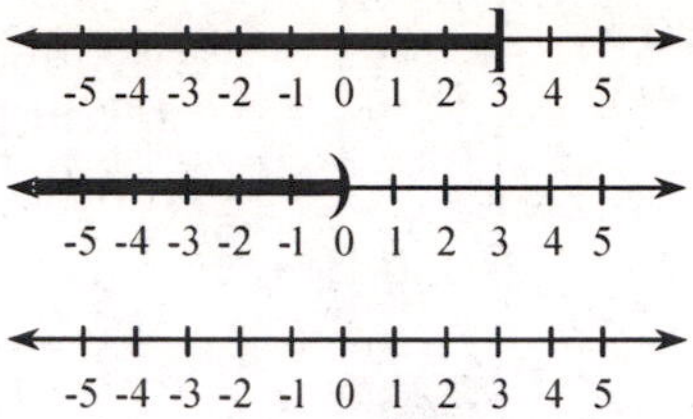

g. A: $[-3,3]$, B: $(-2,2)$, A or B :

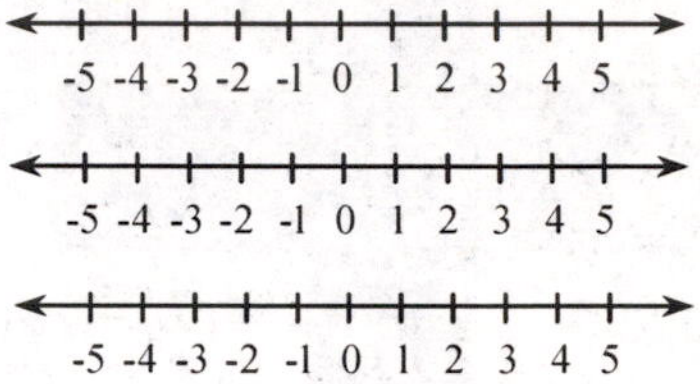

h. A: $[-3,3]$, B: $(-2,2)$, A and B

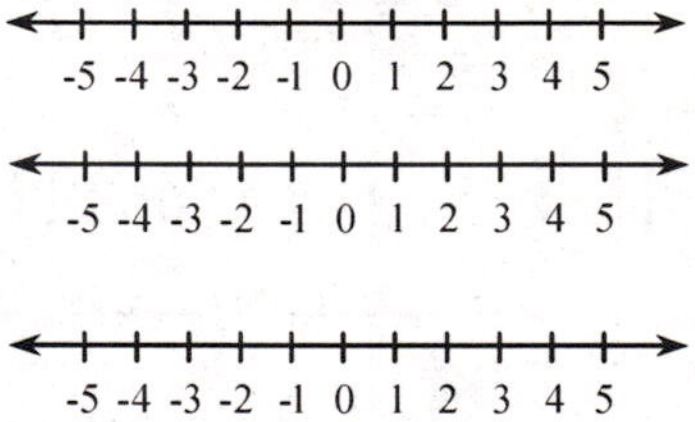

Guided Learning Activity

Solving Compound and Double Inequalities

Solving a Compound Inequality (AND or OR)

1. Solve each inequality separately.
2. Graph each inequality together on a number line. (*use different colors*)
3. Graph the desired intersection (and) or union (or) on a number line. (*use a 3rd color*)
4. Write the interval solution that represents the graphed solution. (*the 3rd color graph*)

Solving a Double Inequality

Apply the properties of inequality to all three of its parts to isolate x in the middle.

---

1. Solve: $3x \leq 15$ and $2x > -6$

Graph the separate pieces:

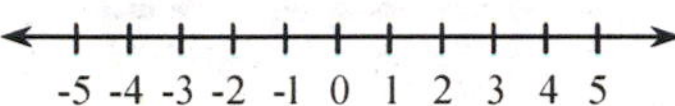

Graph the solution:

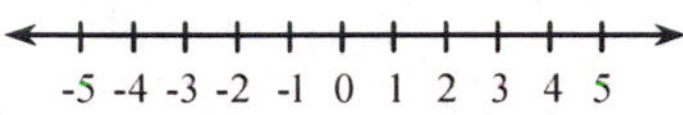

Write the solution: ____________

2. Solve: $3x < 3$ or $-2x < -10$

Graph the separate pieces:

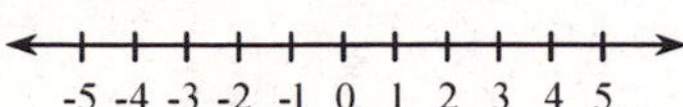

Graph the solution:

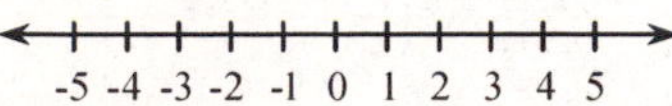

Write the solution: ____________

3. Solve: $5 < 2x + 1 \leq 9$

Graph the solution:

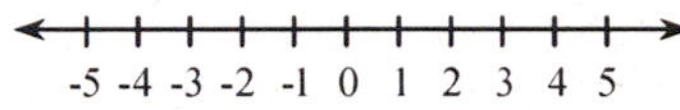

Write the solution: ____________

4. Solve: $2x - 5 < -11$ or $5x + 1 \leq 6$

Graph the separate pieces:

-5 -4 -3 -2 -1 0 1 2 3 4 5

Graph the solution:

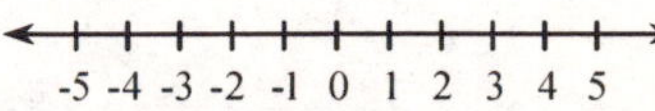

Write the solution: ____________

Student Activity

The Match is Out There

Directions: Solve each compound or double inequality and then match the problem up with the graph of its solution.

___ **1.** Solve: $x+2>4$ and $x-1<4$

___ **2.** Solve: $0 \le x+3 \le 5$

___ **3.** Solve: $2+x \le 2$ or $2-3x \ge -4$

___ **4.** Solve: $4x \le -12$ and $\dfrac{x}{2} \le 1$

___ **5.** Solve: $\dfrac{x}{2}+3 \le 1$ or $\dfrac{5x}{4}-1 > \dfrac{3}{2}$

___ **6.** Solve: $2x-5<5$ or $3x+1>-2$

a. -5 -4 -3 -2 -1 0 1 2 3 4 5

b. -5 -4 -3 -2 -1 0 1 2 3 4 5

c. Empty set
-5 -4 -3 -2 -1 0 1 2 3 4 5

d. -5 -4 -3 -2 -1 0 1 2 3 4 5

e. -5 -4 -3 -2 -1 0 1 2 3 4 5

f. -5 -4 -3 -2 -1 0 1 2 3 4 5

g. All reals
-5 -4 -3 -2 -1 0 1 2 3 4 5

h. -5 -4 -3 -2 -1 0 1 2 3 4 5

i. -5 -4 -3 -2 -1 0 1 2 3 4 5

j. -5 -4 -3 -2 -1 0 1 2 3 4 5

Student Activity

Matching Up the Different Cases

Directions: Begin by isolating the absolute value in each of the equations and inequalities below. Then categorize the problem as one of the four cases or a special case. The first one has been done for you.

Let k be a **positive** constant and let X, X_1, and X_2 be mathematical expressions.

Case 1:	Case 2:	Case 3:	Case 4:	Special Case:
$\lvert X \rvert = k$	$\lvert X_1 \rvert = \lvert X_2 \rvert$	$\lvert X \rvert < k$ $\lvert X \rvert \le k$	$\lvert X \rvert > k$ $\lvert X \rvert \ge k$	If the constant is negative or zero.

$2\lvert x-3 \rvert < 4$ $\lvert x-3 \rvert < 2$ **CASE 3**	$\lvert x-3 \rvert - 4 \ge 2$	$\lvert 2a-3 \rvert + 5 > 3$	$\frac{\lvert x \rvert}{3} - 2 = 4$
$\lvert x+4 \rvert = \lvert 2x-2 \rvert$	$1 = \lvert 2t-4 \rvert + 4$	$5 - \lvert 2x \rvert < 2$	$4\lvert x+3 \rvert - 3 \le 2$
$-2\lvert 4y \rvert < 10$	$\frac{\lvert 3x-4 \rvert}{-6} \le -2$	$1 > \lvert 3-b \rvert - 6$	$\lvert 2x \rvert = \lvert x+3 \rvert$
$5 - \lvert x+4 \rvert = 2$	$6 - 2\lvert x+4 \rvert \le 0$	$\lvert x-4 \rvert + 5 = 5$	$3 < \frac{\lvert d+12 \rvert}{5}$

Guided Learning Activity
Solving Absolute Value Equations and Inequalities

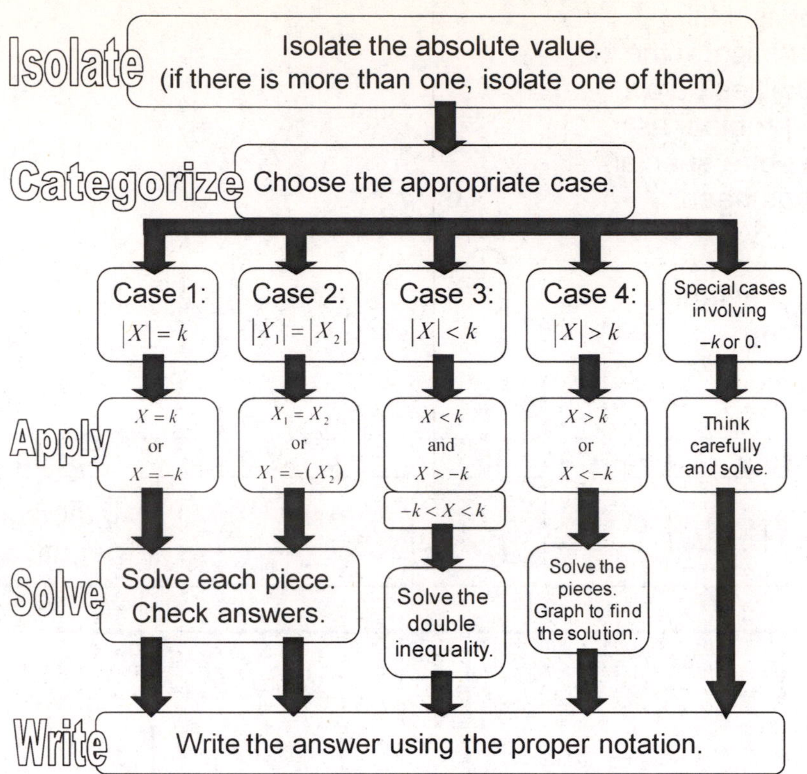

Directions: With the guidance of your instructor, work through the strategy for solving an absolute value equation or inequality with each of the problems below.

1. Solve $|2x| - 3 = 11$

2. Solve $3 + |2x + 1| \le 7$

3. Solve $\frac{|3x+4|}{5} > 1$

5. Solve $|x+4| = |3x-8|$

4. Solve $-4|x+1| < 20$

Student Activity

Connecting Absolute Value Problems to Graphs

The graph of $y = |x|$ gives a V-shaped curve. We can interpret why the absolute value equations and inequalities come out the way they do by looking at graphs that correspond to the equations. For example:

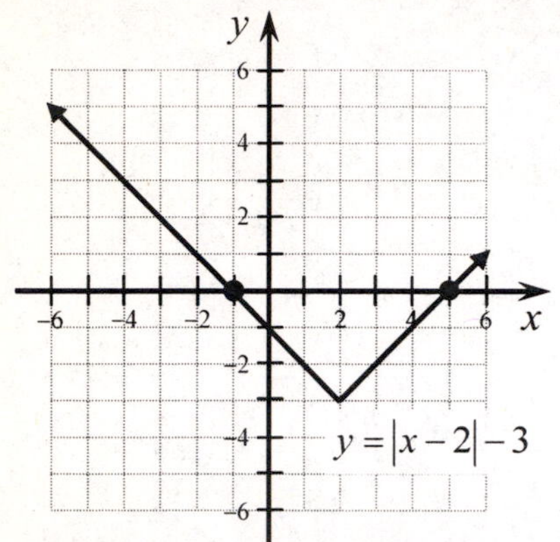

Where is $|x-2|-3=0$?

Where $x=-1$ or $x=5$.

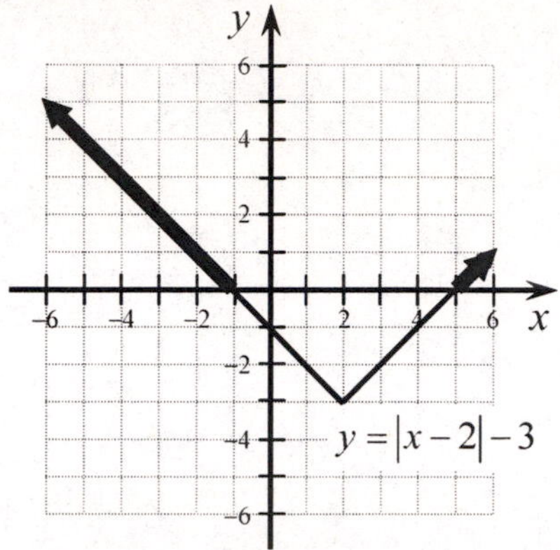

Where is $|x-2|-3>0$?

Where $x<-1$ or $x>5$.

Intervals: $(-\infty,-1)\cup(5,\infty)$

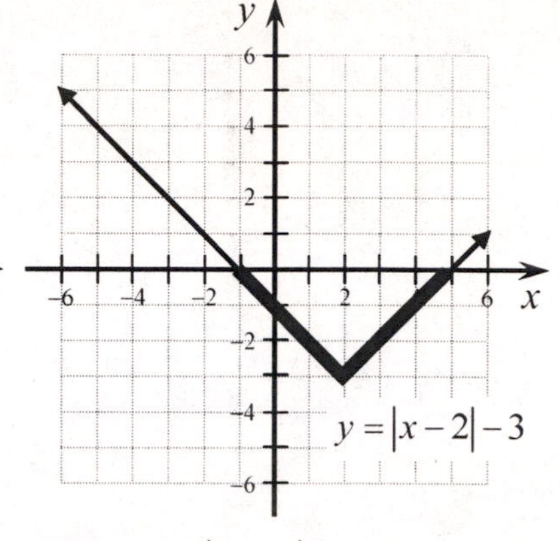

Where is $|x-2|-3<0$?

Where $-1<x<5$.

Interval: $(-1,5)$

Now solve the corresponding equations and inequalities using the techniques we have learned in this section. Remember to isolate the absolute values first!

1. a. Solve: $|x-2|-3=0$ **b.** Solve: $|x-2|-3>0$ **c.** Solve: $|x-2|-3<0$

As you answer the following problems, you might find it helpful to trace each answer on the graph in a different color.

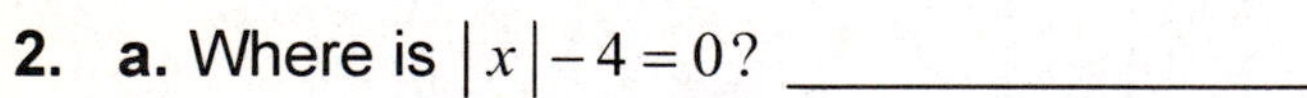

2. a. Where is $|x|-4=0$? ______________

b. Where is $|x|-4<0$? ______________

c. Where is $|x|-4\geq 0$? ______________

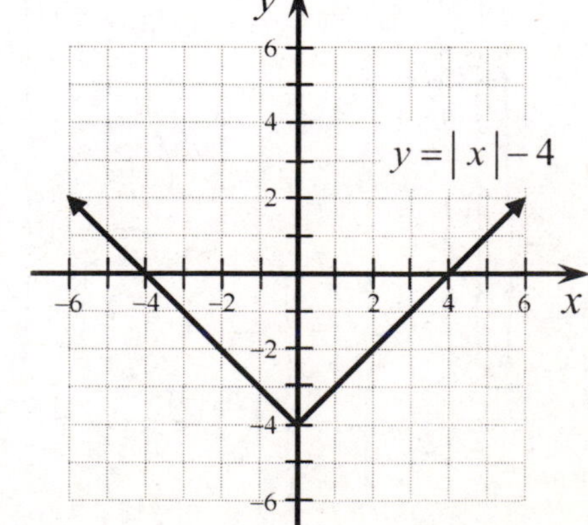

3. **a.** Where is $|x+1|-4=0$? ______________

Solve: $|x+1|-4=0$

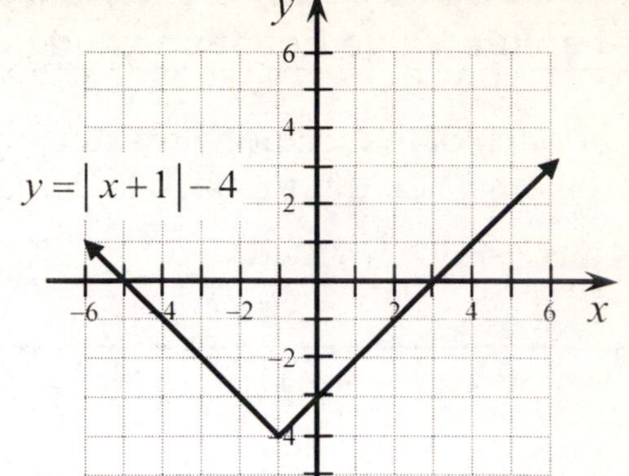

b. Where is $|x+1|-4\le 0$? ______________

Solve: $|x+1|-4\le 0$

---

4. **a.** Where is $|x+3|+1<0$? ______________

Solve: $|x+3|+1<0$

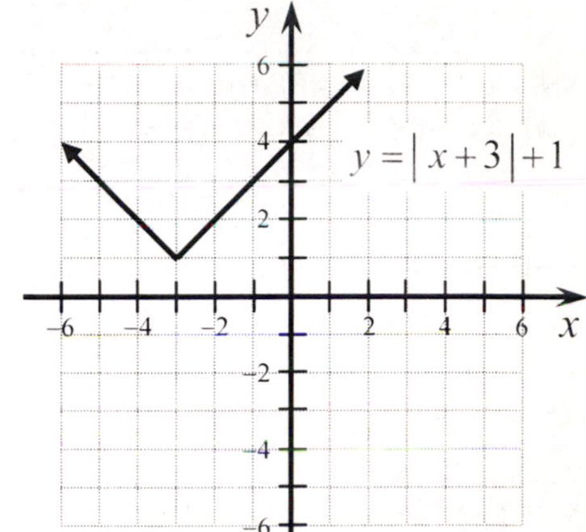

b. Where is $|x+3|+1\ge 0$? ______________

Solve: $|x+3|+1\ge 0$

---

5. **a.** Where is $|x-3|\le 0$? ______________

Solve: $|x-3|\le 0$

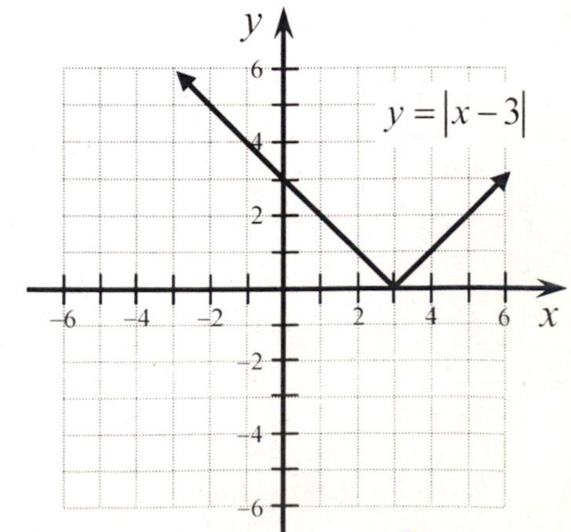

Pay attention to this next one!

b. Where is $|x-3|\ge \mathbf{2}$? ______________

Solve: $|x-3|\ge 2$

Assess Your Understanding

Absolute Value Equations and Inequalities

For each of the following, describe the strategies or key steps that will help you **start** the problem. You do **not** have to complete the problems.

		What will help you to start this problem?
1.	Solve: $\lvert x+4\rvert - 1 < 6$	
2.	Solve: $-13 \le 5x+2 \le 17$	
3.	Solve the compound inequality: $5x+1<11$ or $3x-2>13$	
4.	Solve: $\lvert x+1\rvert = \lvert 2x-1\rvert$	
5.	Let $A=\{-2,0,2,4\}$ and $B=\{1,3,5\}$. Find $A \cup B$.	
6.	Let $A=\{1,3,5,7\}$ and $B=\{5,6,7,8\}$. Find the intersection of A and B.	
7.	Solve: $\lvert x+2\rvert = -4$	
8.	Solve: $\lvert 2x+1\rvert \ge 9$	

Metacognitive Skills

Absolute Value Equations and Inequalities

Metacognitive skills refer to the ability to judge how well you have learned something and to effectively direct your own learning and studying. This is a self-evaluation tool designed to help you focus your studying and to improve your metacognitive skills with regards to this math class.

Fill the 1st column out **before** you begin studying. Fill the 2nd column out after you study for your test.

Go back to this assessment after your test and circle any of the ratings that you would change – this identifies the "disconnects" between what you **thought** you knew well and what you **actually** knew well.

Use the scale below to assign a number to each topic.

5 *I am confident I can do any problems in this category correctly.*
4 *I am confident I can do most of the problems in this category correctly.*
3 *I understand how to do the problems in this category, but I still make a lot of mistakes.*
2 *I feel unsure about how to do these problems.*
1 *I know I don't understand how to do these problems.*

Topic or Skill	**Before Studying**	**After Studying**
Understanding intersection and union, including their relationship to the words "and" and "or" and the symbols $\cup$ and $\cap$.		
Solving a compound inequality (one that uses **and** or **or**).		
Solving a double inequality (like $-3 < x+1 < 5$).		
Isolating the absolute value in an equation or inequality.		
Determining the type of absolute value equation or inequality (Case 1, 2, 3, 4, or special case).		
Remembering the proper procedure after you determine the category of absolute value equation or inequality.		
Reasoning out the answer to a "special case" absolute value equation or inequality.		

LINE: Lines and More

Student Activity

Take Me to the Point

Directions: Capital letters are often used to name specific points. For each point that is graphed below, write down the letter and coordinates of the point in the appropriate category in the table below. Point A has been done for you.

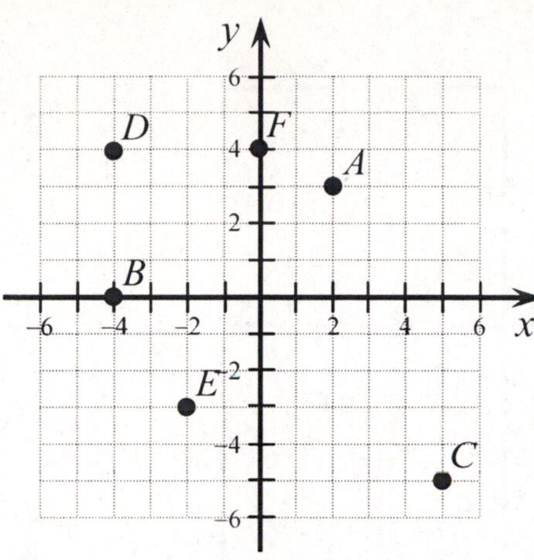

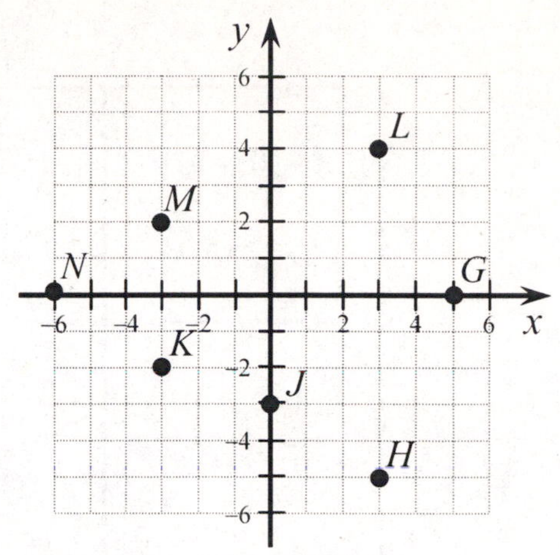

Quadrant I	Quadrant II	Quadrant III	Quadrant IV	x-axis	y-axis
$A(2,3)$					

Directions: Now plot these points on the axis provided. Indicate each point with its letter on the graph.

$P(2,3)$ $G(-3,4)$ $A(0,0)$ $R(-2,2)$

$U(0,-3)$ $N(3,-4)$ $F(-3,-4)$ $H(4,5)$

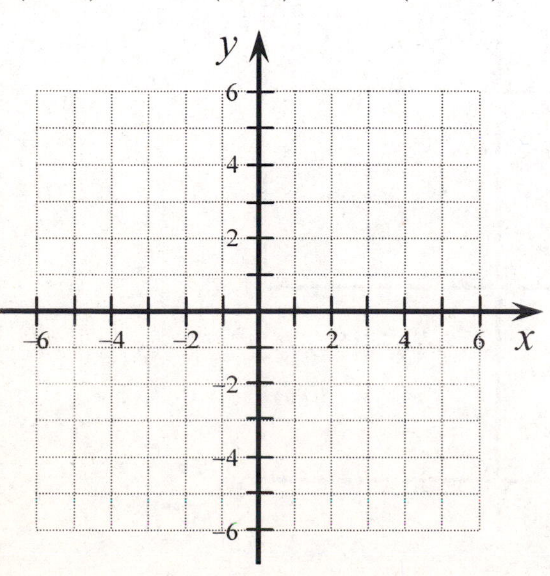

Student Activity

Tic-Tac-Toe with Ordered Pairs

Directions for Game #1: If the ordered pair in the square **is** a solution to the equation $y = 2x - 5$, then circle it (thus placing an **O** on the square). If the ordered pair is **not** a solution, then put an **X** over it.

$(-10,-30)$	$(-10,-25)$	$\left(\frac{2}{3},-\frac{11}{3}\right)$
$(26,47)$	$(0,-5)$	$(0,0)$
$(5,0)$	$\left(-\frac{1}{2},-4\right)$	$(1,3)$

Directions for Game #2: If the ordered pair in the square **is** a solution to the equation $4x - 6y = 12$, then circle it (thus placing an **O** on the square). If the ordered pair is **not** a solution, then put an **X** over it.

$(-9,-8)$	$\left(1,-\frac{4}{3}\right)$	$(3,0)$
$(-2,0)$	$(0,0)$	$\left(-\frac{3}{2},-3\right)$
$(1,1)$	$\left(\frac{2}{3},-\frac{14}{9}\right)$	$(9,-4)$

Guided Learning Activity

Graphing Linear Equations

Example 1: Fill in the missing values in the table of solutions for the linear equation, $y=-x+4$, plot the points, and then draw the line through the points.

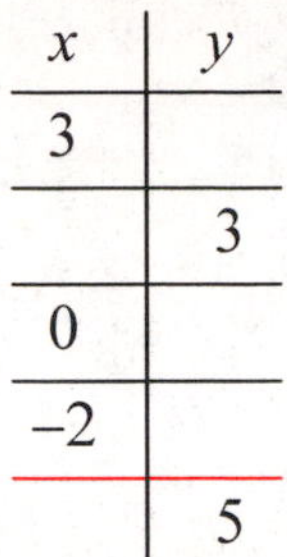

x	y
3	
	3
0	
-2	
	5

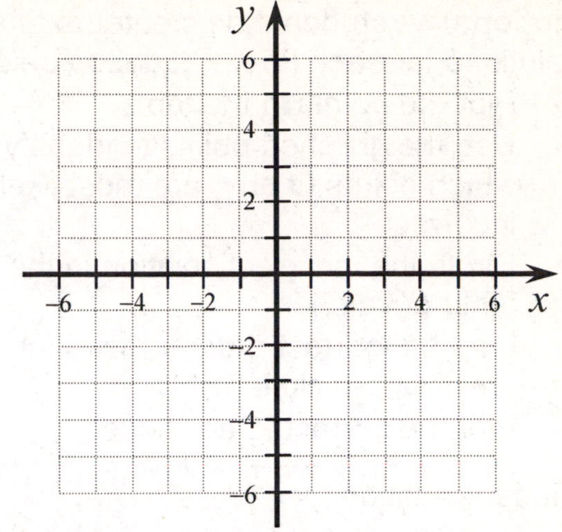

If $x=3$, find y.	If $y=3$, find x.	If $x=0$, find y.	If $x=-2$, find y.	If $y=5$, find x.
$y=-(\quad)+4$	$(\quad)=-x+4$	$y=-(\quad)+4$	$y=-(\quad)+4$	$(\quad)=-x+4$

Example 2: Fill in the missing values in the table of solutions for the linear equation, $3x-y=-2$, plot the points, and then draw the line through the points.

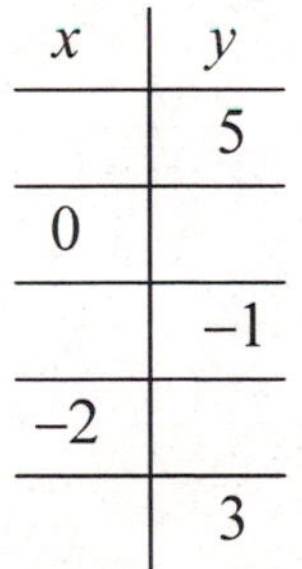

x	y
	5
0	
	-1
-2	
	3

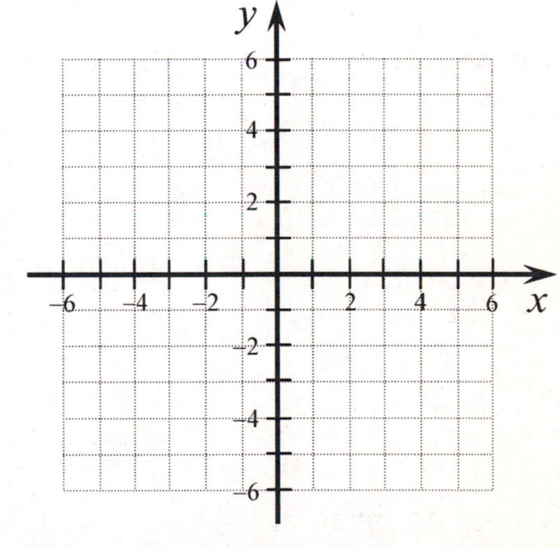

If $y=5$, find x.	If $x=0$, find y.	If $y=-1$, find x.	If $x=-2$, find y.	If $y=3$, find x.
$3x-(\quad)=-2$	$3(\quad)-y=-2$	$3x-(\quad)=-2$	$3(\quad)-y=-2$	$3x-(\quad)=-2$

Student Activity

Out-of-line Suspects

Directions: A student has created a table of solutions for each **linear** equation below.

- Plot the points in the table.
- Use the graphed points to identify which points (if any) are most likely incorrect.
- Circle the incorrect solution in the table of values.
- Find the correct ordered pairs for any that are circled.
- Plot the corrected solutions.

1. Linear equation: $y = 3x - 1$

x	y
-1	2
-2	3
0	-1
$\frac{1}{3}$	0
2	5

2. Linear equation: $y = \frac{1}{2}x + 2$

x	y
-2	0
-1	$\frac{3}{2}$
2	5
0	2
4	4

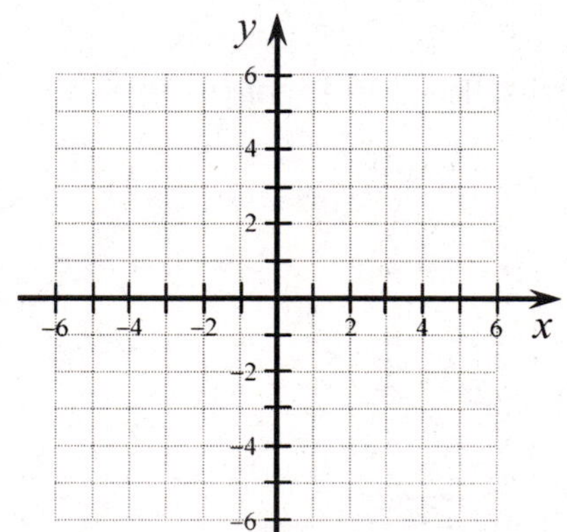

3. Linear equation: $y = 4 - x$

x	y
-2	6
-6	1
1	3
2	-4
6	-2

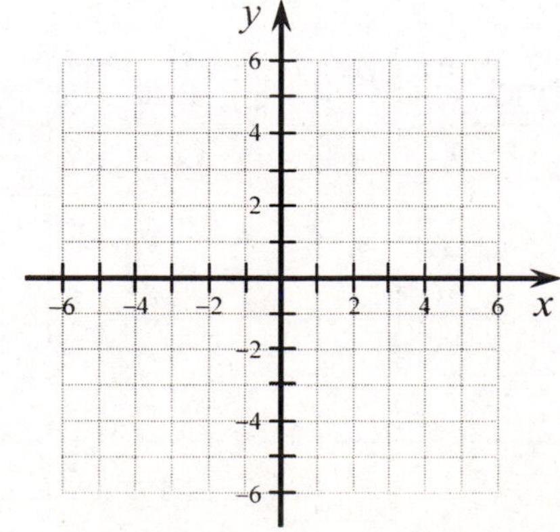

Student Activity

The Cost of College

When you pay for your college education, there are typically three different types of charges:

1) ***Variable cost*** – Tuition, which is paid per credit hour (or unit of study). There may also be fees that vary depending on the number of credit hours (for example, an additional \$6 per credit hour).
2) ***Fixed cost*** – Student fees, which are paid regardless of how many or what types of classes you take, (for example, a \$25 Registration Fee). These are fixed costs.
3) Course fees or lab fees, which are paid for certain courses only (we will ignore these in this model). Your instructor may tell you to ignore some other fee or tuition oddities to allow a linear model to be used.

Our linear model of college costs will only consider the first two types of charges. Your instructor will help you locate the information necessary to build the linear model – perhaps they have asked you to bring in a copy of your tuition bill!

How much do you pay in **fixed** costs for your enrollment at your college? __________

Per credit hour, how much do you pay in tuition (and **variable** fees)? ______

Fill in the table below for a student at your college:

Credits (or units) taken by student, *n*	Fixed student fees	Tuition and other variable costs for this number of credits (or units)	Total cost for this number of credits, *C*
4			
8			
12			
16			

Create a graph containing the data points above.

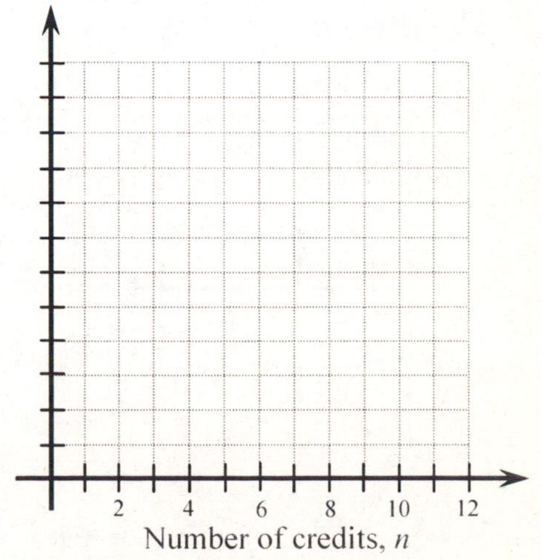

Write a linear equation to model the cost C, for taking a certain number of credits n.

Show how you could use the linear equation you just wrote to estimate the cost of taking 9 credits (or units).

Student Activity

ID the Intercepts

Directions: For each graph, identify the x- and y-intercepts (if they exist) using **both** coordinates of the point.

1. x-intercept: (,)

y-intercept: (,)

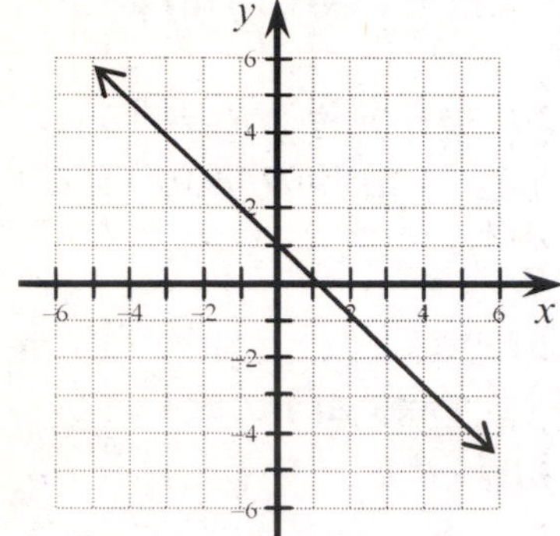

2. x-intercept: (,)

y-intercept: (,)

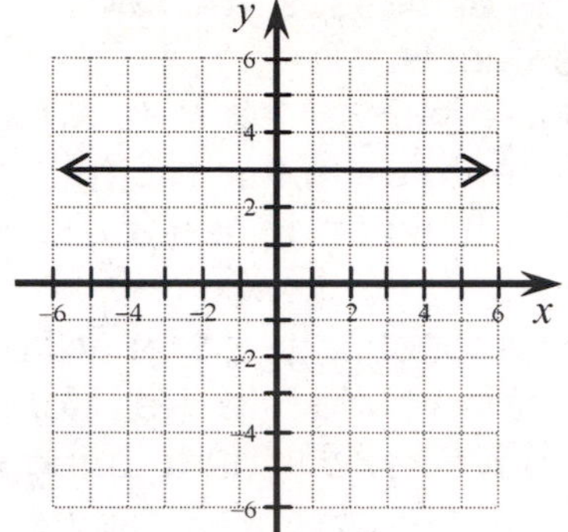

3. x-intercept: (,)

y-intercept: (,)

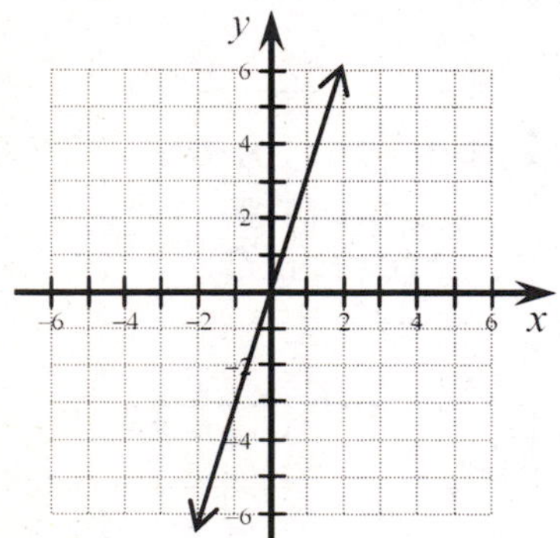

4. x-intercept: (,)

y-intercept: (,)

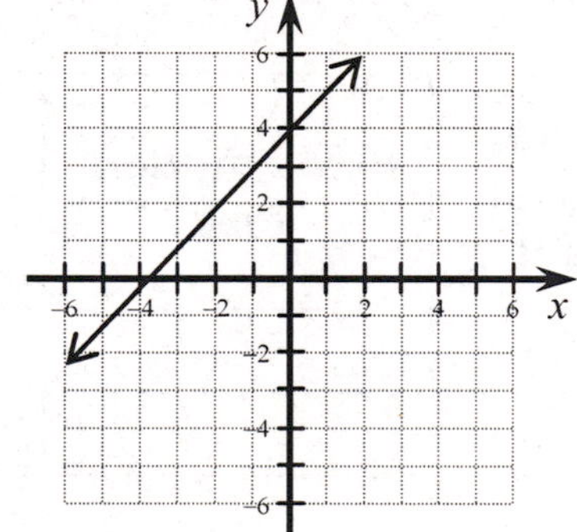

5. x-intercept: (,)

y-intercept: (,)

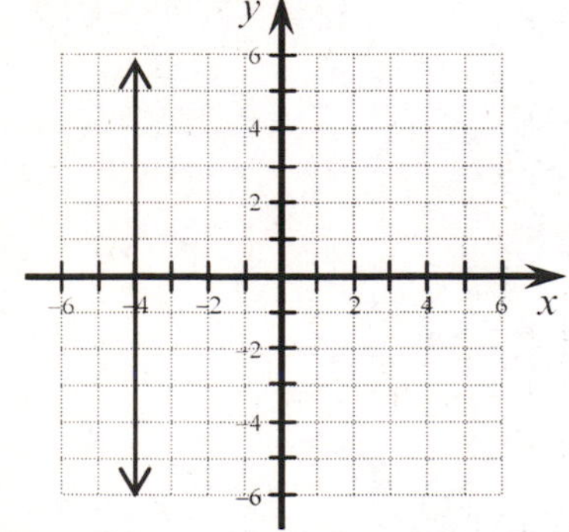

6. x-intercept: (,)

y-intercept: (,)

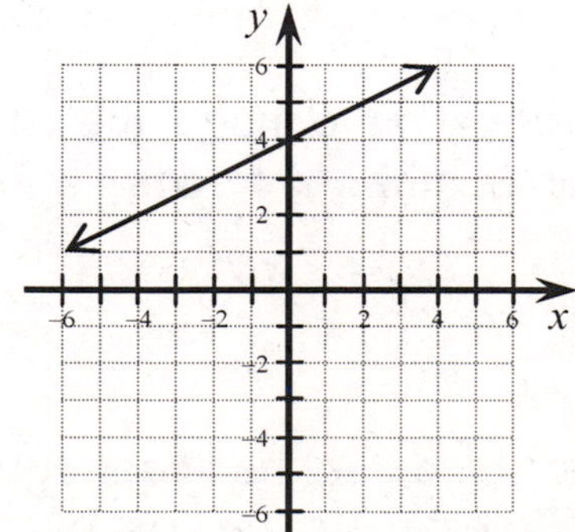

Student Activity
A Web of Lines

Directions: First, use the method of intercepts to find the x- and y-intercepts for each equation below. **Then** draw each of the lines on the graph provided using a straightedge. (Use different colors for each line if you have colored pencils.)

$6x + y = 6$

x-int: (,)

y-int: (,)

$5x + 2y = 10$

x-int: (,)

y-int: (,)

$4x + 3y = 12$

x-int: (,)

y-int: (,)

$3x + 4y = 12$

x-int: (,)

y-int: (,)

$2x + 5y = 10$

x-int: (,)

y-int: (,)

$x + 6y = 6$

x-int: (,)

y-int: (,)

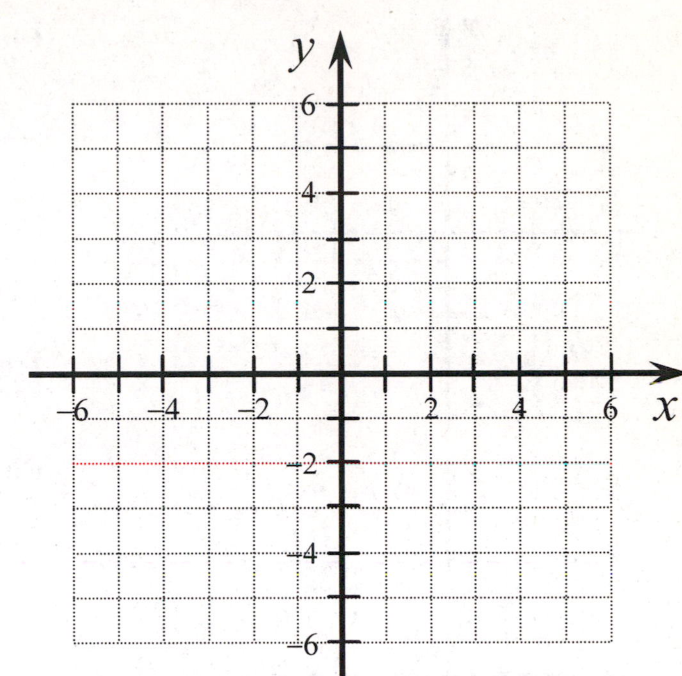

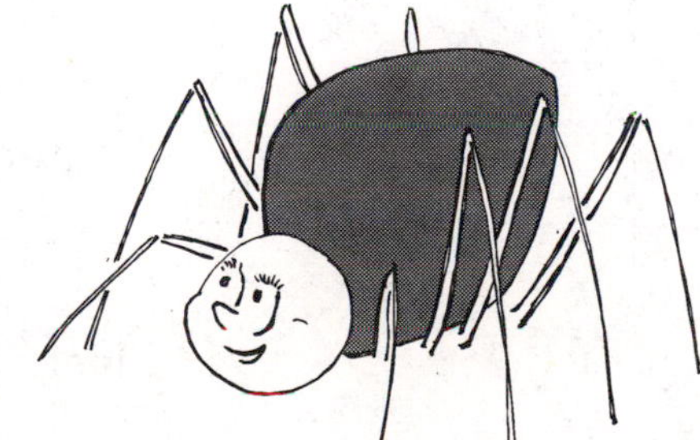

Challenge: Now, if you're really a good math "detective" you can construct and graph a set of your own linear equations to make a similar "web" in **Quadrant III**. Look for patterns in the equations above to help you.

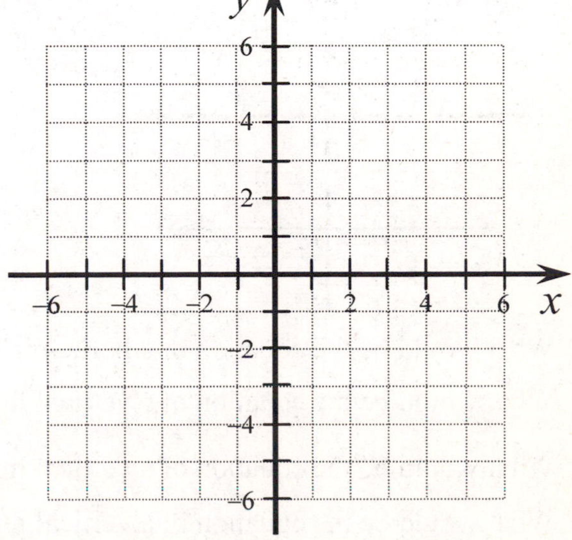

Guided Learning Activity

Clues to the Equation

Directions: For each graphed line, construct a possible table of values that could accompany the line. Then, together with your class, you will write a linear equation to describe the line mathematically.

1.

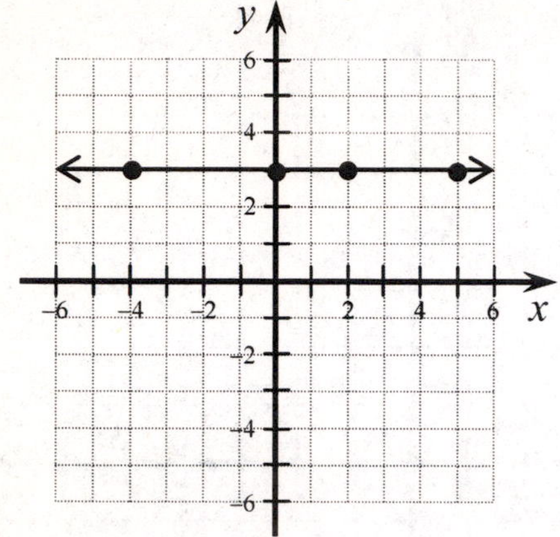

Clue: Table of Solutions

x	y

Equation: ____________

2.

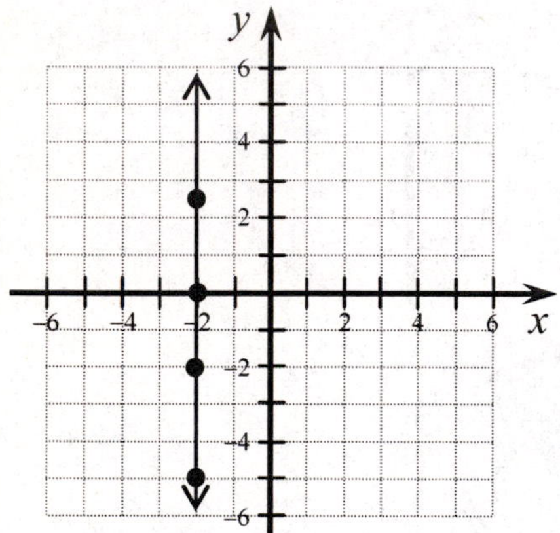

Clue: Table of Solutions

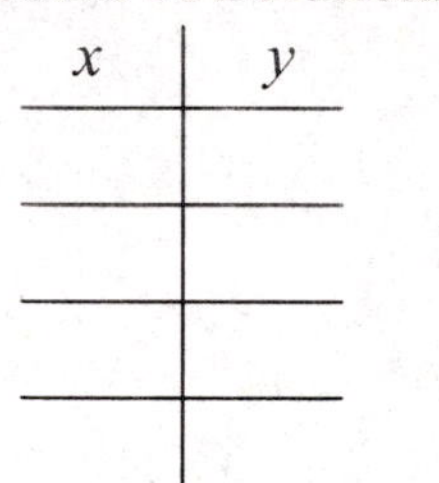

Equation: ____________

3.

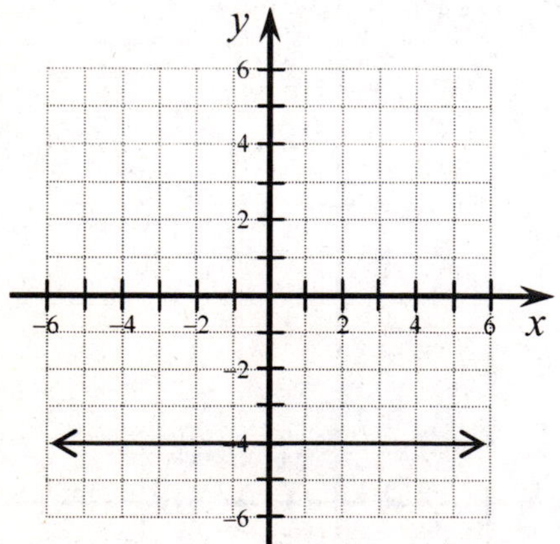

Clue: Table of Solutions

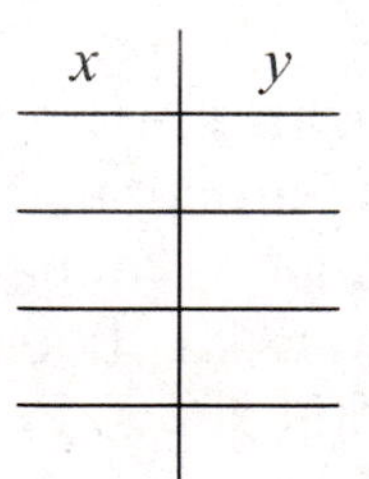

Equation: ____________

4. What would be the equation of a horizontal line that passes through $(2,3)$? _______

5. What would be the equation of a vertical line that passes through $(-2,4)$? _______

6. What would be the equation of a vertical line with an x-intercept of 5? ______

7. What would be the equation of a vertical line with an x-intercept of 0? ______

Student Activity

LINE-11

Graphing Linear Equations with a Calculator

You can use your graphing calculator to graph linear equations. Each model of calculator will have a different set of keystrokes that will allow you to graph these equations. Before you start, it is a good idea to guess what your graph should look like (in case you enter something into your calculator incorrectly). The standard viewing window on most graphing calculators goes from -10 to 10 on the x-axis and from -10 to 10 on the y-axis.

1. Sketch a graph of the equation $2x+3y=3$ on the axes below.

Graphing calculators can usually only accept equations beginning with $y=$.

2. Solve the equation $2x+3y=3$ for y.

3. Now find the key on your calculator that takes you to the $y=$ screen. It will usually look like [Y =]. It may be a function above another key, so you may have to press the *function* or [2nd] key first. Below, draw the keys you need to press.

Once you reach the $y =$ screen, you should see your cursor next to $Y_1 =$. **Enter the equation you found in problem 2.** Note that you do not need to type $y =$ since it is already there.

4. Next, find your graph key. It may say GRAPH or it may be a function above another key. Select graph. Draw the keys you need to press.

5. Often times, it is convenient to view a graph in a **standard** viewing window. Find your calculator's zoom menu. It may be a key that says ZOOM or it may be a function above another key. If you have it as an option, select **ZoomStd** or **Zoom Standard**. Draw the keys you need to press to do this.

Hopefully your calculator's graph looks like the one you sketched in problem 1.

6. You can change the size of the viewing window by using other options in the zoom menu. You can also change the window to a size of your choice using the WINDOW key. Note that WINDOW may be found *above* a key on some calculators. Draw the keys you need to press to get into the window menu.

Now change the window to these settings:

$\text{xmin} = 0$	$\text{ymin} = 0$
$\text{xmax} = 5$	$\text{ymax} = 5$
$\text{xscale (or xscl)} = 1$	$\text{yscale (or yscl)} = 1$

Using the GRAPH key or function, look at the graph again.

7. What has happened to the view of the graph on the calculator screen?

8. What window settings would show you only the second Quadrant?

$\text{xmin} =$	$\text{ymin} =$
$\text{xmax} =$	$\text{ymax} =$
$\text{xscale (or xscl)} =$	$\text{yscale (or yscl)} =$

Guided Learning Activity

Hit the Slopes

Directions: For each line that is graphed below, determine the slope of the line using either a slope triangle or the slope formula.

1.

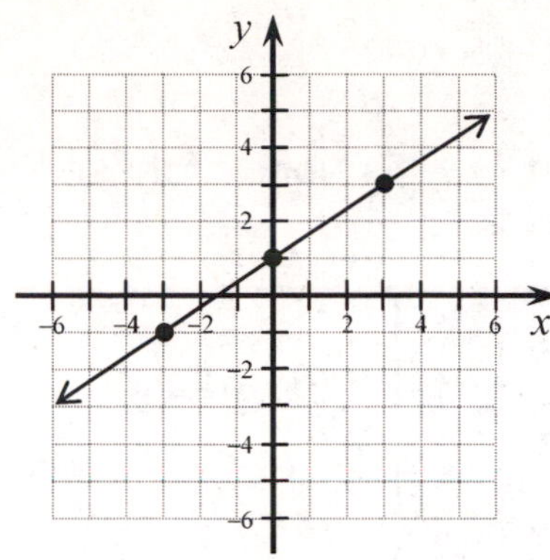

2.

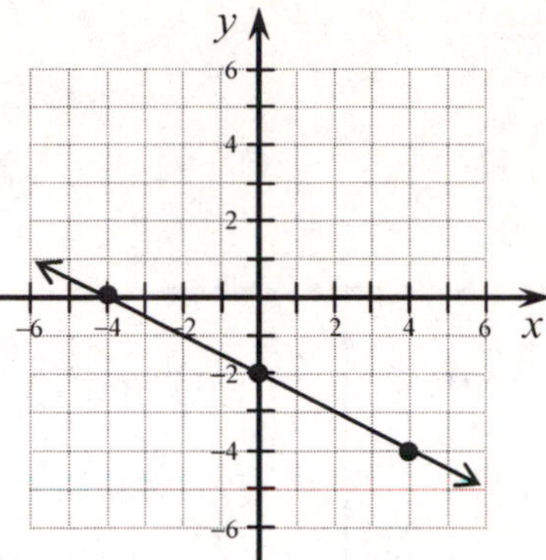

3.

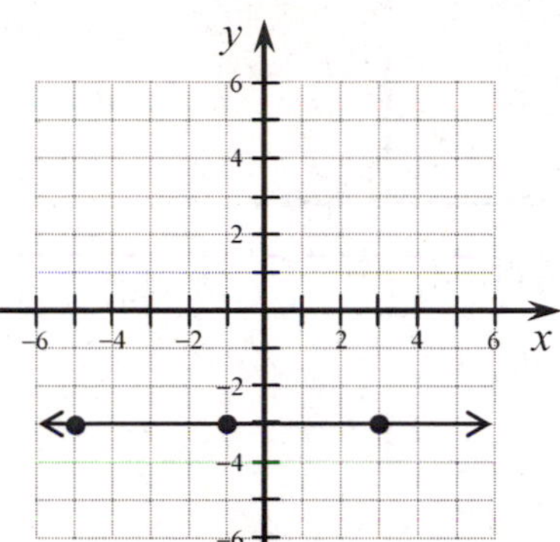

4.

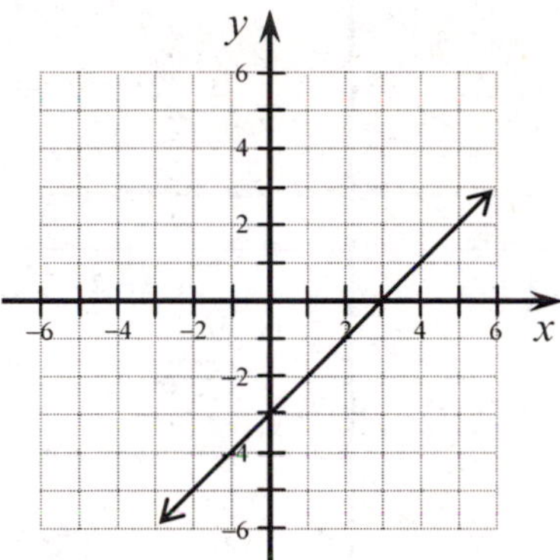

5.

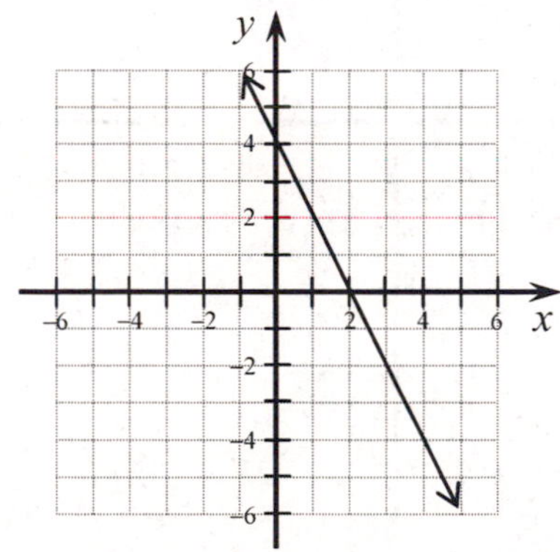

6.

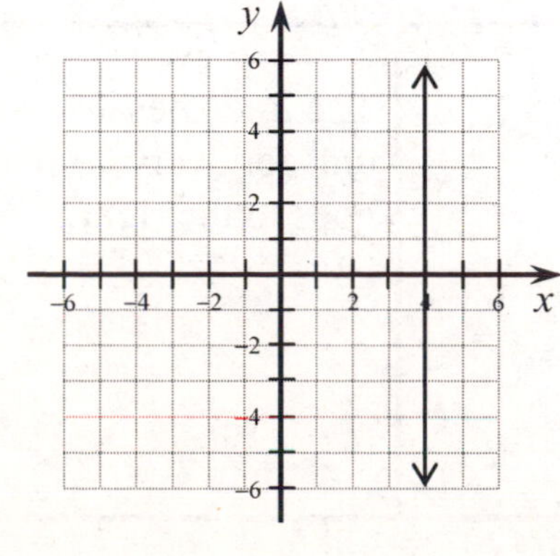

Student Activity

Match Up on Slopes

Match-up: In each box of the grid below, you will find either a pair of points that is on a line or the description of a line in words. Match each line that is described with its slope in the choices A through E. If the slope is not found in A through E, then choose F (none of these).

A 2 **B** $\frac{1}{2}$ **C** -1 **D** 0 **E** undefined **F** None of these

$(0,0)$ and $(3,0)$	$(4,4)$ and $(-4,0)$	$\left(\frac{3}{2},\frac{1}{2}\right)$ and $\left(\frac{3}{2},\frac{10}{2}\right)$	$(-3,-6)$ and $(3,-4)$
$(2,5)$ and $(3,4)$	$\left(\frac{7}{6},\frac{1}{9}\right)$ and $\left(\frac{1}{6},\frac{10}{9}\right)$	$(3.14, 2.72)$ and $(1.14, 1.72)$	$(-6,-1)$ and $(-8,0)$
$(-100,300)$ and $(-100,50)$	$(9,7)$ and $(7,9)$	$(1,1)$ and $(-5,1)$	$(8,8)$ and $(7,6)$
The line is a vertical line.	The line goes to the right 1 unit for every 2 units it goes up.	The line is a horizontal line.	The line goes down 1 unit for every 1 unit it goes to the right.

Student Activity

Midpoints as Averages

1. Plot the two given points on each graph provided. Connect the two points with a line **segment** using a straight-edge. Then estimate the midpoint of the line segment.

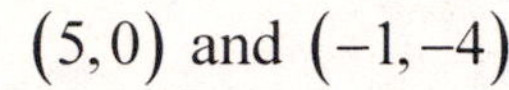

$(5,0)$ and $(-1,-4)$

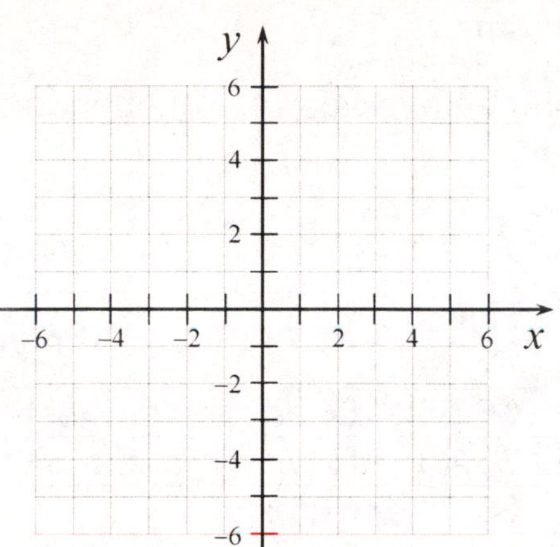

Midpoint: (____,____)

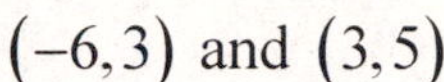

$(-6,3)$ and $(3,5)$

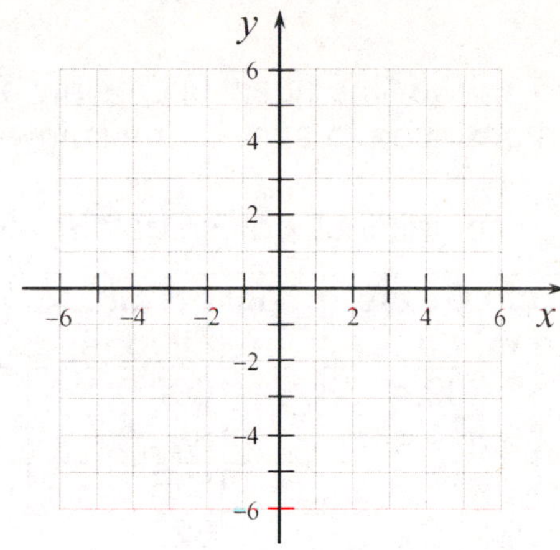

Midpoint: (____,____)

2. We find an average (or mean) of several values by finding the sum of the values, and then dividing by the number of values. Find the average of 5, 12, 3, and 4.

Average: $\dfrac{\square+\square+\square+\square}{\square}=$

3. We find the midpoint of two points by averaging the *x*-coordinates and averaging the *y*-coordinates. Find the midpoint for $(5,0)$ and $(-1,-4)$ using this technique.

x-coordinate: $\dfrac{\square+\square}{\square}=$ *y*-coordinate: $\dfrac{\square+\square}{\square}=$

Midpoint: (____,____) Is this consistent with the midpoint you found in #1?

4. Repeat this process for $(-6,3)$ and $(3,5)$

Midpoint: $\left(\dfrac{\square+\square}{\square},\dfrac{\square+\square}{\square}\right)=($____,____$)$

5. See if you can find "third-points" for the points $(-2,-6)$ and $(1,6)$. These would be two points that split the line segment into thirds.

Student Activity

Entertaining Rates of Change

It is often helpful to summarize raw data with a rate describing the amount of change in one quantity with respect to the amount of change in another. This is called an **average rate of change**, and is often used to describe change that occurs *over time*.

$$\text{Average Rate of Change} = \frac{\text{change in quantity}}{\text{change in time}}$$

Directions: Answer the questions below by calculating an average rate of change and using this information to answer the follow-up questions.

1. This table contains box office data from 2000 – 2006. (Source: www.boxofficemojo.com)

Year	Total Gross (in millions)	Tickets Sold (in millions)	# of Pics	Ticket Price	#1 Picture of the year
2006	$9,209.4	1400.0	606	$6.58	Dead Man's Chest
2005	$8,840.4	1381.3	547	$6.40	Revenge of the Sith
2004	$9,418.3	1516.6	551	$6.21	Shrek 2
2003	$9,185.9	1523.3	508	$6.03	Return of the King
2002	$9,167.0	1578.0	467	$5.81	Spider-Man
2001	$8,412.5	1487.3	482	$5.66	Harry Potter
2000	$7,661.0	1420.8	478	$5.30	The Grinch

a. What is the average annual rate of change for the total box office gross between 2000 and 2006? Show your calculation and round your answer to the nearest tenth.

b. Complete the statement: *On average, total box office gross increased* ___________ *dollars per year over the 6-year period between 2000 and 2006. The total increase in box office gross during this period was* ____________.

c. What is the average annual rate of change for the total box office gross between 2003 and 2006? Show your calculation and round your answer to the nearest tenth.

d. Complete the statement: *On average, total box office gross increased* ___________ *dollars per year over the 3-year period between 2003 and 2006. The total increase in box office gross during this period was* ____________.

e. In general, Is the average annual rate of change for total box office gross increasing or decreasing? ______________

The following tables contain U.S. Music purchasing data for 2005 and 2006.

Units sold (in millions):	**2005**	**2006**
Overall Music Sales (Albums, singles, music video, digital tracks)	1,003	1,198
Total Album Sales (Includes CD, CS, LP, Digital albums)	618.9	588.2
Digital Track Sales	352.7	581.9
Overall Album Sales (Includes all albums & track equivalent albums)	654.1	646.4
Internet Album Sales (Physical album purchases via e-commerce sites)	24.7	29.4
Digital Album Sales	16.2	32.6

(Source: www.businesswire.com)

Top Ten Selling Albums of 2006	**Units Sold**
Soundtrack/High School Musical	3,719,071
Me and My Gang/Rascal Flatts	3,479,994
Some Hearts/Carrie Underwood	3,015,950
All the Right Reasons/Nickelback	2,688,166
Futuresex/Lov.../Justin Timberlake	2,377,127
Back to Bedlam/James Blunt	2,137,142
B'day/Beyonce	2,010,311
Soundtrack/Hannah Montana	1,987,681
Taking the Long Way/Dixie Chicks	1,856,284
Extreme Behavior/Hinder	1,817,350

2. Use the tables above to answer the following questions. Show your calculations and round your answer to the nearest hundredth.

a. What is the average *monthly* rate at which **overall music sales** increased from 2005 to 2006?

b. What is the average *monthly* rate at which **total album sales** decreased from 2005 to 2006?

c. What is the average *monthly* rate at which **digital album sales** increased from 2005 to 2006?

3. Based on your results from parts b and c, does it appear that all the people who topped buying physical albums are now purchasing digital albums? Explain.

4. What is the average *daily* sales rate for the top selling album of 2006? What is the average daily sales rate for the *tenth* top selling album of 2006? (Round your answer to the nearest whole number.)

Student Activity

Match Up on Slope-Intercept Form

Match-up: In each box of the grid below, you will find either the equation of a line or a description of a line. For each, determine the slope **and** the y-intercept and match it with the appropriate letters. If the slope or the y-intercept is not found among the choices or cannot be determined from the information given, then choose E or N respectively.

Slope: **A** 2 **B** $\frac{1}{2}$ **C** -1 **D** 0 **E** None of these or cannot be determined

y**-Intercept:** **J** 3 **K** -2 **L** 0 **M** 1 **N** None of these or cannot be determined

$y=\frac{1}{2}x-3$	$y=-x+3$	$x+y=0$	The line is a vertical line passing through $(-2,3)$.
$6x-3y=6$	$y=1-3x$	The line passes through $(0,3)$ and $(2,1)$.	The line passes through $(0,-2)$ and $(2,-1)$.
The line is a vertical line passing through $(2,1)$.	The line is a horizontal line passing through $(-2,3)$.	The line has intercepts $(0,1)$ and $(-2,0)$.	The line is parallel to a horizontal line and passes through $(0,1)$.
The line is perpendicular to a line with a slope of -2 and passes through $(0,3)$.	$10x-5y=10$	$4y-2x=0$	The line is parallel to a line with a slope of 2 and passes through the origin.

Guided Learning Activity

Graphing with Slope-Intercept Form

Slope-Intercept Form: $y = mx + b$ where m is the slope and $(0, b)$ is the y-intercept.

To graph using slope-intercept form:

1. Graph a point (the y-intercept).
2. Use $m = \frac{\text{rise}}{\text{run}}$ to move from that point to locate another point on the line.

1. Graph: $y = \frac{1}{2}x + 3$

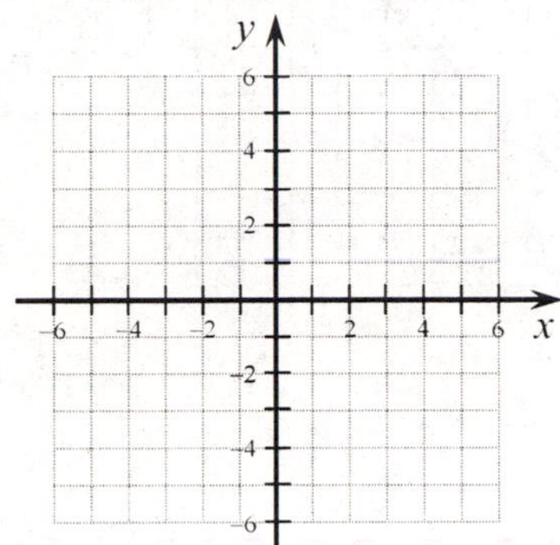

2. Graph: $y = 3x - 2$

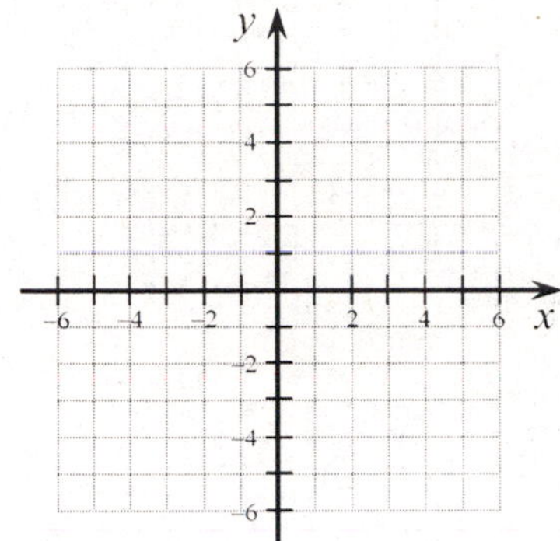

3. Graph: $y = -\frac{3}{2}x + 3$

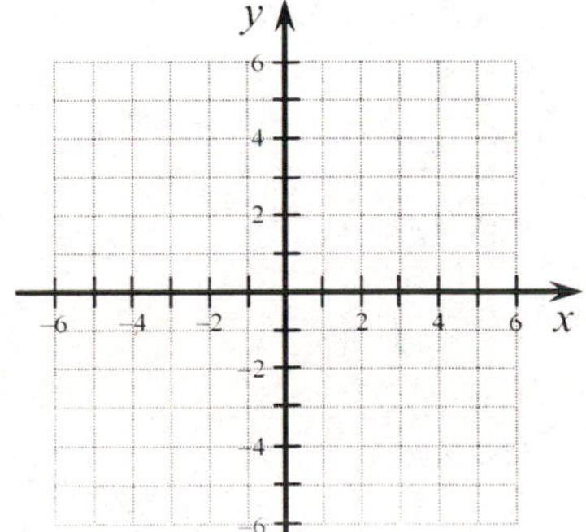

4. Graph: $y = -2x$

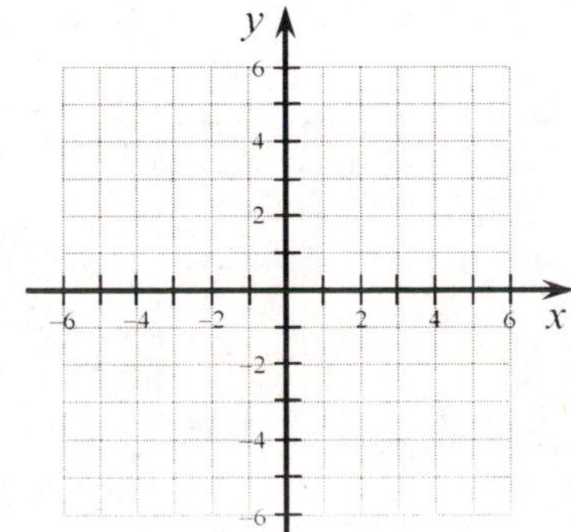

5. Graph: $3x - 4y = -4$

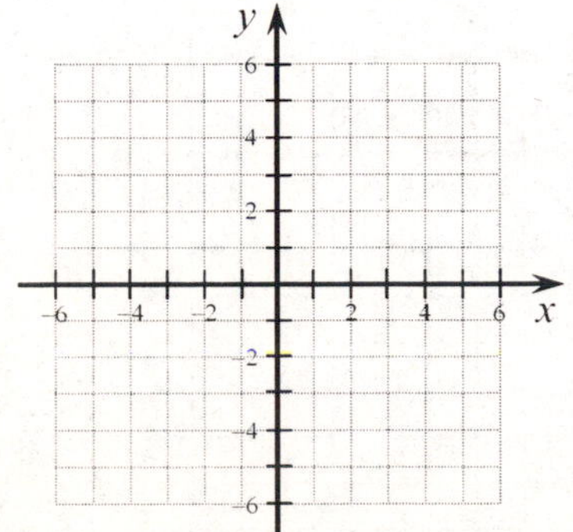

6. Graph: $y - x = 2$

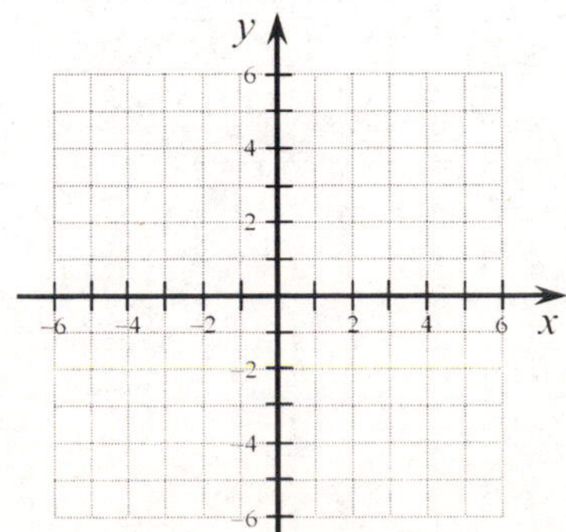

Student Activity
Evidence from the Graph

Directions: For each line that is graphed below, determine the equation of the line and write it in slope-intercept form.

1.

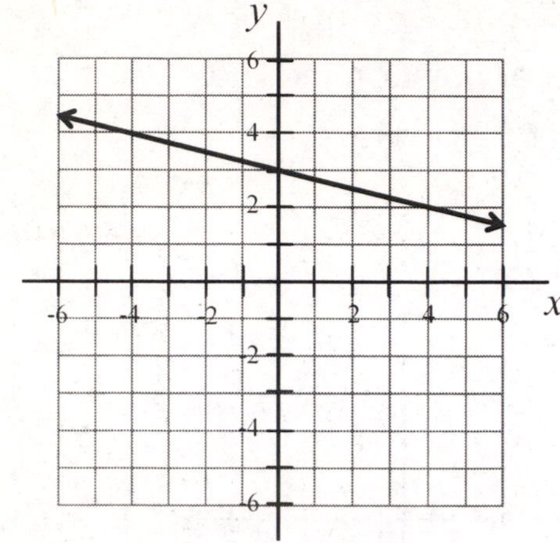

2.

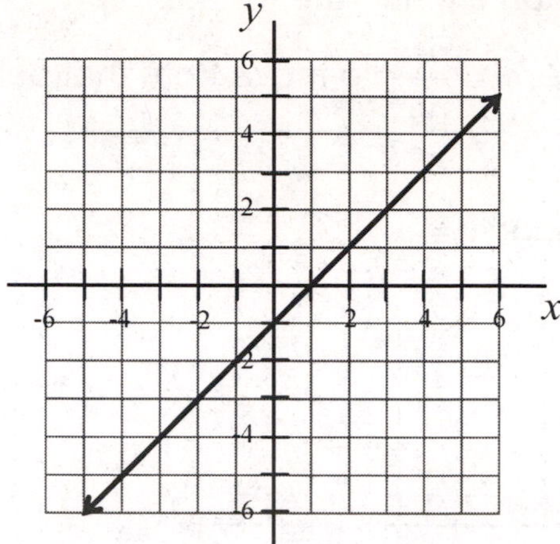

3.

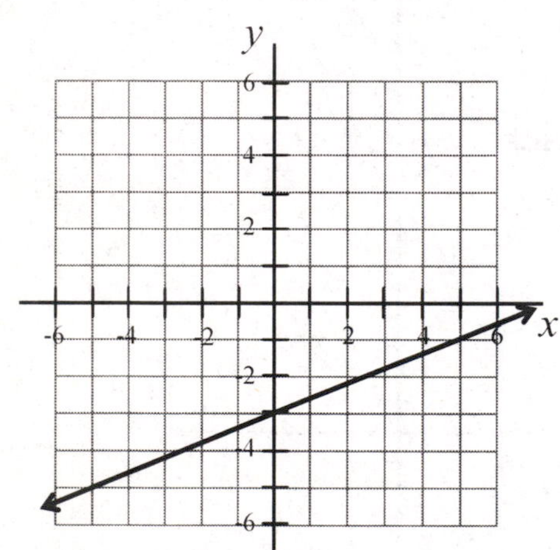

4.

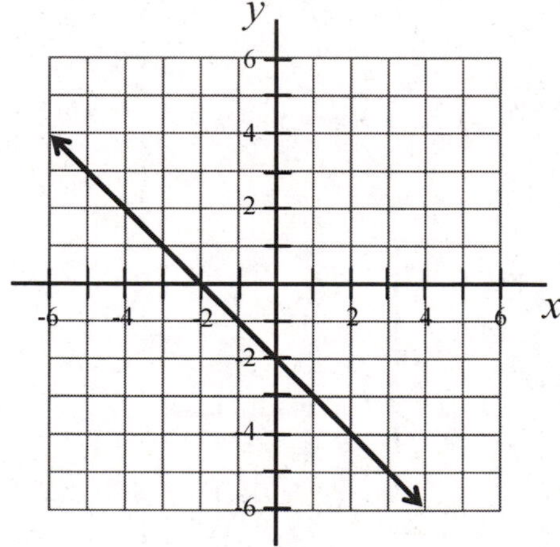

5.

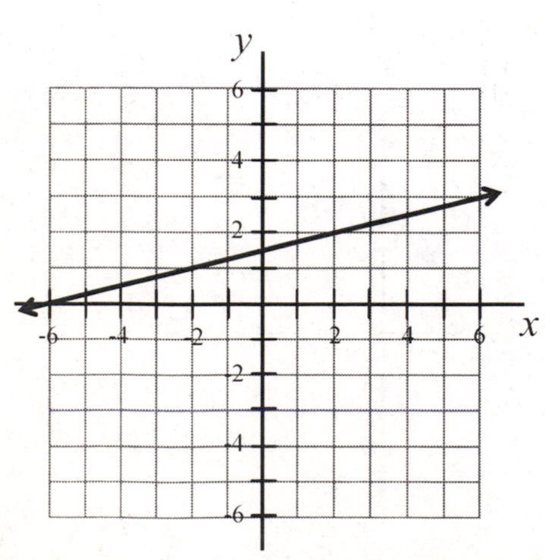

6.

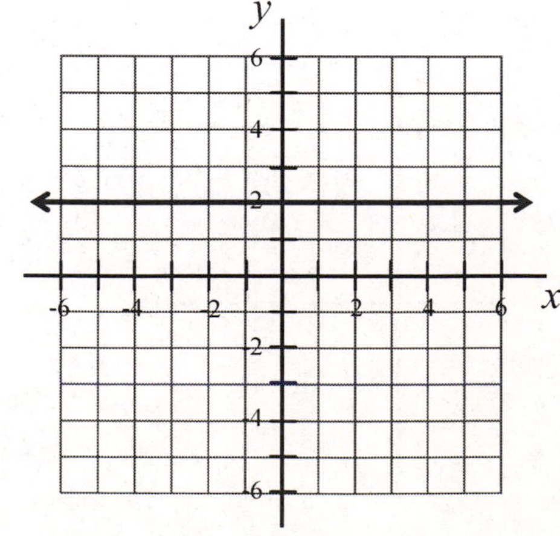

Student Activity

What's the Verdict, Parallel or Perpendicular?

Directions: Each of the "accused" pairs of lines is going to trial. Your task is to graph each pair of lines, and then assign each pair a verdict: parallel or perpendicular. If the pair of lines is neither parallel nor perpendicular, you may declare a mistrial.

1. $y=\frac{1}{2}x+3$; $y=\frac{1}{2}x+1$

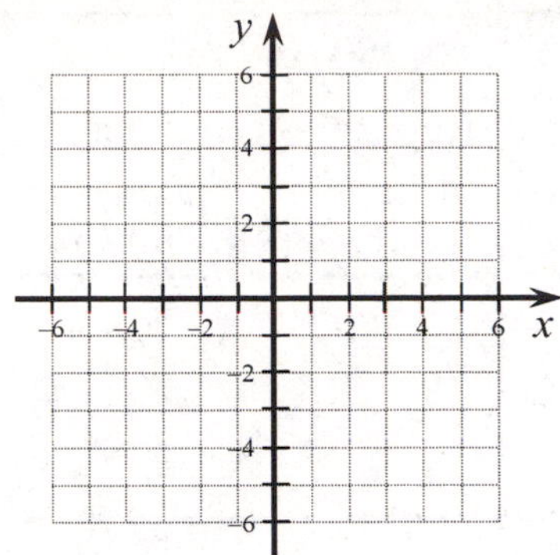

2. $y=\frac{1}{3}x-1$; $y=3x$

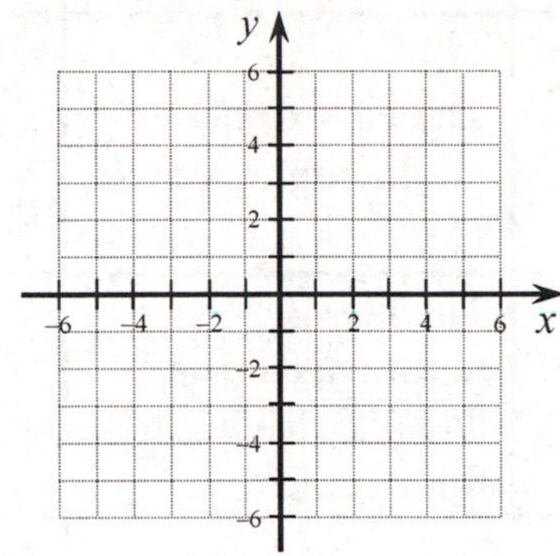

3. $y=\frac{1}{2}x+1$; $y=-\frac{1}{2}x+2$

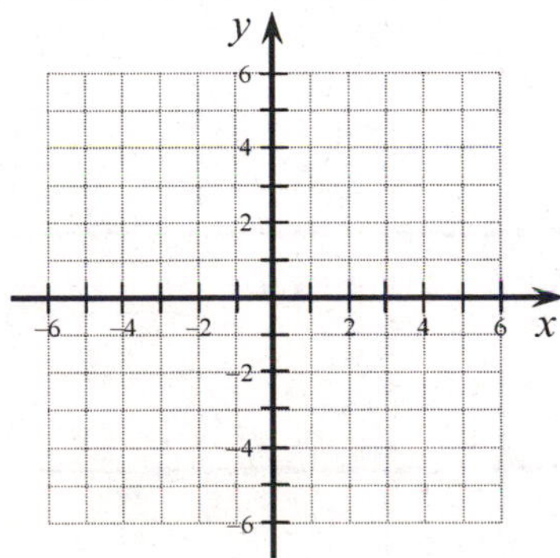

4. $y=2x-3$; $y=-\frac{1}{2}x+4$

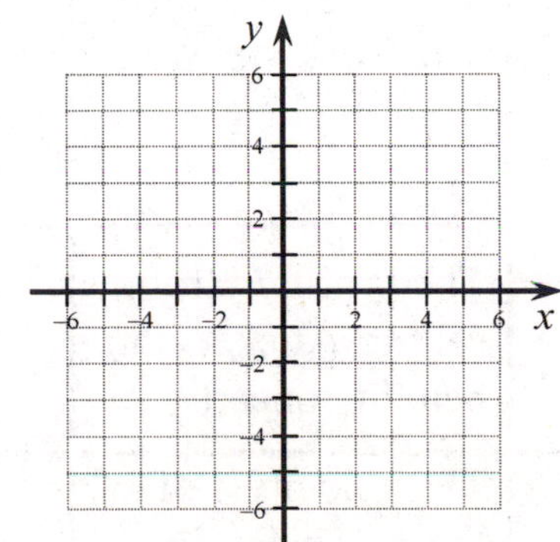

5. $y=3$; $y=-1$

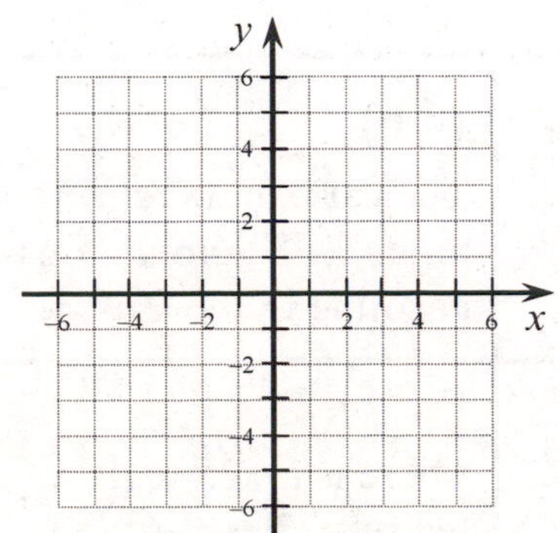

6. $y=-x+4$; $y=x-2$

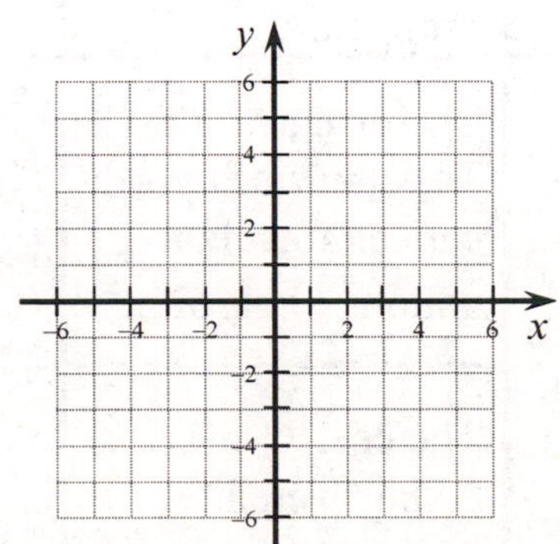

Student Activity

Modeling Data with Linear Equations

Directions: Fill in the missing information in the table below. The first one has been done for you.

Mathematical model	Description	Slope and meaning of the slope	y-intercept and meaning of the y-intercept
$C = 4.5x + 250$	The cost C of producing x items.	$m = 4.5$ It costs \$4.50 to produce each item.	$(0, 250)$ Regardless of the number of items produced, the fixed costs are \$250.
$C = 0.10n + 25$	The cost C of renting a car for a day and driving it for n miles.		
$R = 8.5t$	The revenue R from selling t movie tickets.		
$P = 10h$	The weekly pay P from working h hours.		
$P = 25t - 500$	The profit P from selling t tickets to a benefit concert.		
$v = 32t$	The velocity v of a falling object (in feet per second) t seconds after it is dropped.		
	The distance d a Toyota Prius hybrid car can travel on g gallons of gasoline.	$m = 55$ A Toyota Prius gets 55 miles per gallon of gasoline. (Source: www.toyota.com/prius/specs.html)	$(0, 0)$ The car can travel zero miles with zero gallons of gasoline in the tank.
	The cost C of taking n credit hours at a community college.	$m = 66$ The cost is \$66 per credit hour.	$(0, 25)$ There is a \$25 registration fee regardless of the number of credit hours taken.

Guided Learning Activity

Writing Equations that Model Data

Example 1: In the first year of a research study, the population of flamingoes at a zoo was 25. In the fifth year of the study the population of flamingoes was 45. Find a linear equation to model the population of flamingoes, *P*, in the year *t*.

Declare the variables: t = year of study, P = number of flamingoes

In terms of these variables, what form do ordered pairs have? (t, P)

Rewrite the point-slope form using the new variables: $P - P_1 = m(t - t_1)$

What information are you given in the problem?

Ordered pairs(s): $(1, 25)$ and $(5, 45)$ **Slope:** (not given)

Strategy to find an equation of the line: Find the slope and then use a point and the slope in the point-slope form.

Finish:

Example 2: A factory produces ethanol at a rate of 4000 liters every 5 hours. After the factory had run for 8 hours, the factory manager noted that there were 10,000 liters of ethanol in the storage tank. Write a linear equation relating the time *t* in hours since the factory began production and the number of liters *L* of ethanol in the storage tank.

Declare the variables: $t =$ ____________________

$L =$ ____________________

In terms of these variables, what form do ordered pairs have? (,)

Rewrite the point-slope form using the new variables: ____________________

What information are you given in the problem?

Ordered pair(s): ____________________ **Slope:** ____________

Strategy to find an equation of the line: ____________________

____

Finish:

Student Activity

Tic-Tac-Toe on Inequality Solutions

Directions for Game #1: If the ordered pair in the square **IS** a solution of the inequality, then circle the ordered pair (thus putting an **O** on the square). If it **IS NOT** a solution, then put an **X** over the ordered pair in the square.

$x+y<5$ $(2,3)$	$y<x+6$ $(9,3)$	$y>0$ $(2,-1)$
$2x-3y\le 0$ $\left(\frac{1}{2},\frac{1}{3}\right)$	$x>y+4$ $(4,0)$	$x\le 5$ $(4,9)$
$3x+3\ge 4y+4$ $(2,1)$	$3-x<-4$ $(8,1)$	$y-2x\le 3x-y$ $(1,5)$

Directions for Game #2: If the ordered pair in the square **IS** a solution of the graphed inequality, then circle the ordered pair (thus putting an **O** on the square). If it **IS NOT** a solution, then put an **X** over the ordered pair in the square.

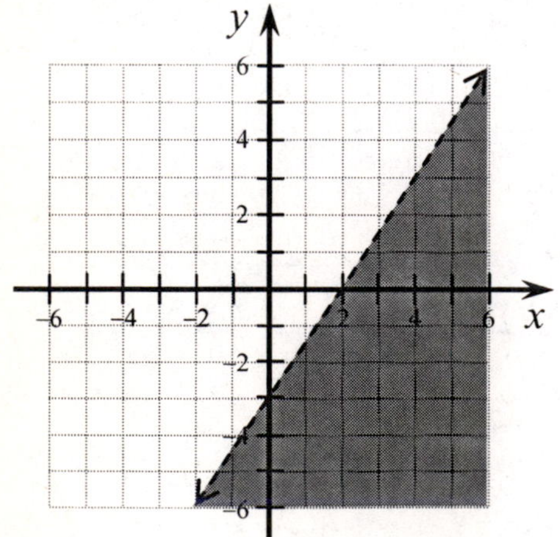

$(4,1)$	$(4,3)$	$(-2,-4)$
$(1,-3)$	$(-0.5,-6)$	$(0,0)$
$(-1,-4)$	$(-3,-6)$	$(7,-7)$

Guided Learning Activity

Shady Inequalities

Graphing an Inequality in two variables:

1. Draw the boundary line (remember to draw it either dashed or solid).
2. Determine which side of the boundary line to shade.
3. Shade the half-plane that shows the solution of the inequality.

1. Graph: $y \le \frac{1}{2}x - 1$

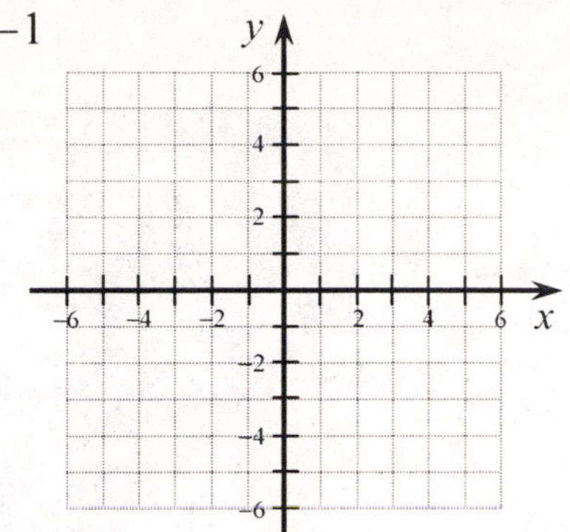

2. Graph: $x + y > 3$

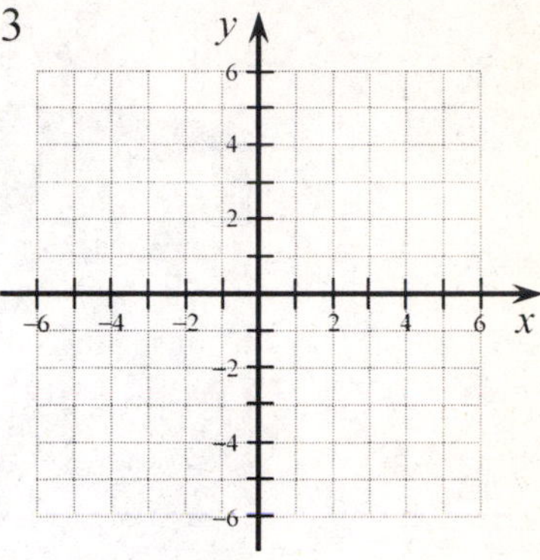

3. Graph: $2y + 3x < 6$

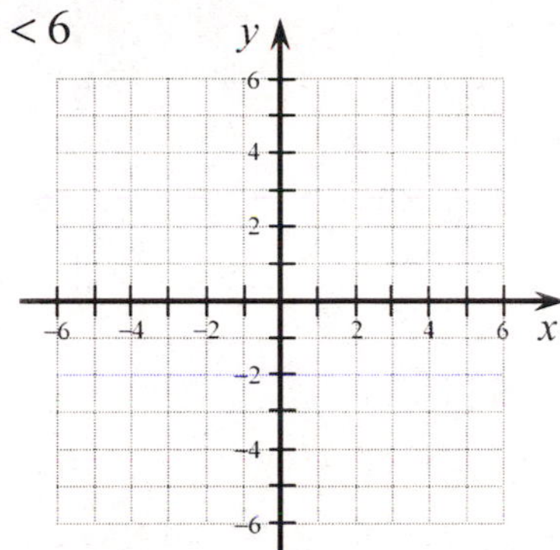

4. Graph: $y \ge 4$

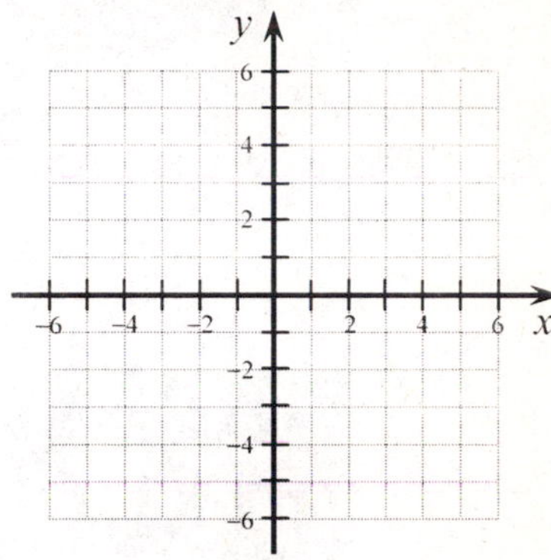

5. Graph: $x < -1$

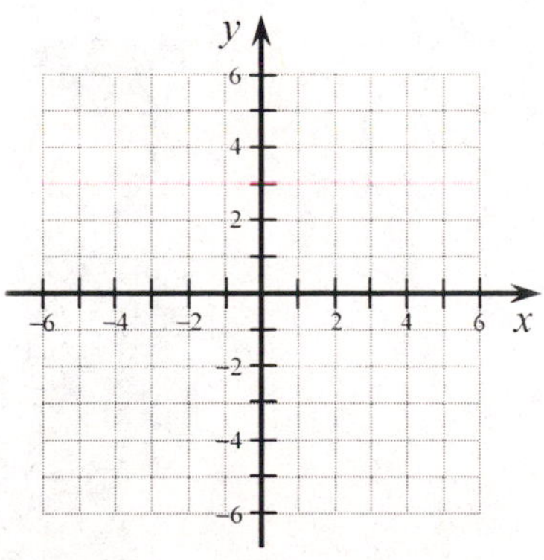

6. Graph: $x - 2y > 4$

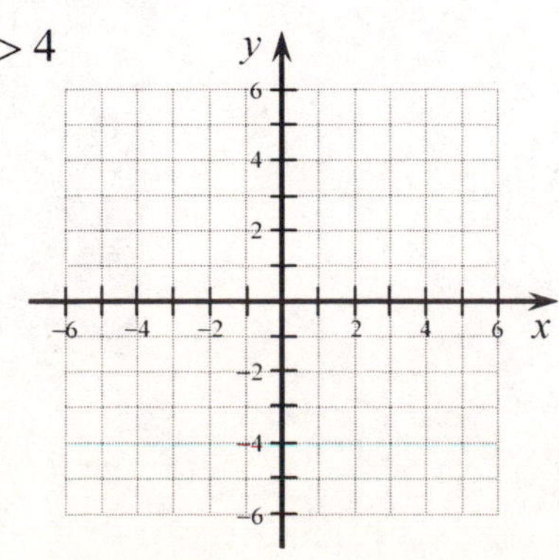

Student Activity

Following the Clues Back to the Inequality

Directions: In each “crime-scene” below, you are shown the graph of an inequality. Use your mathematical powers of reasoning (and detective skills) to determine what the inequality must have been to result in this graph.

1.

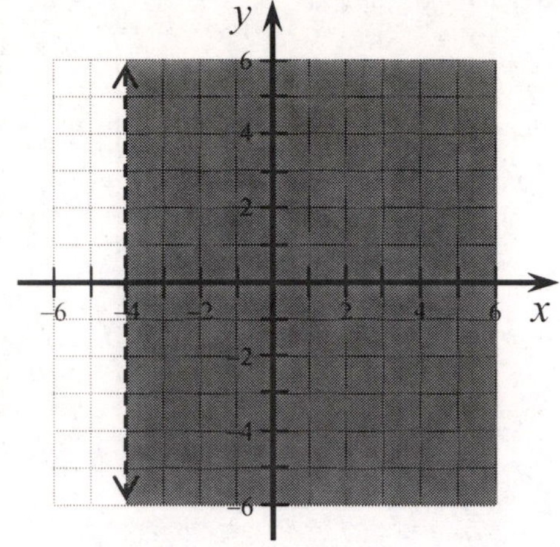

3.

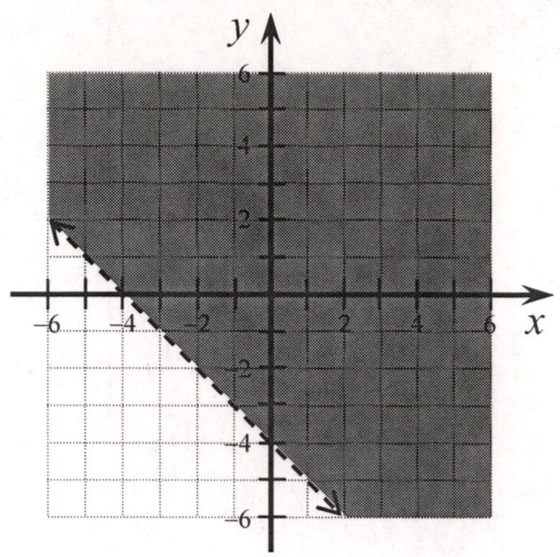

2.

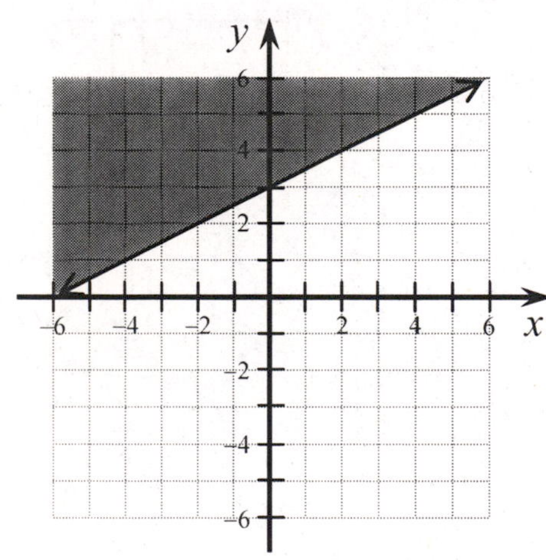

Assess Your Understanding

Lines and More

For each of the following, describe the strategies or key steps that will help you **start** the problem. You do **not** have to complete the problems.

		What will help you to start this problem?
1.	Find the intercepts of the graph of $2x-8y=8$.	
2.	Find the slope of a line perpendicular to the one that passes through $(0,0)$ and $(-3,4)$.	
3.	Find the equation of a vertical line that passes through $(3,7)$.	
4.	Graph the solution of $y<2x-3$.	
5.	Find the slope between $(-2,3)$ and $(1,-5)$.	
6.	Graph $2x+4y=8$ by creating a table of values.	
7.	Graph $3x-y=6$ by using its intercepts.	
8.	Is $(6,1)$ a solution of $x-2y\le 4$?	
9.	Graph the line that has slope 1 and passes through $(-2,-3)$.	
10.	Find the intercept(s) of the horizontal line $y=2$.	

Metacognitive Skills

Lines and More

Metacognitive skills refer to the ability to judge how well you have learned something and to effectively direct your own learning and studying. This is a self-evaluation tool designed to help you focus your studying and to improve your metacognitive skills with regards to this math class.

Fill the 1st column out **before** you begin studying. Fill the 2nd column out after you study for your test.

Go back to this assessment after your test and circle any of the ratings that you would change – this identifies the "disconnects" between what you **thought** you knew well and what you **actually** knew well.

Use the scale below to assign a number to each topic.

5 *I am confident I can do any problems in this category correctly.*
4 *I am confident I can do most of the problems in this category correctly.*
3 *I understand how to do the problems in this category, but I still make a lot of mistakes.*
2 *I feel unsure about how to do these problems.*
1 *I know I don't understand how to do these problems.*

Topic or Skill	**Before Studying**	**After Studying**
Plotting points or identifying points on a graph using an ordered pair.		
Identifying the quadrant or axis that a point is located in/on.		
Creating a line graph given a set of data.		
Deciding whether an ordered pair is a solution of a linear equation.		
Deciding whether an ordered pair is a solution of a linear inequality.		
Constructing a table of solutions for a linear equation.		
Graphing a linear equation by plotting points.		
Finding the x- and y-intercepts of a linear equation (if they exist).		
Graphing a linear equation by using the x- and y-intercepts.		
Interpreting a graph or answering questions based on the graph of a linear equation.		
Understanding the equations and graphs of vertical and horizontal lines.		
Finding the slope of a line by looking at its graph.		
Finding the slope of a line if you are given two points.		
Understanding the slopes of horizontal and vertical lines.		
Calculating a rate of change (including the units).		
Understanding the special properties of slope for parallel and perpendicular lines.		
Finding the slope and y-intercept of a line given in slope-intercept form.		
Rearranging a linear equation so that it is written in slope-intercept form.		
Using the slope and any point on the line to graph a linear equation.		

Continued on next page.

Use the scale below to assign a number to each topic.

5 *I am confident I can do any problems in this category correctly.*
4 *I am confident I can do most of the problems in this category correctly.*
3 *I understand how to do the problems in this category, but I still make a lot of mistakes.*
2 *I feel unsure about how to do these problems.*
1 *I know I don't understand how to do these problems.*

Topic or Skill	**Before Studying**	**After Studying**
Writing an equation of a line if you are given the slope and *y*-intercept.		
Using the slope and *y*-intercept to graph a linear equation.		
Using the point-slope form to write an equation of a line.		
Writing an equation for a line if you are given two points on the line.		
Writing an equation of a horizontal or a vertical line.		
Knowing the formulas for slope, slope-intercept form, and point-slope form.		
Applying the formulas for linear equations to write an equation to model data.		
Solving application problems that involve linear equations.		
Graphing a linear inequality.		
Understanding whether the boundary line in a linear inequality is dashed or solid and which side of the boundary line to shade.		
Solving for *y* in a linear inequality (including the special rule that applies to dividing or multiplying by a negative number when solving inequalities)		
Solving application problems that involve inequalities.		

FUNC: Functions and More

Student Activity
Everyday Functions

Case #1: During an eight hour shift, hospital patients in the ICU are assigned to a nurse at a ratio of 2:1. If the ICU has eight patients, the nursing assignments would be as follows:

Patient	Nurse
Patient 1	Nurse A
Patient 2	
Patient 3	Nurse B
Patient 4	
Patient 5	Nurse C
Patient 6	
Patient 7	Nurse D
Patient 8	

1. Consider the set of patients to be the domain and the set of nurses to be the range. The situation described here **is** a function. Why?

2. Describe the practical implications if this relation were **not** a function.

3. If the domain was the set of nurses and the range was the set of patients, would the relation still be a function? Why or why not?

Case #2: Suppose the exchange rate for the US Dollar to Mexican Peso is $1=10.880 pesos. The table below shows the conversion of several dollar amounts to pesos.

US Dollars	Mexican Pesos
5	54.4
24	261.12
78	848.64
110	1,196.8
500	5,440
1250	1,360

4. Let the domain be US Dollars and the range be Mexican Pesos. This relation **is** a function. Why?

5. Describe the practical implications if this relation were **not** a function.

6. If the domain and range were reversed, would it still be a function? Why or why not?

Case #3: These ordered pairs represent the identification numbers of commercial fishing boats and the profit (or loss) per boat in thousands of dollars.

$(1,37)$ $(2,43)$ $(5,37)$ $(7,13)$ $(8,-5)$ $(10,37)$ $(12,-117)$

7. Let the domain be ______________________________ and the range be ______________________.

8. Is this a function? Why or why not?

9. Describe the practical implications if this relation were **not** a function.

10. If the domain and range were reversed, would it still be a function? Why or why not?

Student Activity

Skeleton Crew

Skeleton Method: To evaluate a function that uses function notation, it is helpful to "see" the function without the variables. To make the "skeleton" of the problem, replace the variables with an empty parentheses "skeleton".

For example, $f(x)=x^2-5x+1$ becomes $f(\ )=(\ )^2-5(\ )+1$.

Now if you have to evaluate the expression for $x=-2$, write the -2 inside each parentheses skeleton and evaluate.

So $f(\ )=(\ )^2-5(\ )+1$ becomes $f(-2)=(-2)^2-5(-2)+1$.

Directions: For each row of the table, a function is described in words, in function notation, or as a parentheses skeleton. Complete the table and then evaluate the function for the given value. The first one has been done for you.

Word Description	$f(x)$	Skeleton	$f(-2)$
Multiply by 2 and then subtract 3	$f(x)=2x-3$	$f(\ )=2(\ )-3$	$f(-2)=2(-2)-3$ $=-4-3=-7$
Subtract 3 and then multiply by 2			
	$f(x)=x^3-5$		
Add 3, then square the result			
		$f(\ )=\lvert(\ )-6\rvert$	
			$f(-2)=\dfrac{(-2)+3}{4}=\dfrac{1}{4}$
	$f(x)=9-x$		
Take the reciprocal of it (assume it is not zero)			

Guided Learning Activity

Investigating Functions

Definition of a function: a relation in which each **first component** corresponds to exactly one **second component**.

Vertical Line Test: If a vertical line intersects a graph in more than one point, the graph is **not** the graph of a function.

1. Graph: $f(x) = |x|$

x	$f(x)$

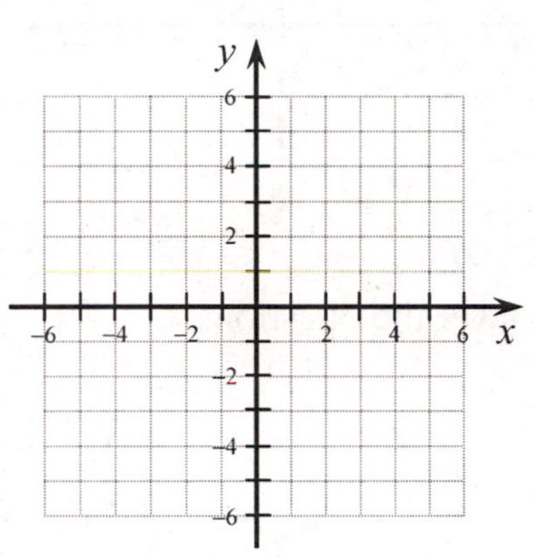

Is it a function? _____

2. Graph: $f(x) = x - 3$

x	$f(x)$

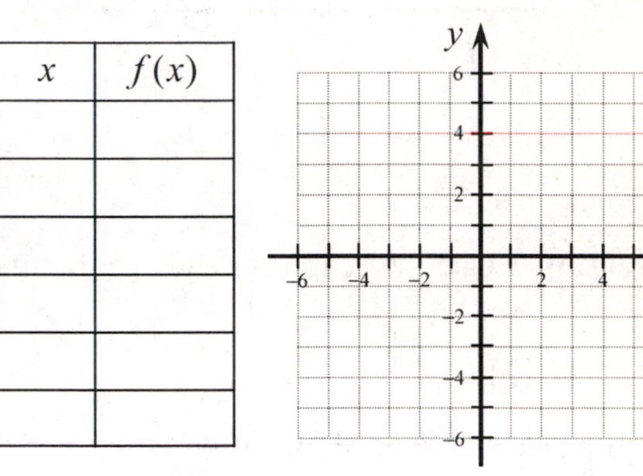

Is it a function? _____

3. Graph: $x = |y|$

x	y
	−2
	−1
	0
	1
	2

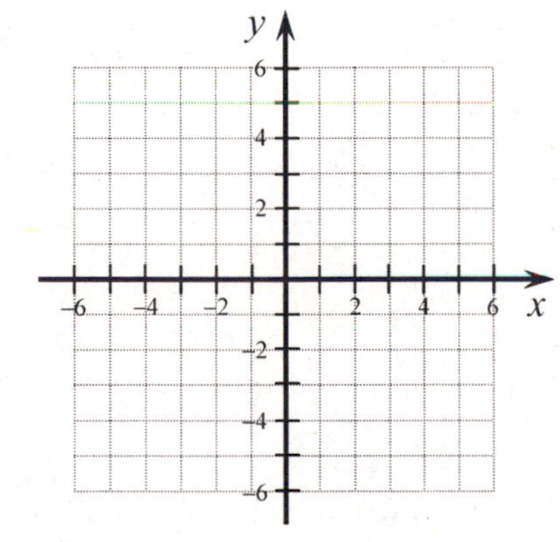

Is it a function? _____

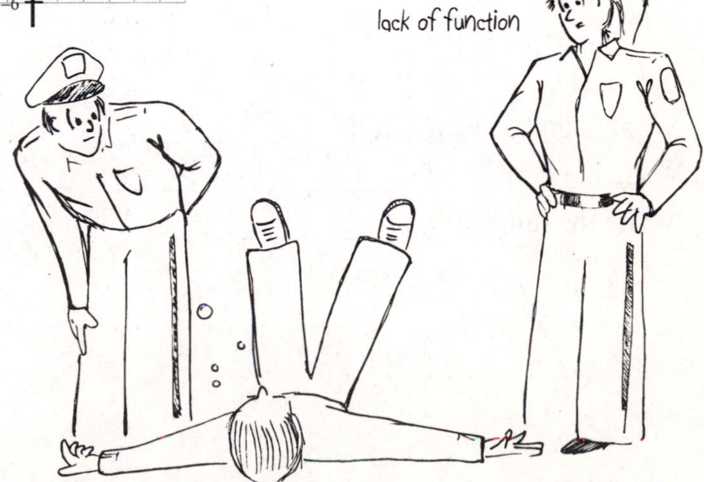

Matt fails the Vertical Line Test

Student Activity

Gleaning Information from Graphs

In general, the value of $f(a)$ is given by the y-coordinate of a point on the graph of f with x-coordinate a.

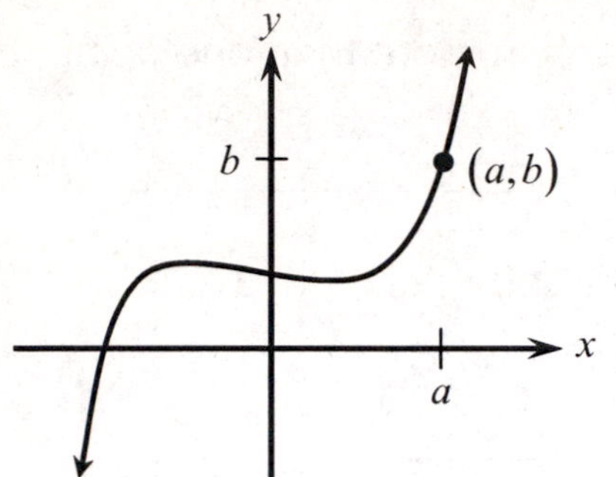

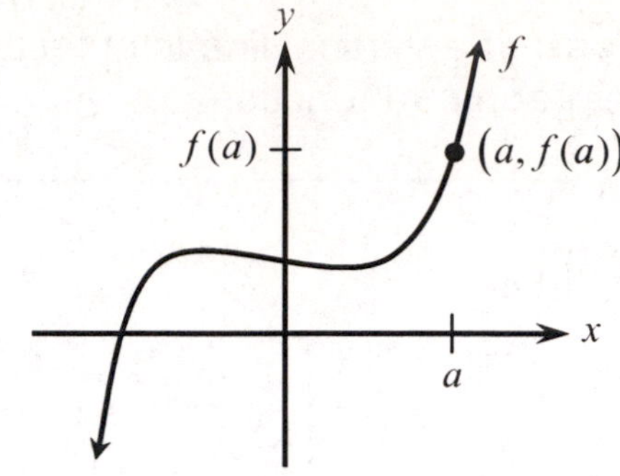

Directions: Use the graph for each problem to answer the questions.

1. a. $f(-2) =$ ____

b. Find the value(s) of x where $f(x) = 1$. ______

c. $f(5) =$ ____

d. $f($___$) = -3$

e. Find the value(s) of x where $f(x) = 2$. ______

f. Write the domain of f : ____________________

g. Write the range of f : ____________________

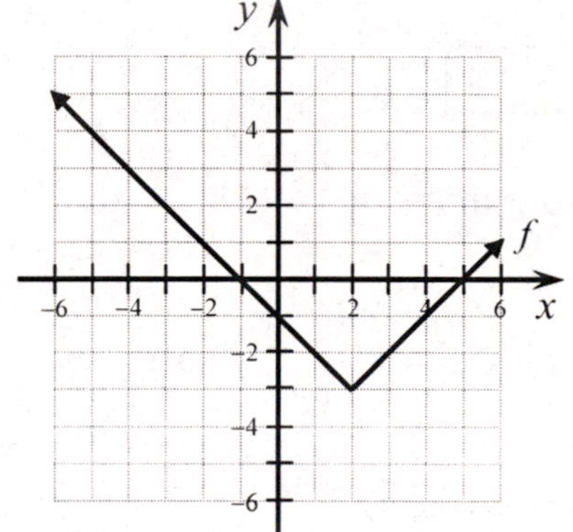

2. a. $g(2) =$ ____

b. Find the value(s) of x where $g(x) = 4$. ______

c. $g(0) =$ ____

d. $g($___$) = 2$

e. What is the slope, m, of this line? _____

f. Write the domain of g : ____________________

g. Write the range of g : ____________________

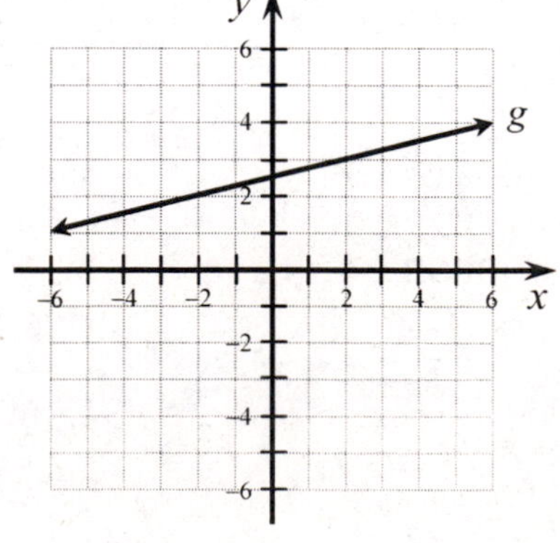

Relationship counseling for function properties.

3. **a.** $h(-1) =$ ____

b. Find the value(s) of x where $h(x) = 0$. ______

c. $h(-4) =$ ____

d. $h($___$) = 2$

e. Find the value(s) of x where $h(x) = -2$. ______

f. Write the domain of h : ____________________

g. Write the range of h : ____________________

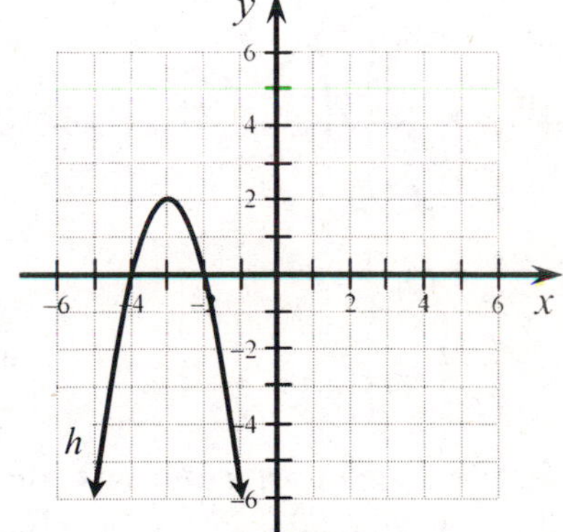

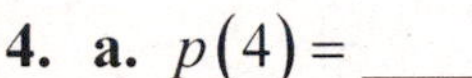

4. **a.** $p(4) =$ ____

b. Find the value(s) of x where $p(x) = 1$. ________

c. $p(-2) =$ ____

d. Find the value(s) of x where $p(x) = 4$. _______

e. What is the slope of p where $x > 2$? _____

f. Write the domain of p : ____________________

g. Write the range of p : ____________________

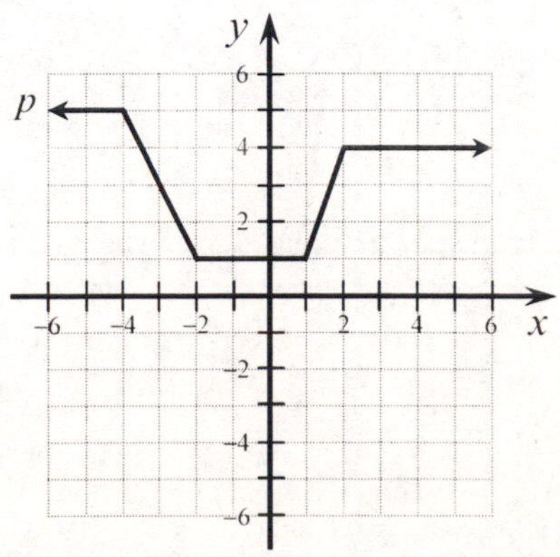

Student Activity
Shapeshifters

Directions: For problems 1-4, a function is drawn using a dashed line. Follow the directions for each problem to draw the graph that is described using a **solid line**.

1. Shift the graph of f down 4 units.

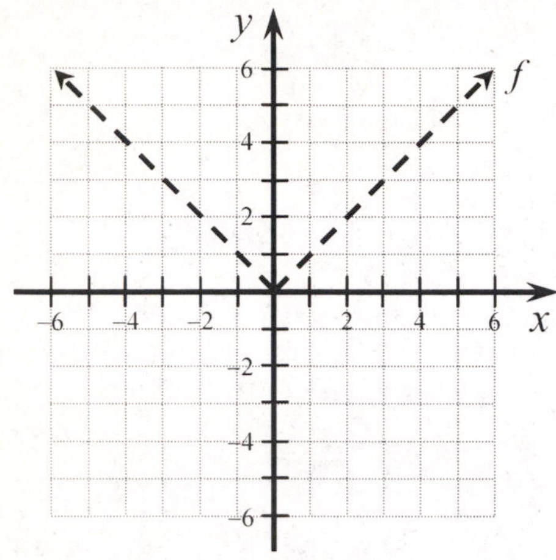

2. Reflect the graph of g over the x-axis.

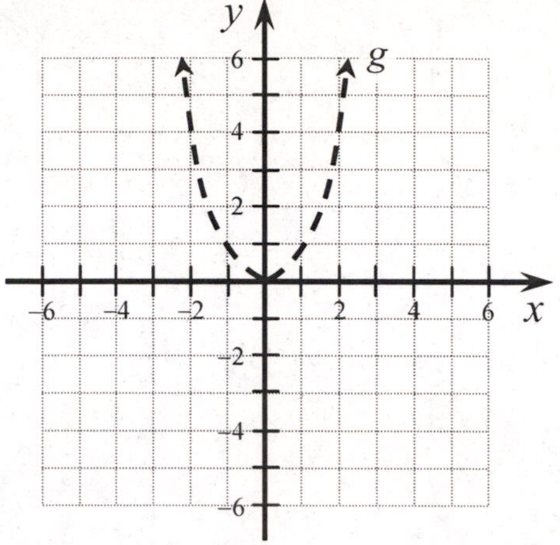

3. Shift the graph of h to the right 2 units.

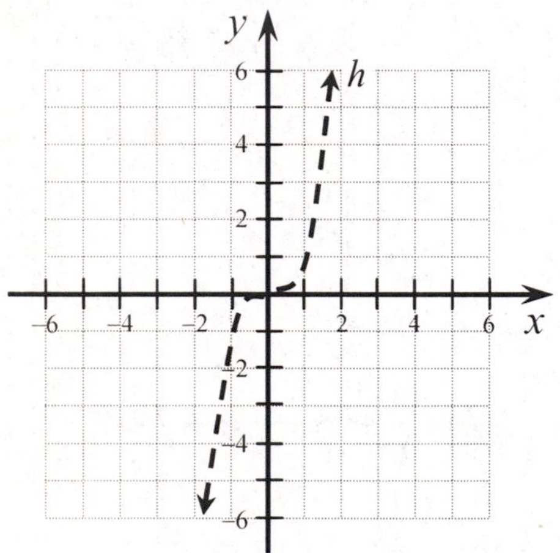

4. Shift the graph of p up 3 units and to the left 5 units.

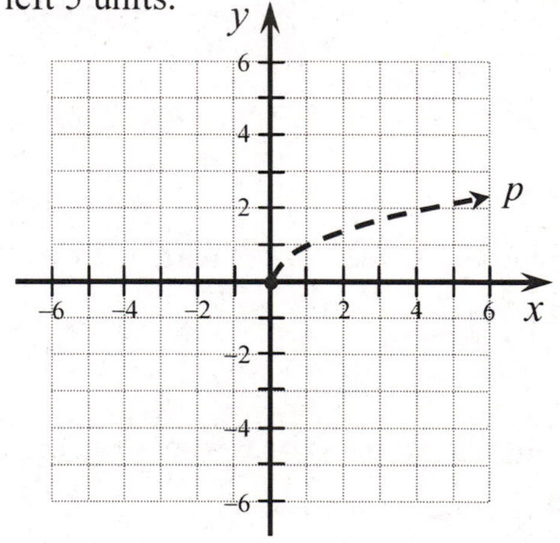

5. Describe the translations and/or reflections of f that must occur to get the graph of g.

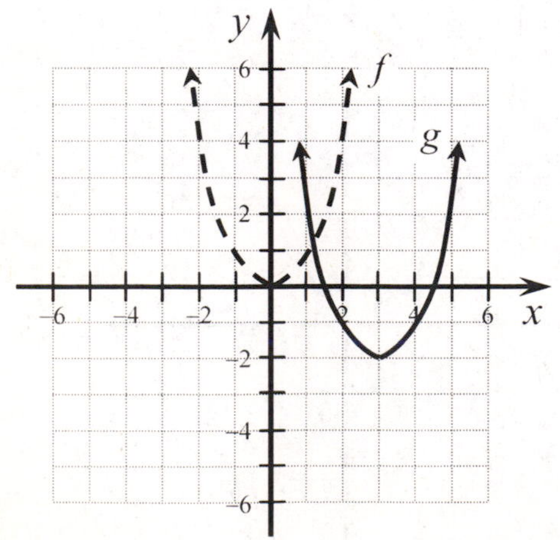

6. Describe the translations and/or reflections of f that must occur to get the graph of g.

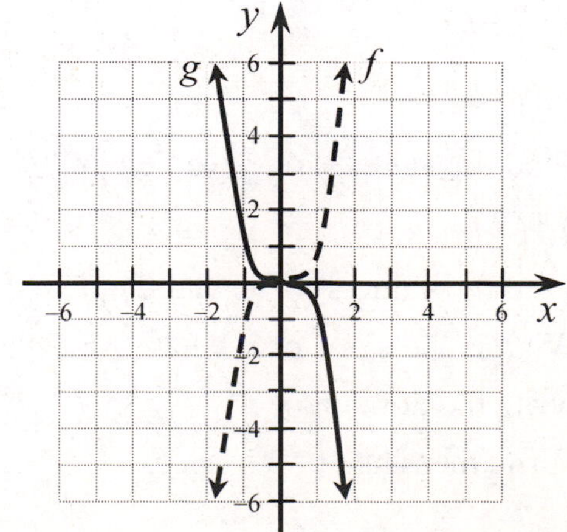

Student Activity

Reflecting on Graph Translations

1. In the graphs below, examine the functions f and g in each picture. How is the change in the mathematical function related to the change between the graph of f and the graph of g?

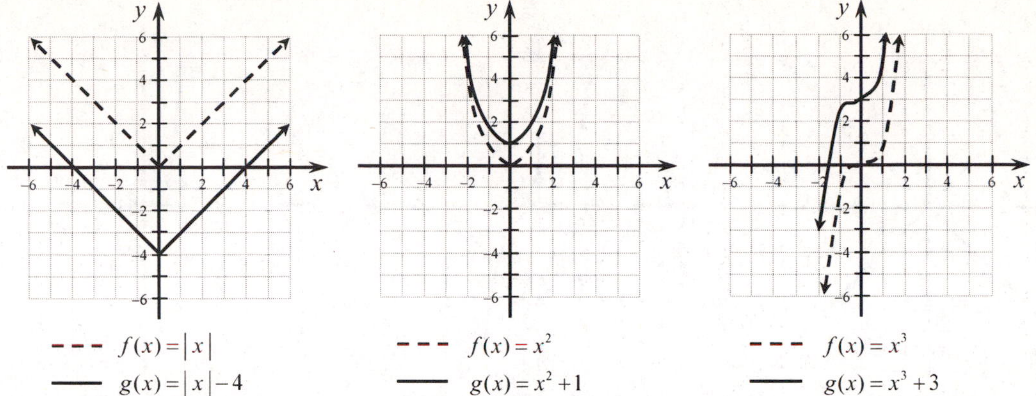

If f is a function and k is a positive number then

a. The graph of $y=f(x)+k$ is identical to the graph of $y=f(x)$ except that it is translated k units _____________.

b. The graph of $y=f(x)-k$ is identical to the graph of $y=f(x)$ except that it is translated k units _____________.

2. In the graphs below, examine the functions f and g in each picture. How is the change in the mathematical function related to the change between the graph of f and the graph of g?

If f is a function and h is a positive number then

a. The graph of $y=f(x-h)$ is identical to the graph of $y=f(x)$ except that it is translated h units _____________.

b. The graph of $y=f(x+h)$ is identical to the graph of $y=f(x)$ except that it is translated h units _____________.

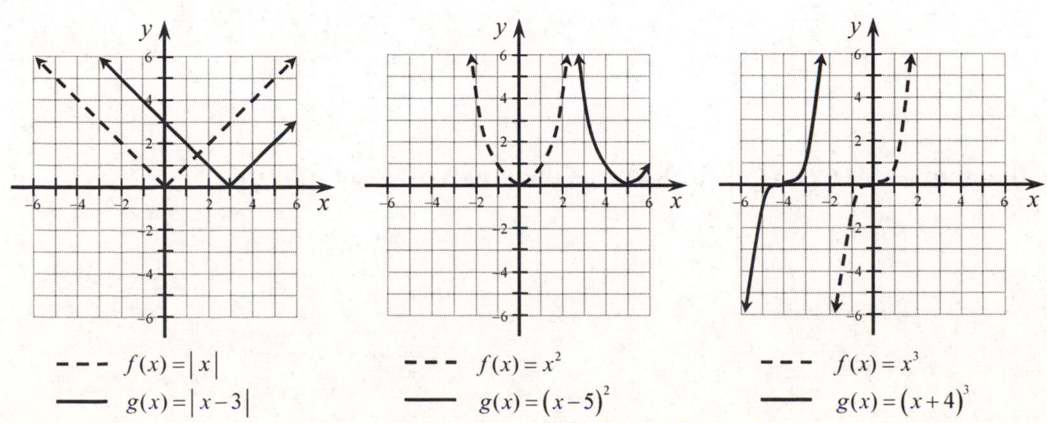

3. In the graphs below, examine the functions f and g in each picture.

A vertical reflection is a reflection about the x-axis. These graphs show vertical reflections:

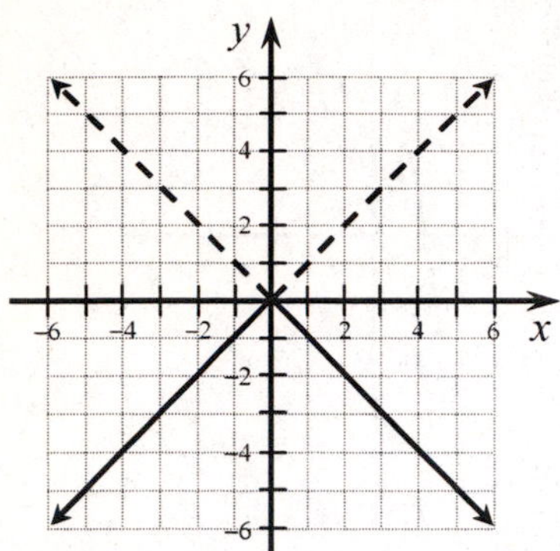
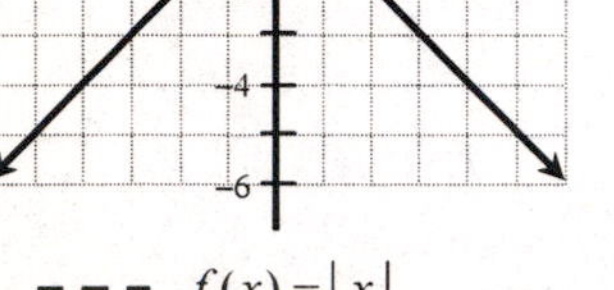

$f(x) = |x|$ (dashed)
$g(x) = -|x|$ (solid)

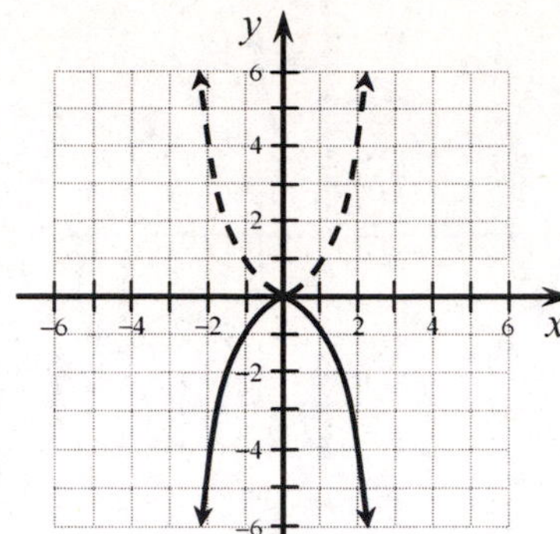
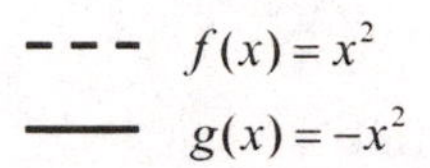

$f(x) = x^2$ (dashed)
$g(x) = -x^2$ (solid)

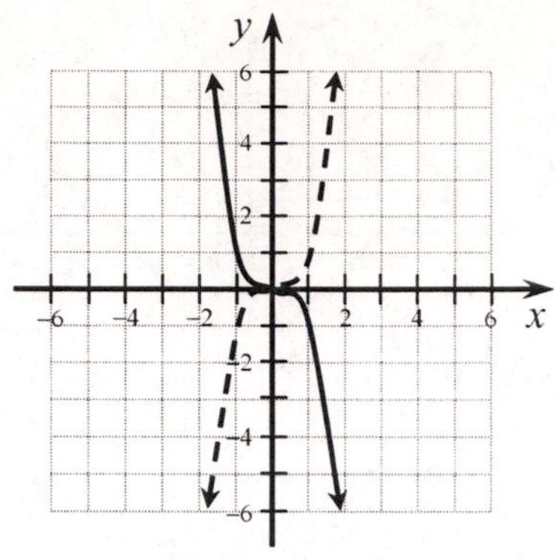

$f(x) = x^3$ (dashed)
$g(x) = -x^3$ (solid)

A horizontal reflection is about the y-axis. These graphs show horizontal reflections:

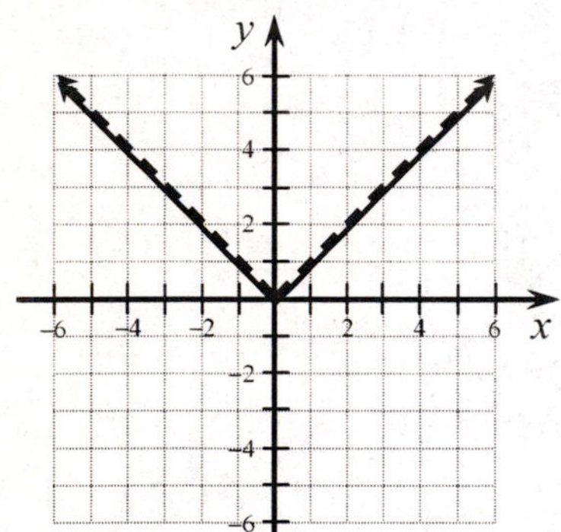
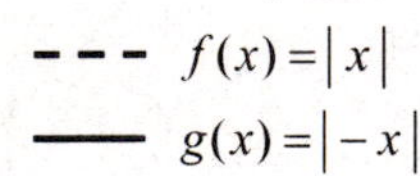

$f(x) = |x|$ (dashed)
$g(x) = |-x|$ (solid)

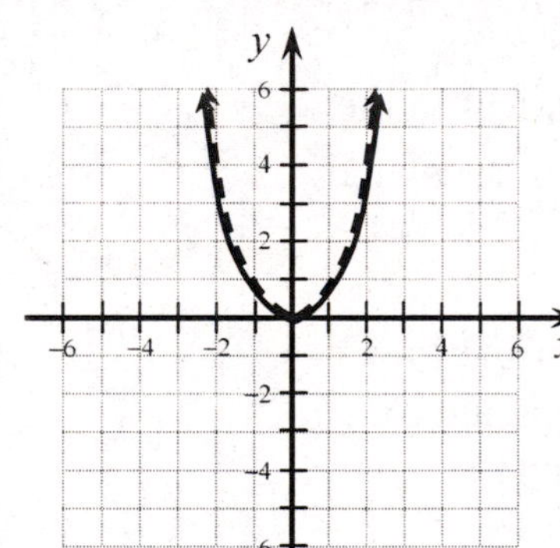
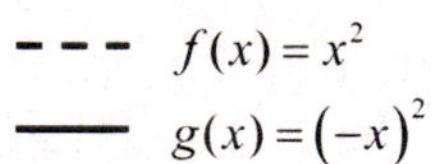

$f(x) = x^2$ (dashed)
$g(x) = (-x)^2$ (solid)

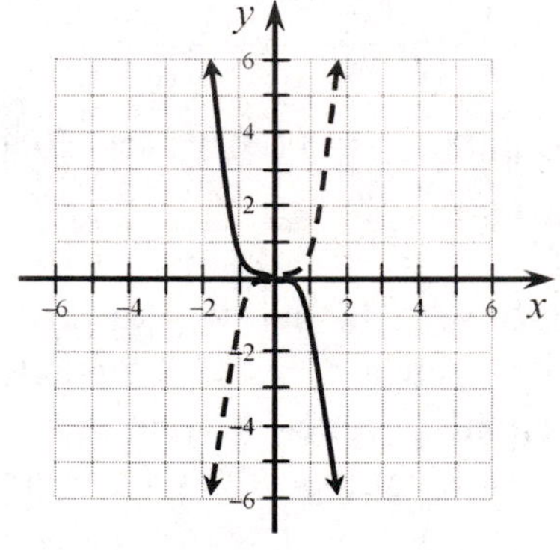

$f(x) = x^3$ (dashed)
$g(x) = (-x)^3$ (solid)

a. The graph of $y = -f(x)$ is the graph of $y = f(x)$ reflected ________________.

b. The graph of $y = f(-x)$ is the graph of $y = f(x)$ reflected ________________.

c. Why is the graph of $g(x) = -x^3$ the same as the graph of $g(x) = (-x)^3$?

d. Why is the graph of $f(x) = |x|$ the same as the graph of $g(x) = |-x|$?

4. Let's recap. Let k and h represent positive numbers.

   a. The graph of $y = f(x+h)$ is a translation of h units ______________ from $f(x)$.

   b. The graph of $y = f(x-h)$ is a translation of h units ______________ from $f(x)$.

   c. The graph of $y = f(-x)$ is a ___________ reflection of $f(x)$.

   d. It seems that when x is replaced by another expression involving x, the general effect on the graph is _____________ (*horizontal or vertical?*).

   e. The graph of $y = f(x)+k$ is a translation of h units ______________ from $f(x)$.

   f. The graph of $y = f(x)-k$ is a translation of h units ______________ from $f(x)$.

   g. The graph of $y = -f(x)$ is a ___________ reflection of $f(x)$.

   h. It seems that when something is added to, subtracted from, or multiplied by $f(x)$, the general effect on the graph is _____________ (*horizontal or vertical?*).

5. a. $g(x) = (x-1)^2$ is the graph of $f(x) =$ ____ shifted ___ units ___________.

   b. $g(x) = |x| + 6$ is the graph of $f(x) =$ ____ shifted ___ units ___________.

   c. $g(x) = (x+2)^3$ is the graph of $f(x) =$ ____ shifted ___ units ___________.

   d. $g(x) = (x+5)^2 - 2$ is the graph of $f(x) =$ ____ shifted 5 units ____________ and 2 units ______________.

   e. $g(x) = -|x|$ is the graph of $f(x) =$ ____ reflected about the _____________.

   f. $g(x) = \sqrt{x} + 3$ is the graph of $f(x) = \sqrt{x}$ shifted ____ units ____________.

   g. $g(x) = \sqrt{-x}$ is the graph of $f(x) = \sqrt{x}$ reflected about the _____________.

6. Suppose we define a new function called the "box" function, mathematically, it is written like this: $f(x) = \boxed{x}$. Based on the graphing rules we just developed, describe how each of the following functions would be graphed, in comparison with the graph of *f*. Think carefully about whether x has been replaced with another expression or whether the $f(x)$ portion of the graph has remained intact.

   a. $g(x) = \boxed{x+2}$ ____

   b. $g(x) = \boxed{x} - 4$ ____

   c. $g(x) = 2 + \boxed{x}$ ____

   d. $g(x) = -\boxed{x}$ ____

   e. $g(x) = \boxed{x-3}$ ____

   f. $g(x) = \boxed{-x}$ ____

   g. $g(x) = \boxed{x+3} - 1$ ____

Student Activity

Algebra of Domains

When we create a new function using the algebra of functions *f* and *g*, the domain of the new function is the set of real numbers that are in the domain of both *f* **and** *g* (as in, the intersection of the two domains). For a quotient *f* / *g*, there is the additional restriction that the new denominator, *g*, not be equal to zero.

Directions: Begin by writing the domain for the functions *f*, *g*, and *h*. Then complete the table below.

$f(x) = 2x - 1$ $\quad$ $g(x) = \sqrt{x+4}$ $\quad$ $h(x) = x + 5$

Domain: _______ $\quad$ Domain: ________ $\quad$ Domain: ________

	Find the new function and simplify it.	What is the domain of new function?
1.	$(f+g)(x)$	
2.	$(f-h)(x)$	
3.	$(f/h)(x)$	
4.	$(f \cdot h)(x)$	
5.	$(h/g)(x)$	
6.	$g(h(x))$	
7.	$g(f(x))$	

Student Activity

Graphical Algebra of Functions

When we perform an algebraic operation on functions (addition, subtraction, multiplication, division), we are performing the operation on the *output* of the functions.

1. Below you see the graphs of two functions, f and g. A table of values has been given for each of the functions. Fill in the table of values for $f+g$ and $f-g$.

x	$f(x)$	$g(x)$	$(f+g)(x)$ $f(x)+g(x)$	$(f-g)(x)$ $f(x)-g(x)$
–5	–5	–2		
–4	–4	3		
–3	–3	3		
–2	–2	3		
–1	–1	1		
0	0	1		
1	1	1		
2	2	3	5	–1
3	3	3		
4	4	–5		
5	5	–5		

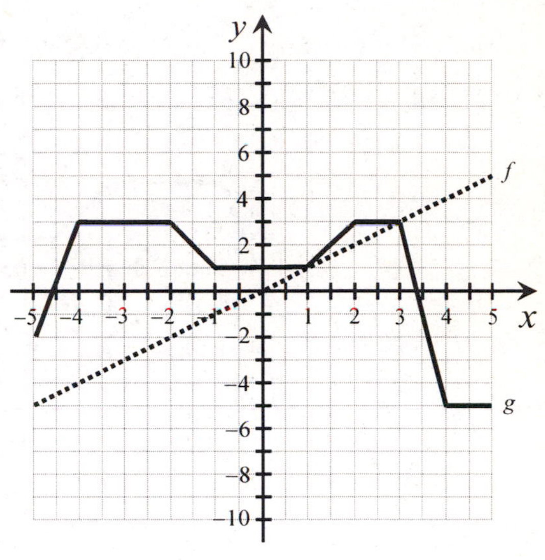

2. Now use the table or the graph to evaluate the expressions in the tables below.

$f(4)$	4
$g(4)$	–5
$f(4)+g(4)$	–1
$(f+g)(4)$	–1
$g(1)$	1
$g(3)$	3
$g(1+3)$	–5
$g(1)+g(3)$	4

$f(-3)$	–3
$g(-3)$	3
$f(-3)-g(-3)$	–6
$(f-g)(-3)$	–6
$g(1)$	1
$g(4)$	–5
$g(1-4)$	3
$g(1)-g(4)$	6

3. Why aren't $g(1+3)$ and $g(1)+g(3)$ equivalent?

4. Now we will graphically perform the addition and subtraction of the functions. For example, at $x=2$, $f(2)=2$ and $g(2)=3$. Adding the values of f and g gives 5, so we plot a point at $(2,5)$ for the addition graph. Subtracting the values at two, $f-g$ gives -1, so we plot a point at $(2,-1)$ for the subtraction graph.

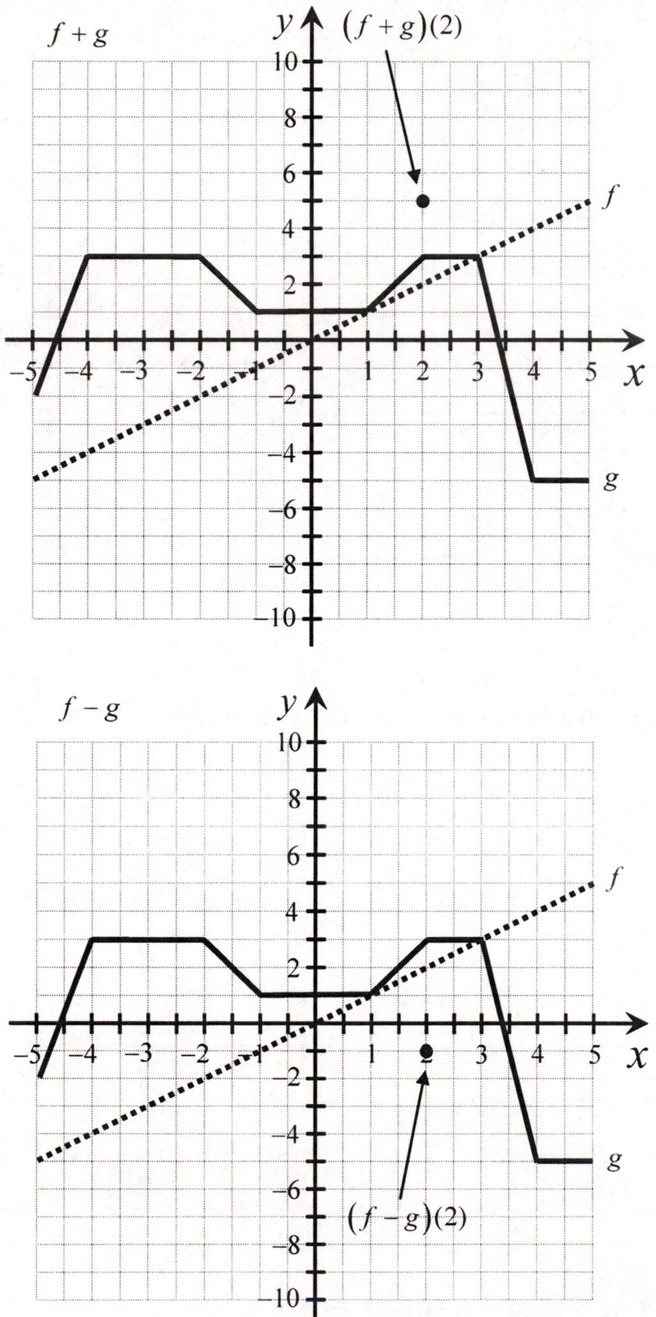

5. Now that you see the graphs of $f+g$ and $f-g$, try to explain what's happened to make the new curve for each one.

Student Activity

Composing Skeletons

Skeleton Method: To evaluate a function that uses function notation, it is helpful to "see" the function without the variables. To make the "skeleton" of the problem, replace the variables with an empty parentheses "skeleton".

For example, $f(x) = x^2 + 4x + 3$ becomes $f(\ ) = (\ )^2 + 4(\ ) + 3$.

For this activity, we define: $f(x) = 3x - 5$ $g(x) = x - 5$ $h(x) = x^2 - 5x$

	Evaluate this expression	Write with parentheses notation	Make a parentheses skeleton of the outer function	Substitute the inner function and simplify
1.	$(f \circ g)(x)$	$f(g(x))$	$f(\quad) = 3(\quad) - 5$	$f(g(x)) = 3(x-5) - 5$ $= 3x - 15 - 5 = 3x - 20$
2.	$(g \circ f)(x)$			
3.		$h(g(x))$		
4.				$g(h(x)) = (x^2 - 5x) - 5$ $= x^2 - 5x - 5$
5.	$(f \circ f)(x)$			
6.				$g(g(x)) = (x-5) - 5$ $= x - 10$
7.		$h(f(x))$		

Student Activity

From Tables and Graphs

The functions *f*, *g*, and *h* are evaluated for the values of *x* in the table below.

For example, we can use the table to see that $f(0)=2$ or $g(-2)=7$.

x	$f(x)$	$g(x)$	$h(x)$
-2	0	7	4
-1	1	5	1
0	2	3	0
1	3	1	1
2	4	-1	4

x	$f(x)$	$g(x)$	$h(x)$
-2	0	7	4
-1	1	5	1
0	2	3	0
1	3	1	1
2	4	-1	4

$g(-2)=7$

$f(0)=2$

Directions: Evaluate each expression using the table of values.

1. $h(-1)=$

2. $(f+g)(2)=f(2)+g(2)=$

3. $(h-g)(1)=$

4. $(g/f)(-1)=$

5. $(f\cdot g)(0)=$

6. $g(3-1)=$

7. $h(f(0))=$

8. $g(h(-1))=$

9. $(g\circ f)(-2)=$

10. $f(f(-2))=$

Directions: Evaluate each expression below using the graphs of *f* and *g*.

11. $(f+g)(2)=f(2)+g(2)=$

12. $(g-f)(-2)=$

13. $(f\cdot g)(0)=$

14. $g(f(-1))=$

15. $(f\circ f)(-1)=$

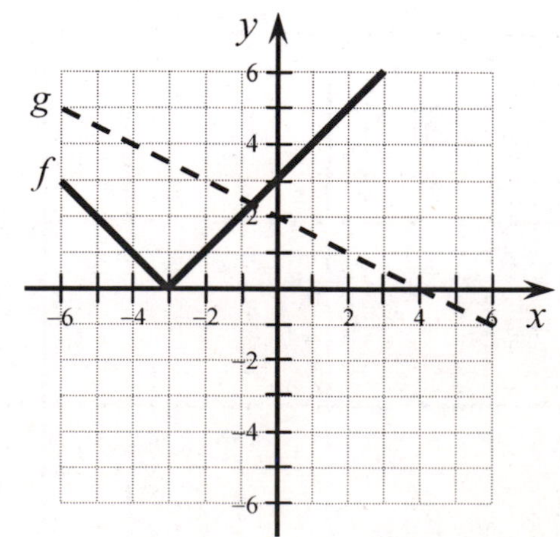

Student Activity

Why is 1-1 Important?

1. Below, you see a graph that represents f. Is f a function? _____
2. Is f one-to-one? _____
3. Make a table of solutions for f.
4. Swap the x- and y-coordinates of each solution to make a second table.
5. Graph the solutions from the second table on the same set of axes as f.
6. Does the new graph represent a function? _____

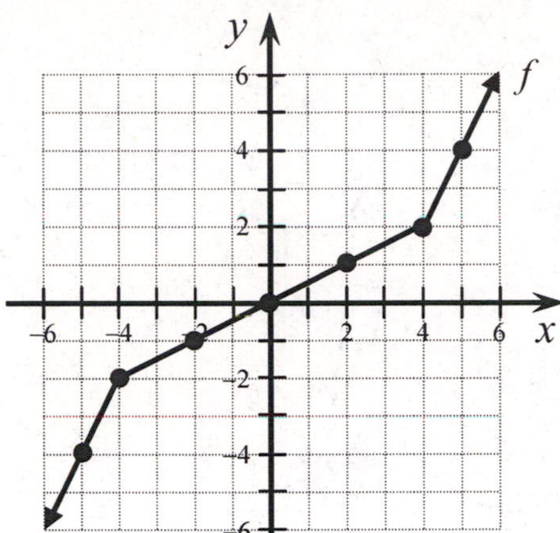

x	$f(x)$
−5	−4
−4	
−2	
0	
2	
4	
5	

x	y
−4	−5

7. Below, you see a graph that represents g. Is g a function? _____
8. Is g one-to-one? _____
9. Make a table of solutions for g.
10. Swap the x- and y-coordinates of each solution to make a second table.
11. Graph the solutions from the second table on the same set of axes as g.
12. Does this new graph represent a function? _____

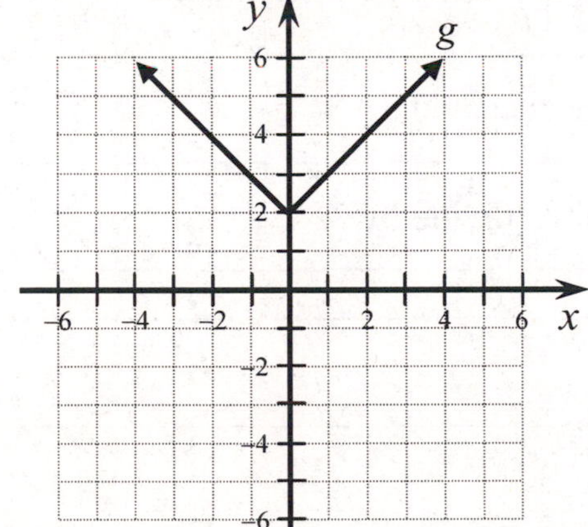

x	$g(x)$
−3	5
−2	
−1	
0	
1	
2	
3	

x	y
5	−3

13. If the original function is one-to-one, what does that mean for the inverse?

14. If the original function is NOT one-to-one, what does that mean for the inverse?

Guided Learning Activity

The Magic of Inverse Functions

Operation	Inverse Operation
Add 5.	Subtract 5.
Multiply by 3.	
Subtract 7.	
Divide by -2.	
Cube it.	
Take a square root.	

We reverse any process by walking through the steps completely backwards.

f : **Putting on socks & shoes.**	f says: Put on a sock, put on a shoe, tie the shoe.
f^{-1} : **Removing socks & shoes.**	f^{-1} says: Untie the shoe, take off the shoe, take off the sock.

We can write inverse mathematical functions by reversing the function process, in the same way.

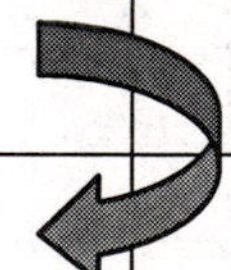

$f(x) = 2x + 3$	f says: Multiply by 2, then add 3.
$f^{-1}(x) = \dfrac{x-3}{2}$	f^{-1} says: Subtract 3, then divide by 2.

$f(x) = \dfrac{x-5}{3}$	f says:
$f^{-1}(x) =$	f^{-1} says:

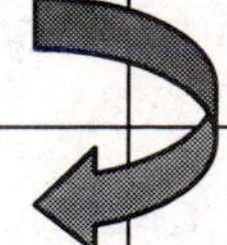

$f(x) = \sqrt[3]{x+4}$	f says:
$f^{-1}(x) =$	f^{-1} says:

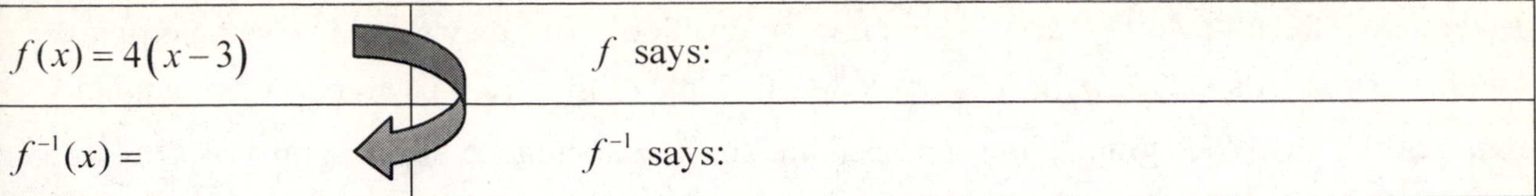

$f(x) = 4(x-3)$	f says:
$f^{-1}(x) =$	f^{-1} says:

Inverse functions have a special property that $f\left(f^{-1}(x)\right) = x$ and $f^{-1}\left(f(x)\right) = x$.
Why is that? Let's input a "value" into our Socks & Shoes function and see.

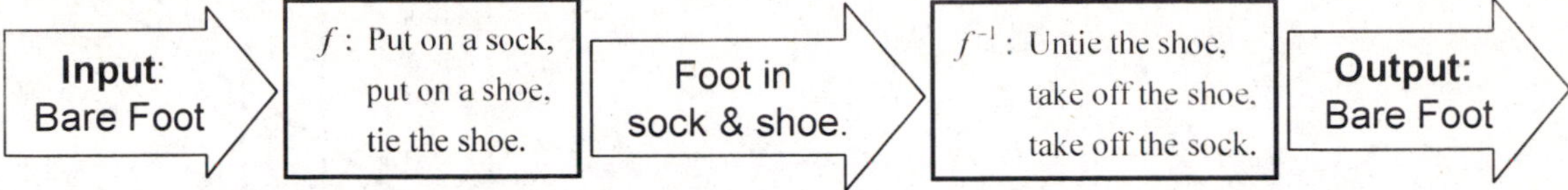

This represents the composition $f^{-1}\left(f(\text{bare foot})\right) = \text{bare foot}$.

Let's try a few mathematical compositions with numerical inputs.

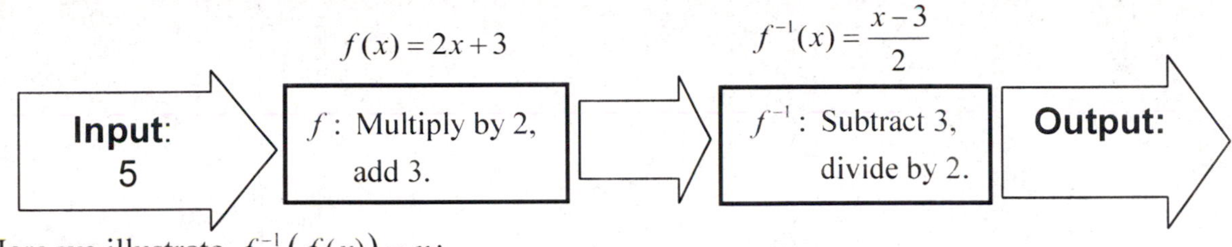

Here we illustrate $f^{-1}\left(f(x)\right) = x$:

Input	Apply the function	Output of function = Input to inverse	Apply the inverse function	Output
4	$f(x) = 5x-2$		$f^{-1}(x) = \dfrac{x+2}{5}$	
3	$f(x) = \sqrt[3]{x+5}$		$f^{-1}(x) = x^3 - 5$	
8	$f(x) = \dfrac{x+4}{3}$		$f^{-1}(x) = 3x-4$	

And here we illustrate $f\left(f^{-1}(x)\right) = x$:

Input	Apply the inverse function	Output of inverse = Input to function	Apply the function	Output
13	$f^{-1}(x) = \dfrac{x+2}{5}$		$f(x) = 5x-2$	
2	$f^{-1}(x) = x^3 - 4$		$f(x) = \sqrt[3]{x+4}$	
5	$f^{-1}(x) = 3x-4$		$f(x) = \dfrac{x+4}{3}$	

Student Activity

Mirror Images

Directions: Recall that if f and f^{-1} are inverse functions, then they should be symmetric over the line $y = x$. Also, if (a,b) is a point on the graph of f, then (b,a) will be a point on the graph of f^{-1}. On each graph, draw a dashed line for $y = x$ and then use the graph of f to graph f^{-1} on the same axes. The first one has been started for you.

1.

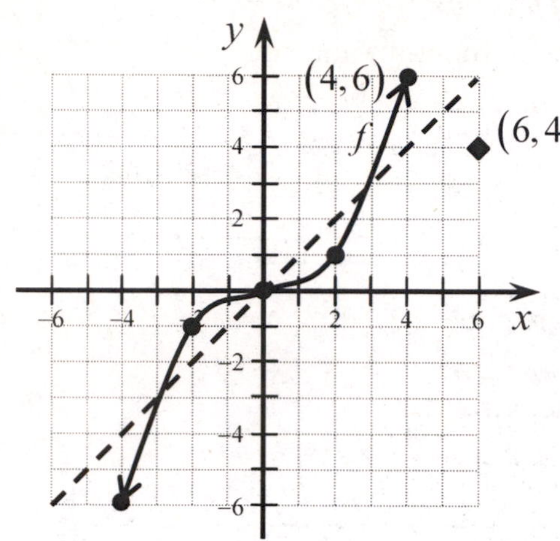

2.

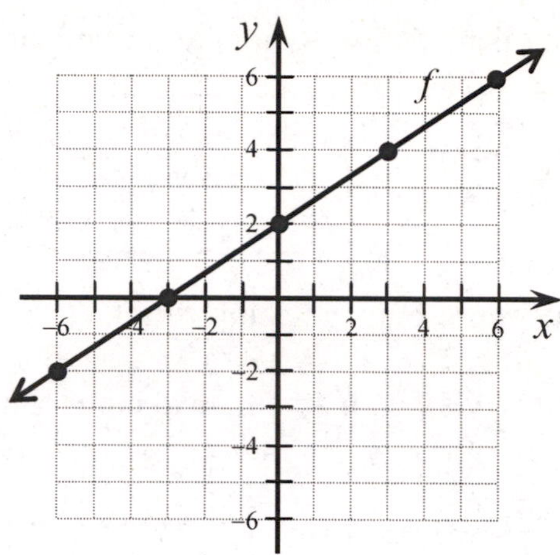

3.

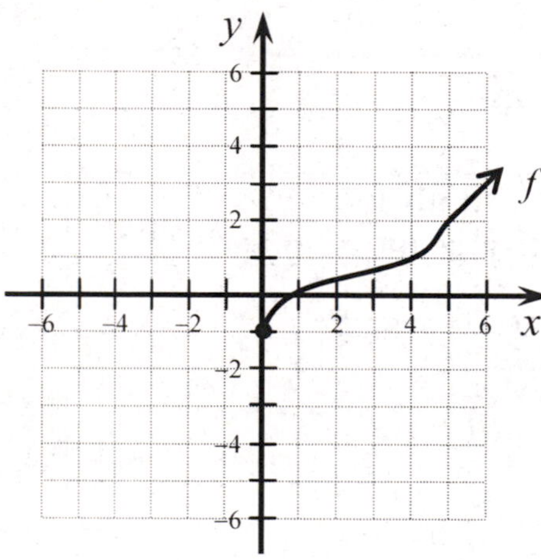

4.

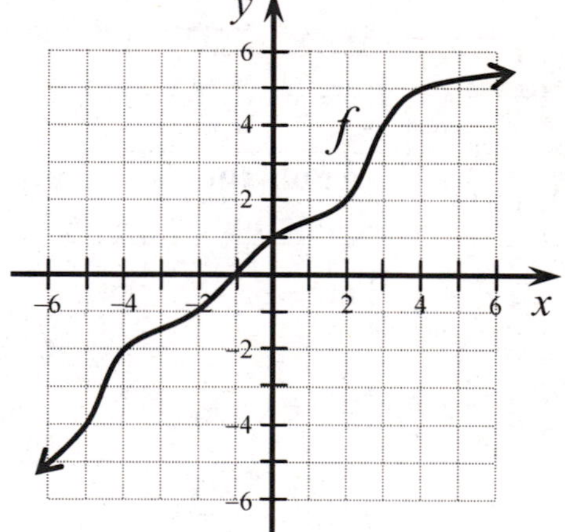

5. Now draw your own one-to-one function, and graph its inverse.

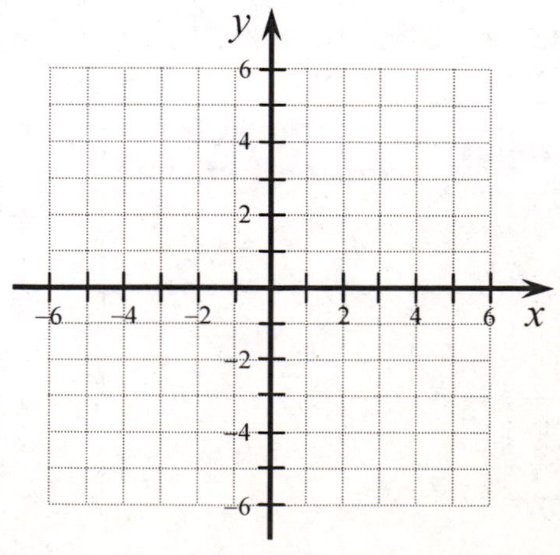

Student Activity

Put it All Together

If f and f^{-1} are inverse functions of each other then the following are true:

- The domain of f is the range of f^{-1} and the range of f is the domain of f^{-1}.
- Let (a,b) be a point on the graph of f, then (b,a) is a point on f^{-1}.
- The graphs of f and f^{-1} are symmetric over the line $y = x$.
- $f\left(f^{-1}(x)\right) = x$ and $f^{-1}\left(f(x)\right) = x$

Problem A: Let $f(x) = \dfrac{3}{2}x + 3$.

1. The point $(2,6)$ is on the graph of f. What point must be on the graph of f^{-1}? (__,__)

2. Find $f^{-1}(x)$. Check this step with someone else before you go on.

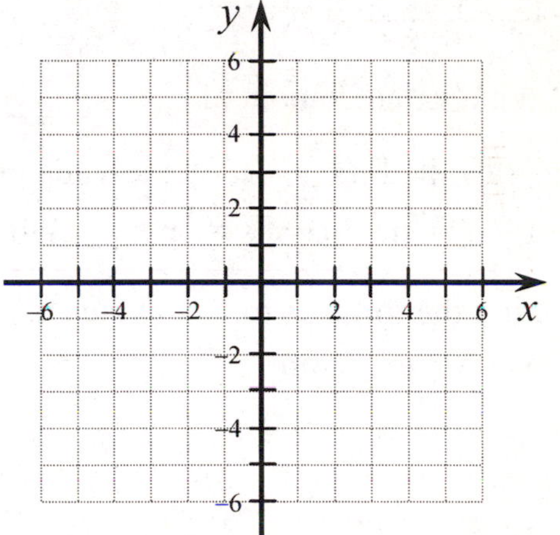

3. Graph f and f^{-1} on the axes to the right. Label which is which.

4. Draw the line $y = x$ on the same graph using a dashed line.

5. Fill in the domain and range for each:

$f(x)$ Domain: ______________
Range: ______________

$f^{-1}(x)$ Domain: ______________
Range: ______________

6. Find and simplify $f\left(f^{-1}(x)\right)$.

7. Find and simplify $f^{-1}\left(f(x)\right)$.

Problem B: Let $g(x) = \sqrt{x+3} + 1$.

1. Graph g on the axes below. Use the table of solutions to assist you. Label the graph with g.

x	$g(x)$
–5	
–4	
–3	
–2	
1	
6	

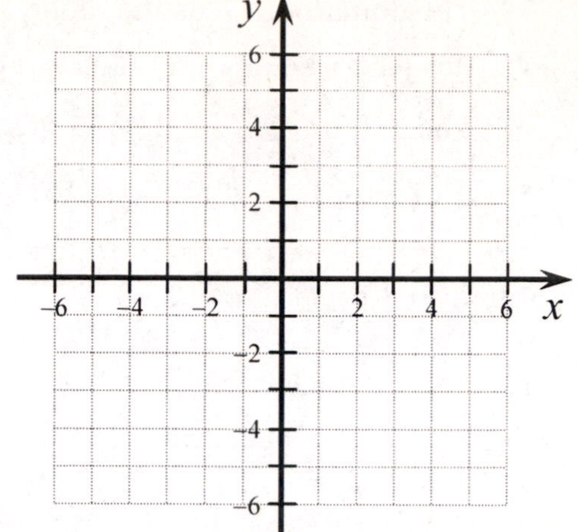

2. What is the domain of g? _______ What is the range? _______

3. Draw the line $y = x$ on the same graph using a dashed line. Draw a reflection of the graph of g over the line of symmetry. Label this g^{-1}.

4. Find the formula for $g^{-1}(x)$ using the formula for $g(x)$. Check this step with someone else before you go on.

5. Based on the domain and range of g, what is the domain of g^{-1}? _________ What is the range? _________

6. Fill in the table of solutions for g^{-1} using the formula you just found **and** the domain for g^{-1}. Then graph this function on the new axes provided.

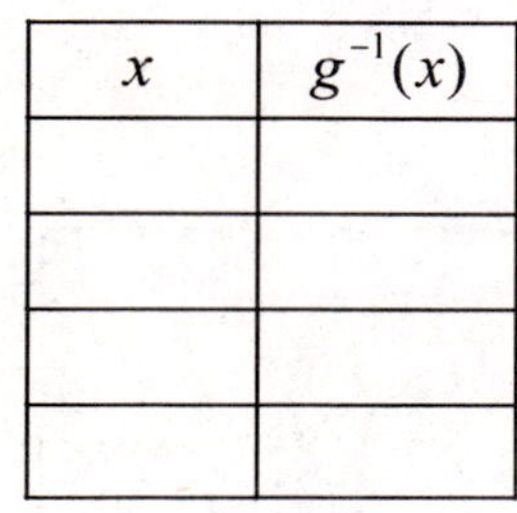

x	$g^{-1}(x)$

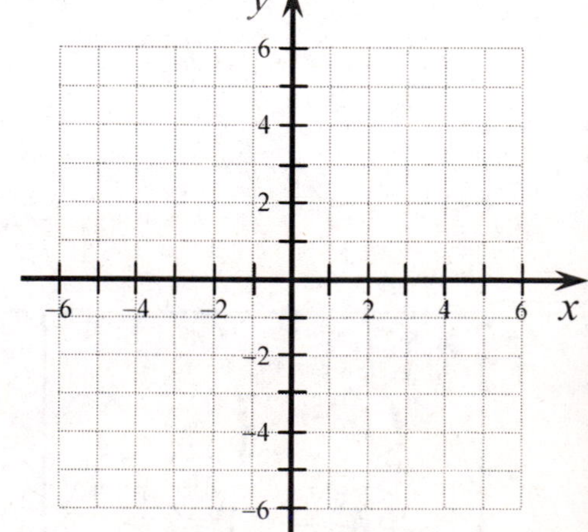

7. Include a condition with the formula for g^{-1} so that $g^{-1}(x)$ gives the true inverse of g. $g^{-1}(x) =$ , $x \geq$ ____.

Assess Your Understanding

Functions and More

For each of the following, describe the strategies or key steps that will help you **start** the problem. You do **not** have to complete the problems.

		What will help you to start this problem?
1.	If $f(x)=-3x+1$, find $f(4)$.	
2.	Is the relation $\{(2,3),(4,1),(3,4)\}$ a function?	
3.	Let $f(x)=2x-7$. Find $f^{-1}(x)$.	
4.	Graph $g(x)=(x-1)^2$ and write the domain and range.	
5.	Describe how to graph $f(x)=\|x-2\|$ given the graph of $y=\|x\|$.	
6.	Describe how to graph $g(x)=-x^3+4$ given the graph of $y=x^3$.	
7.	Let $f(x)=x^2-4x$ and $g(x)=2x-1$. Find $(f\circ g)(x)$.	
8.	Let $f(x)=2x+3$ and $g(x)=x-4$. Find $(f/g)(x)$ and state the domain of the new function.	
9.	Is $y=x^2$ a function? Is it one-to-one?	
10.	Let $f(x)=\dfrac{x+1}{x-3}$. Find $f^{-1}(x)$.	

Metacognitive Skills

Functions and More

Metacognitive skills refer to the ability to judge how well you have learned something and to effectively direct your own learning and studying. This is a self-evaluation tool designed to help you focus your studying and to improve your metacognitive skills with regards to this math class.

Fill the 1st column out **before** you begin studying. Fill the 2nd column out after you study for your test.

Go back to this assessment after your test and circle any of the ratings that you would change – this identifies the "disconnects" between what you **thought** you knew well and what you **actually** knew well.

Use the scale below to assign a number to each topic.

5 *I am confident I can do any problems in this category correctly.*
4 *I am confident I can do most of the problems in this category correctly.*
3 *I understand how to do the problems in this category, but I still make a lot of mistakes.*
2 *I feel unsure about how to do these problems.*
1 *I know I don't understand how to do these problems.*

Topic or Skill	Before Studying	After Studying
Understand the difference between a relation and a function.		
Be able to use function notation and evaluate for a specific value.		
Identify whether a relation written as a set is a function.		
Apply the vertical line test to decide if a graph represents a function.		
Write the description of a function using function notation.		
Create a table of solutions and sketch the graph of an equation.		
Identify information on graphs using function notation.		
Find the domain and range of a graph.		
Graph the basic functions (identity, squaring, cubing, absolute value).		
Given a graph, draw various translations, and transformations.		
Identify how a change to an equation will change its graph.		
Determine the domain of a function.		
Perform operations on functions (sum, difference, product, quotient).		
Find the domain of a function created from a sum, difference, product, or quotient of two functions.		
Understand what it means graphically to add or subtract functions.		
Interpret and evaluate of function compositions.		
Perform algebra or composition of functions using algebra.		
Perform algebra or composition of functions using a table of values.		
Perform algebra or composition of functions using a graph.		
Identify when a relation, function, or graph is one-to-one.		
Apply the horizontal line test to determine whether a graph is 1-1.		
Know the four properties of functions and their inverses.		
Know how the property of 1-1 relates to finding an inverse function.		
Given the equation for a 1-1 function, find its inverse.		
Given the graph of a relation, draw its inverse.		
Explain (in words) what an inverse is.		

SYS1: Solving Systems

Student Activity

Targeting Solutions

Directions: In each problem below, test the ordered pairs that are provided to see if they are possible solutions to either of the equations in the system of equations. After testing, place the ordered pairs into the appropriate region of the "target" for each problem.

Outer target region	The ordered pair IS a solution of the first equation, but not the second equation.
Middle target region	The ordered pair is not a solution of the first equation, but IS a solution of the second equation.
Bulls-eye (center target region)	The ordered pair is a solution of both equations.

1. $\begin{cases} y < 5x + 2 \\ y > -x + 8 \end{cases}$

$(1, 6)$

$(4, 25)$

$(2, 10)$

$(-10, 20)$

$(-3, -20)$

$(\frac{1}{2}, 9)$

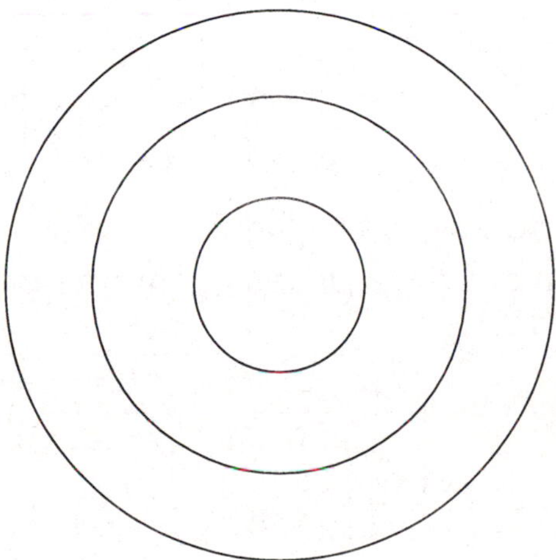

2. $\begin{cases} 4x + 2y = 8 \\ -2x + 4y = 1 \end{cases}$

$(5, -6)$

$(7, \frac{15}{4})$

$(\frac{3}{2}, 1)$

$(-4, -\frac{7}{4})$

$(0, 4)$

$(-3, 10)$

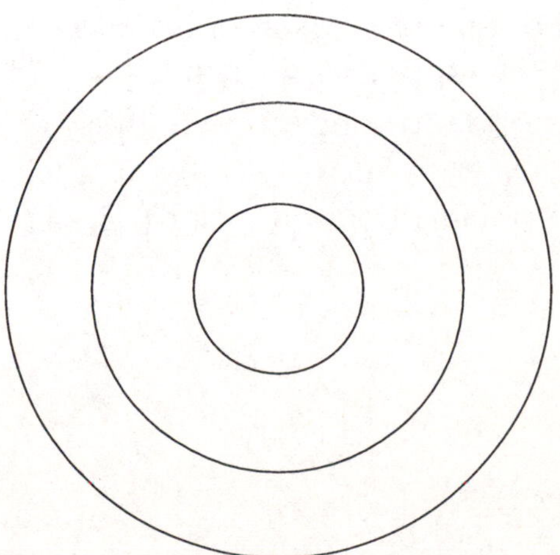

Guided Learning Activity

Graphing to Solve Systems of Equations

Example 1: Write some ordered-pair solutions of the equation $x + y = 5$.

(,) (,) (,) (,) (,) (,)

Write some ordered-pair solutions of the equation $x - y = -3$.

(,) (,) (,) (,) (,) (,)

When we write the two equations like this: $\begin{cases} x + y = 5 \\ x - y = -3 \end{cases}$

we call this a *system of equations*. When we *solve* a system of two equations, we are looking for all of the ordered pairs (x, y) that satisfy *both* equations.

Maybe you found an ordered pair that was a solution of both equations? If so, circle it.

Example 2: What does it mean to be a solution to a system?

Consider the system of equations given by $\begin{cases} y = 2x + 2 \\ y = x - 1 \end{cases}$ and the ordered pair $(-3, -4)$.

You can see by evaluating each equation for $x = -3$ and $y = -4$ that this ordered pair is a solution of both equations. This means it is a solution of the system of equations.

$y = 2x + 2$	$y = x - 1$
$y = 2(\ ) + 2$	$y = (\ ) - 1$
$-4 \stackrel{?}{=} 2(-3) + 2$	$-4 \stackrel{?}{=} (-3) - 1$
$-4 = -4$	$-4 = -4$

So what does the solution of a system of equations look like? Graph $y = 2x + 2$ and $y = x - 1$ on the graphing grid provided to the right.

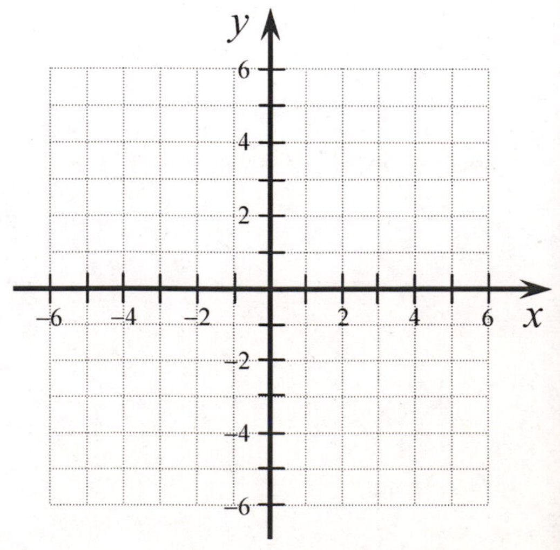

What is special about the point $(-3, -4)$ on the graph?

Now solve these systems of equations by graphing.

1. $\begin{cases} y = -3x + 6 \\ y = 2x - 4 \end{cases}$

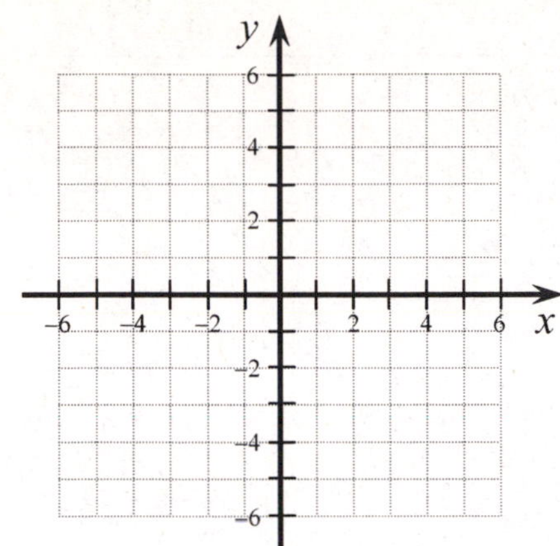

4. $\begin{cases} y = -3x - 2 \\ y = 2x + 3 \end{cases}$

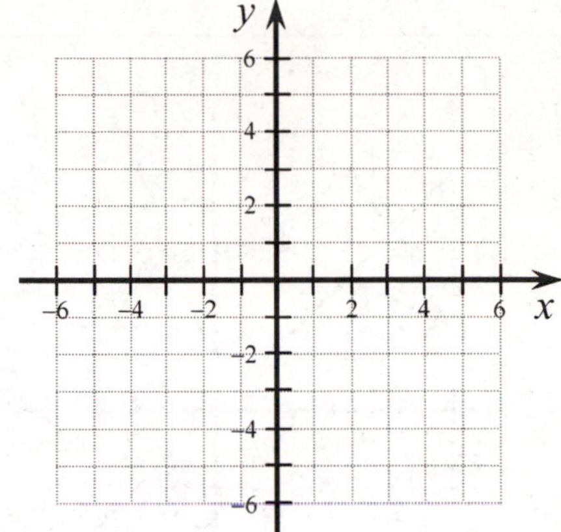

2. $\begin{cases} 5x - y = 6 \\ 3x - 3y = 6 \end{cases}$

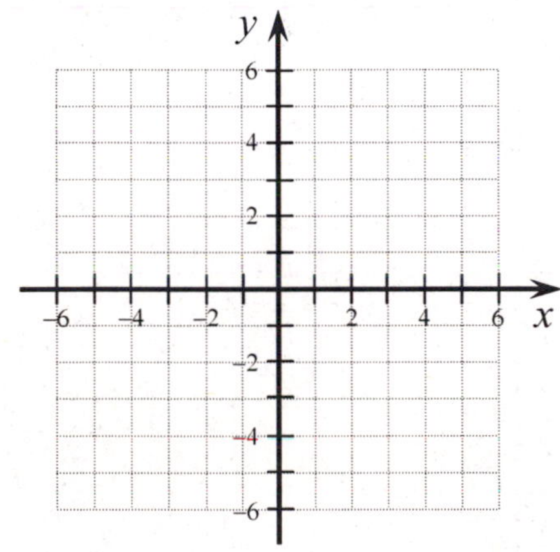

5. $\begin{cases} -2x - 4y = -2 \\ 4x + 8y = 4 \end{cases}$

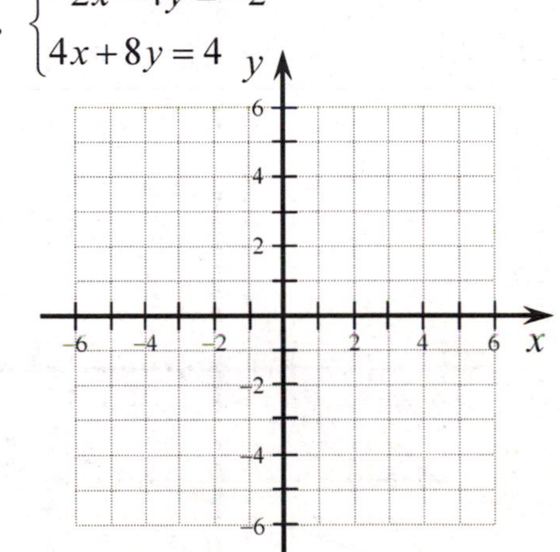

3. $\begin{cases} -4x + 2y = 8 \\ 6x - 3y = 3 \end{cases}$

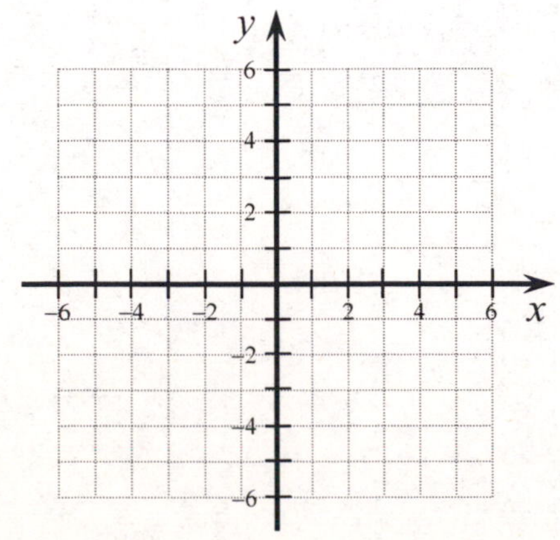

6. $\begin{cases} y = \frac{3}{4}x - 1 \\ y = -\frac{4}{3}x - 1 \end{cases}$

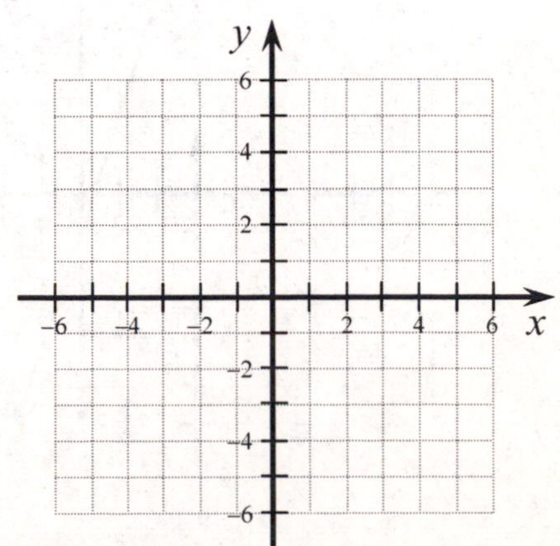

Student Activity

SYS1-4

Following the Clues Back to the System of Equations

Directions: In each "crime-scene" below, you are shown the graph of a system of equations. Use your mathematical powers of reasoning (and detective skills) to determine what the system of equations must have been to result in this graph.

1.

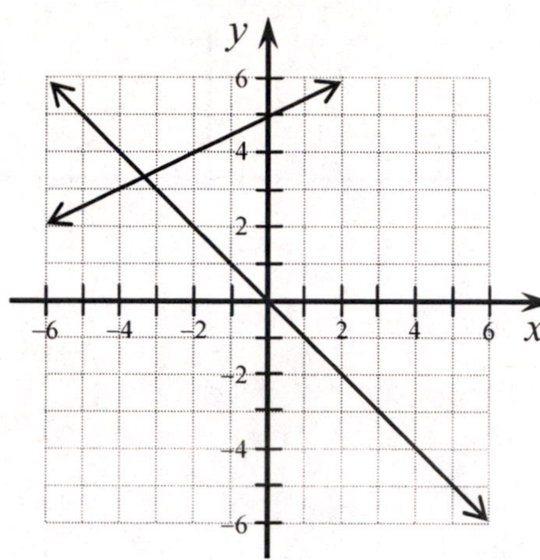

2.

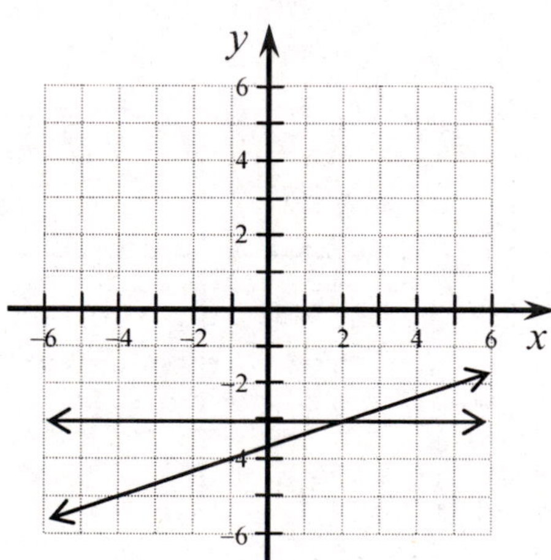

3.

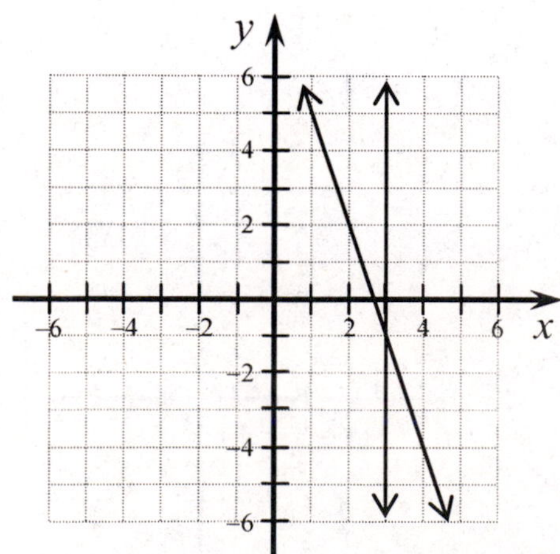

Student Activity

Graphing Systems of Equations with a Calculator

You can use your graphing calculator to solve systems of linear equations by graphing. Each model of calculator will use a different set of keystrokes to graph these equations. Before you start to graph any system of equations, it is a good idea to guess what your graph should look like (in case you type something in your calculator incorrectly). The standard zoom on most graphing calculators goes from -10 to 10 on the x-axis and from -10 to 10 on the y-axis.

1. Without using a calculator, sketch a graph of the system of equations $\begin{cases} y = 5x - 5 \\ y = -3x + 3 \end{cases}$ on the axes provided, and label the solution point.

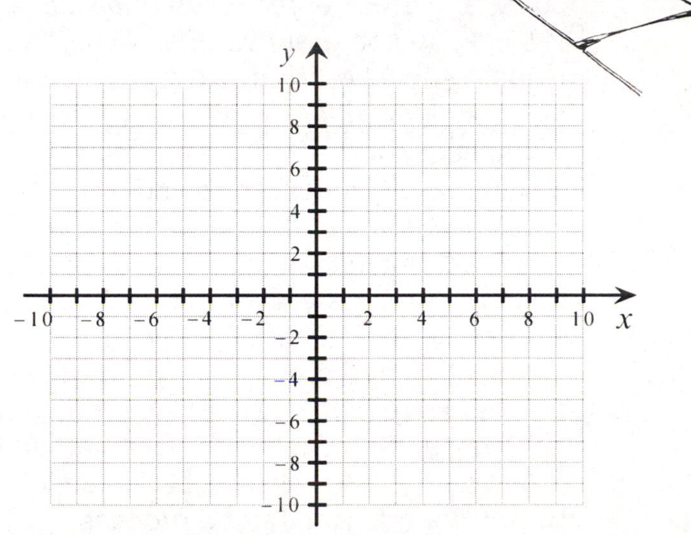

2. Find the $y =$ screen on your calculator. Place your cursor next to $Y_1 =$ and enter the first equation in your system. Then move your cursor to $Y_2 =$ and enter the second equation. Make sure you are in the standard viewing window and then view the graph. Hopefully this looks like the graph you produced above.

3. In order to find the solution to this system of equations on the calculator, we need to find the point of intersection. **From your graph screen**, find the *Math* or *Calculate* menu. You'll know you have found it when you see a menu that contains functions like *zero* (or *root*), *max*, *min*, and *intersect* (it will also have a few others).

Draw the keys or write the steps that were necessary to find this menu.

4. Select *intersect* (or *intersection*) from the menu you just found. Your calculator will give you a series of prompts. At the end of all of the prompts, your calculator should display numerical values labeled $x =$ and $y =$. This should match the solution you found at the beginning of the activity.

Calculator Prompt	What does it mean? What should you do?	Possible Errors
First Curve?	The calculator is asking you to tell it which curve is the first curve in the system of equations. Place your cursor on one of the lines you graphed using the arrow keys and press *enter.* You will then be asked for the second curve. Move the cursor to the other curve and press *enter. Why does it ask this? Well, if you had graphed four lines, it would want to know which lines (curves) were the two that you want to find the intersection of.*	If you give the calculator the **same** curve for both the first and second prompts, you will likely get an error.
Lower Bound?	The calculator is asking you to give it a smaller region to look in to find the intersection point. It is asking you for a value to the left of the point of intersection (lower bound) first. Use your arrow keys to place your cursor to the left of the point of intersection and press *enter* or use the number keypad to enter a value that you know is to the left of the intersection point, and then press *enter.* You will then be asked for the Upper Bound. Repeat the process for a value to the right of the intersection point.	The upper bound must be **above** (to the right of) the lower bound. If you try to give the calculator an upper bound value that is below the lower bound value, you will get an error.
Guess?	It is easier for the calculator to find the exact intersection point if you give it a guess point near the intersection point. Just position your cursor near the intersection point and press enter. *Why does it ask for a guess? If you were graphing more complicated curves (for example polar curves), there might be more than one intersection point in the same lower and upper boundary. In this case, the calculator would want to know which point you wanted to find out exactly.*	You must give the calculator a guess that is **between** the lower and upper bounds. If your guess is outside the region where you told the calculator to look, it will be (understandably) unhappy with you.

Note: If you do not see the intersection values, it could be that you pressed enter one too many times when going through the prompts (inadvertently skipping the solutions). If this happens, repeat the steps a little more slowly, making sure to only press *enter* once for each prompt.

Student Activity

Which Egg is Easier to Crack?

When solving systems of equations by substitution, we begin by solving one of the equations for x or y ; this is called the *substitution equation*. Sometimes one of the variables is easier to solve for than the other (that is, it will take less steps to solve for one than the other).

Directions: Look at the given equations and find the substitution equation where solving for the variable requires **the least number of steps**. Then place the substitution equation in the egg that corresponds to the variable that was solved for. For example, in the first equation, it would be much easier to solve for x, resulting in the equation $x = 2 - 3y$.

$x + 3y = 2$	$7x + y = 14$	$3y = 2x + 2$	$-2y + x = 18$
$x - 2y = 9$	$15 = -5y + x$	$16x = 16y + 16$	$\frac{2}{3}x + y = \frac{1}{3}$
$25x = 225y + 225$	$12x - y = 0$	$3y = 3x + 6$	$5x + y = 10$
$18x = 9y + 36$	$\frac{1}{10}x + y = 1$	$12x + y = 2$	$4x = 2y + 6$

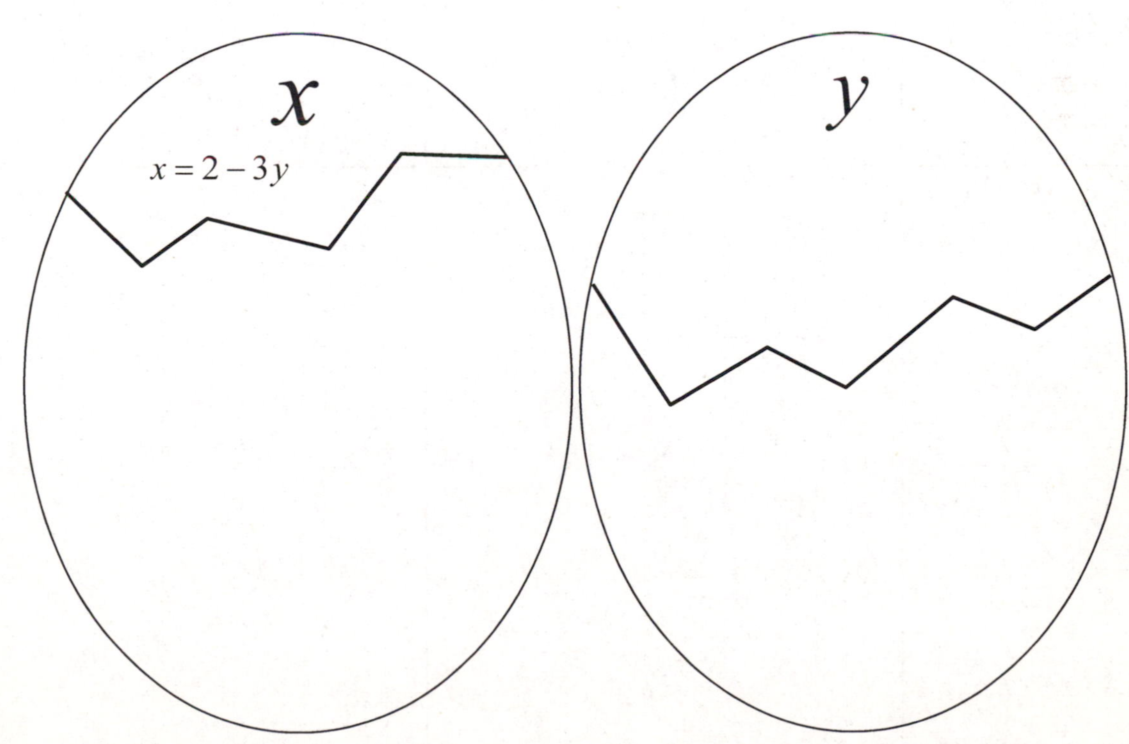

Student Activity
Double Trouble on Substitution

Directions: When you solve a system of equations by substitution, there are *at least* two ways to tackle the problem. For example, in the first system, you could easily solve for y in the first equation, or for x in the second equation to find your substitution equation. **Some choices are easier than others!** For each of the systems below, solve the system of equations two ways as indicated. Your ordered pair solution should be the same for both methods. If they are not, you'll have to go back and look for a mistake. The first one has been started for you.

	System of Equations	Solve for a variable in the FIRST equation, then substitute and solve.	Solve for a variable in the SECOND equation, then substitute and solve.
1.	$\begin{cases} 2x+y=5 \\ x-4y=7 \end{cases}$	Solve for y in the 1st equation: $y=-2x+5$ Substitute into 2nd equation and solve: $x-4(-2x+5)=7$ $x+8x-20=7$ $9x-20=7$ $9x=27$ $x=3$ Find the missing coordinate of the ordered pair: $y=-2(3)+5$ $=-6+5=-1$ $(3,-1)$	
2.	$\begin{cases} x+8y=20 \\ 4x-y=14 \end{cases}$		

	System of Equations	Solve for a variable in the FIRST equation, then substitute and solve.	Solve for a variable in the SECOND equation, then substitute and solve.
3.	$\begin{cases} 3x - y = -2 \\ -12x + 4y = 16 \end{cases}$		
4.	$\begin{cases} 3x + 4y = 0 \\ \frac{3}{4}x + \frac{8}{3}y = -\frac{5}{12} \end{cases}$		
5.	$\begin{cases} x - 7y = -1 \\ -\frac{1}{7}x + y = \frac{1}{7} \end{cases}$		

Student Activity

Forced Elimination

Directions: In order to develop some intuition about which variable is easier to eliminate, we will **start** a few problems by trying both of the possible eliminations. Use the elimination method to solve for a variable. **Then circle the elimination that was easier.** The first one has been started for you.

1.

Eliminate x	**Eliminate y**
$\begin{cases} x+3y=7 \\ 2x-3y=-4 \end{cases}$ Multiply the first equation by -2. $\begin{cases} -2x-6y=-14 \\ 2x-3y=-4 \end{cases}$ $-9y=-18$ $y=2$	$\begin{cases} x+3y=7 \\ 2x-3y=-4 \end{cases}$

2.

Eliminate x	**Eliminate y**
$\begin{cases} -10x+y=-5 \\ 10x-2y=0 \end{cases}$	$\begin{cases} -10x+y=-5 \\ 10x-2y=0 \end{cases}$

3.

Eliminate x	**Eliminate y**
$\begin{cases} 5x+7y=12 \\ 10x-3y=7 \end{cases}$	$\begin{cases} 5x+7y=12 \\ 10x-3y=7 \end{cases}$

4.

Eliminate x	**Eliminate y**
$\begin{cases} 3x+y=2 \\ 4x-2y=6 \end{cases}$	$\begin{cases} 3x+y=2 \\ 4x-2y=6 \end{cases}$

Now inspect these systems of equations. For each system, decide which variable will be easier to eliminate and describe the steps that are necessary to achieve this. Problem 5 has been done for you.

5. $\begin{cases} 3x+y=1 \\ 2x+3y=-11 \end{cases}$ Eliminate y.
Multiply the first equation by -3 and then add the equations.

6. $\begin{cases} 5x-y=-5 \\ x+y=5 \end{cases}$

7. $\begin{cases} x+y=12 \\ -x+y=0 \end{cases}$

8. $\begin{cases} -4x+2y=2 \\ 4x+6y=22 \end{cases}$

9. $\begin{cases} 3x+14y=-7 \\ 2x+2y=12 \end{cases}$

10. $\begin{cases} 3x+8y=2 \\ -6x-7y=5 \end{cases}$

Student Activity

Triple the Fun on Systems of Equations

Directions: For each of the problems below, solve the system of equations by graphing, by substitution, and by elimination – your solution should be the same for all three methods. If they are not, go back and look for a mistake in your work.

1.

System of Equations:

$$\begin{cases} 10x+5y=15 \\ x+2y=-6 \end{cases}$$

Solve by Graphing:

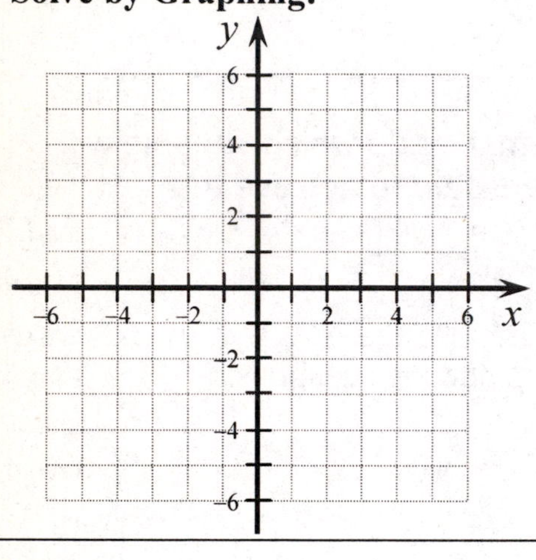

Solve by Substitution

Solve by Elimination

2.

System of Equations:

$$\begin{cases} \frac{1}{4}x-y=-\frac{1}{3} \\ 3x-12y=-12 \end{cases}$$

Solve by Graphing:

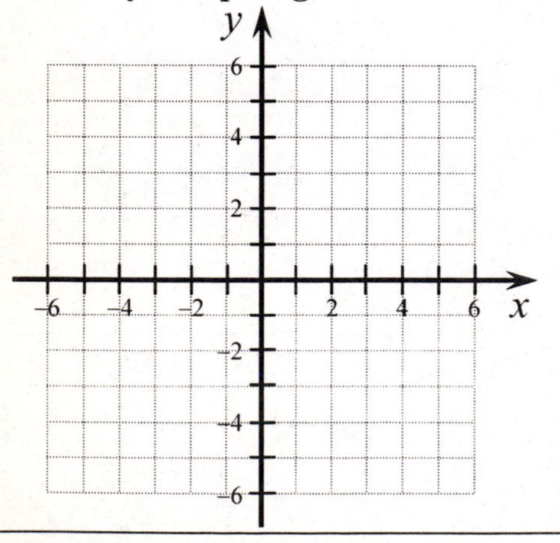

Solve by Substitution

Solve by Elimination

3.

System of Equations:

$$\begin{cases} 3y - 4x = 2 \\ 3x - 4y = -5 \end{cases}$$

Solve by Graphing:

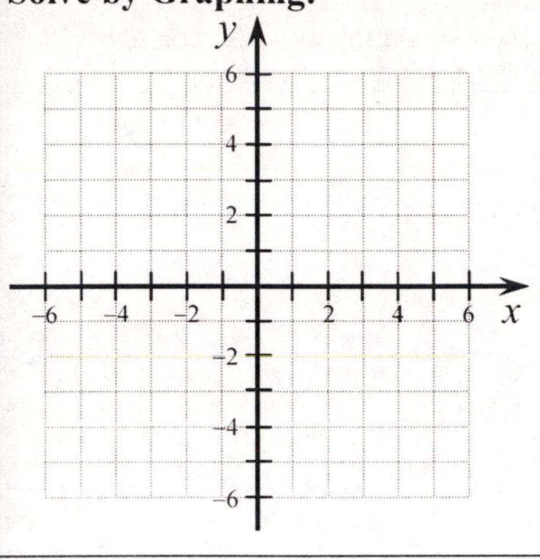

Solve by Substitution

Solve by Elimination

4.

System of Equations:

$$\begin{cases} -4x + y = -5 \\ 12x - 3y = 15 \end{cases}$$

Solve by Graphing:

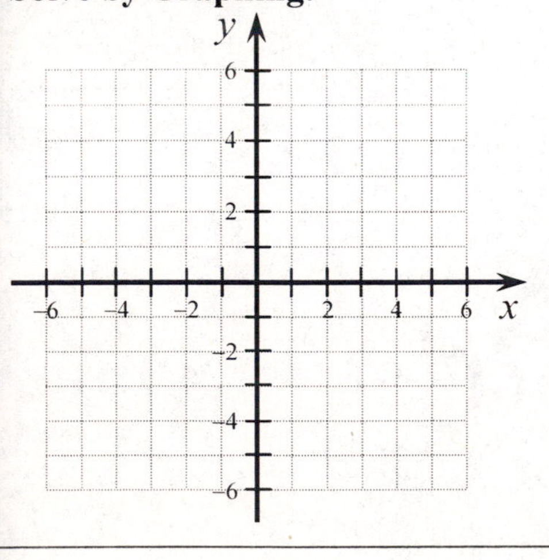

Solve by Substitution

Solve by Elimination

Student Activity

Choose Your Tactic

Directions: For each system of equations below, decide whether it will be easier to solve using substitution or elimination (choose a tactic). Then describe your tactical plan: For **substitution,** which variable in which equation will you solve for? For **elimination** – which variable will you eliminate and how?

	System of Equations	Tactic Substitution or Elimination	Tactical Plan Substitution: Which variable will you solve for? Elimination: Which variable will you eliminate?
1.	$\begin{cases} 5x+3y=-9 \\ y=2x+8 \end{cases}$		
2.	$\begin{cases} 2x+y=9 \\ 5x+3y=26 \end{cases}$		
3.	$\begin{cases} y=-x \\ 6x+6y=0 \end{cases}$		
4.	$\begin{cases} 4x+11y=7 \\ 4x+3y=-1 \end{cases}$		
5.	$\begin{cases} 16x-2y=16 \\ 4x=2y-8 \end{cases}$		
6.	$\begin{cases} 0.02x+0.01y=0.1 \\ -2x+3y=-18 \end{cases}$		
7.	$\begin{cases} -\frac{x}{5}-\frac{y}{3}=2 \\ -\frac{3x}{10}+\frac{2y}{10}=\frac{9}{10} \end{cases}$		
8.	$\begin{cases} x=12y-7 \\ 12y-x=12 \end{cases}$		

Student Activity

The Last Sentence

Directions: In each example below you are given **the last sentence** of an application problem. In many cases, it is possible to declare the variables in the problem just by looking at this sentence. Take your best guess about what the variables should be based on the sentences you are given. The first one has been done for you.

1. Find the length and width of the painting.

   Let L = the length of the painting.
   Let W = the width of the painting.

2. Determine much Maria spent on the rental car and on the hotel.

3. How much money was invested in each stock?

4. How many pounds of peanuts and cashews are required to make the snack mix?

5. Find the speed of the plane and the speed of the wind.

6. Determine how many liters of acid and how many liters of water are required to make the required batch.

7. What base and height measurements give the triangle an area of 12 ft^2?

8. Find the cost of each adult and child ticket.

Guided Learning Activity

Using the Problem Solving Strategy

Problem A: At a scrapbooking party, customers can purchase two special packages, the vacation scrapbook package and the baby album package. Shaude purchases 2 vacation packages and one baby album package for a total of $93.50. Stephan purchases one vacation package and three baby album packages for a total of $138. Find the cost of each of the special packages that were offered at the party

1. **Analyze.** Draw diagrams, tables, and charts to help make sense of the problem.
2. **Declare.** Declare the variables and the quantitative givens in the problem.
3. **Formulate.** Write equations to represent the problem.
4. **Solve.** Solve the equation or system of equations.
5. **Validate.** Check the result. Does it make sense? Is it reasonable?
6. **Conclude.** State the conclusion. Don't forget units.

Problem B: Jasper invests a total of 250,000 L$ (Linden dollars). He splits his investment between a 30-day Linden Bank CD that pays 7% interest at the end of the term and a 30-day Second Life real estate investment scheme organized by his brother that ends up losing 4% of its original value at the end of 30 days. At the end of the 30 days, Jasper makes 2,100 L$. How much did Jasper invest in the CD and how much did he invest in the real estate scheme?

1. **Analyze.** Draw diagrams, tables, and charts to help make sense of the problem.
2. **Declare.** Declare the variables and the quantitative givens in the problem.
3. **Formulate.** Write equations to represent the problem.
4. **Solve.** Solve the equation or system of equations.
5. **Validate.** Check the result. Does it make sense? Is it reasonable?
6. **Conclude.** State the conclusion. Don't forget units

Problem C: Technical grade hydrogen peroxide is 35% H_2O_2. Common household hydrogen peroxide is only 3% H_2O_2. Imogen wants to make a 10% H_2O_2 solution to use in removing a toxic mold in a city building. What amount of the technical grade solution and what amount of the household solution (to the nearest 0.1 mL) should she mix to create the 500 mL of solution she desires?

1. **Analyze.** Draw diagrams, tables, and charts to help make sense of the problem.
2. **Declare.** Declare the variables and the quantitative givens in the problem.
3. **Formulate.** Write equations to represent the problem.
4. **Solve.** Solve the equation or system of equations.
5. **Validate.** Check the result. Does it make sense? Is it reasonable?
6. **Conclude.** State the conclusion. Don't forget units

Student Activity

Setting Up the System

Directions: Read each problem carefully.
a) **Declare** the variables you would use to solve the problem.
b) **Formulate** a system of two equations that could be used to solve each problem. You do not need to solve the application problems.

1. The elevation in Denver, Colorado, is 13 times greater than that of Waikapu, Hawaii. The sum of their elevations is 5,600 feet. Find the elevation of each city.

2. The owner of a pet food store wants to mix birdseed that costs \$1.25 per pound with sunflower seeds that cost \$0.75 per pound to make 50 pounds of a mixture that costs \$1.00 per pound. How many pounds of each type of seed should he use?

3. An airplane can fly 500 miles in 2 hours when flying with the wind. The same journey takes 4 hours when flying against the wind. Find the speed of the wind and the speed of the plane.

4. Barb invested a total of \$25,000 in two accounts. One account has an annual interest rate of 3.5%, while the other has an annual rate of 6%. She earned \$1,257 in interest in one year. How much did she invest in each account?

5. Two angles are supplementary. The measure of the first angle is 10 degrees more than three times the second angle. Find the measure of each angle.

Student Activity
The Moving Walkway Problem

A moving walkway is like an escalator (transporting you from one place to another), only it is a flat belt that you can walk on. Moving walkways can commonly be found now in airports, museums, amusement parks, or zoos. When you walk onto a moving walkway going in your direction and continue to walk, your walking speed is increased by the speed of the walkway.

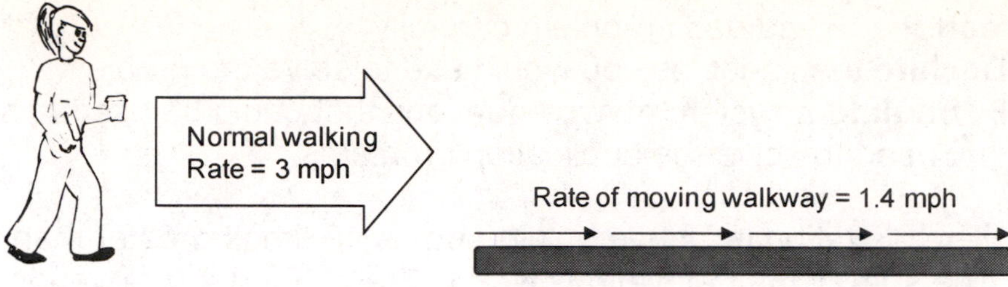

1. Using the information from the diagram above, fill in the missing rates.

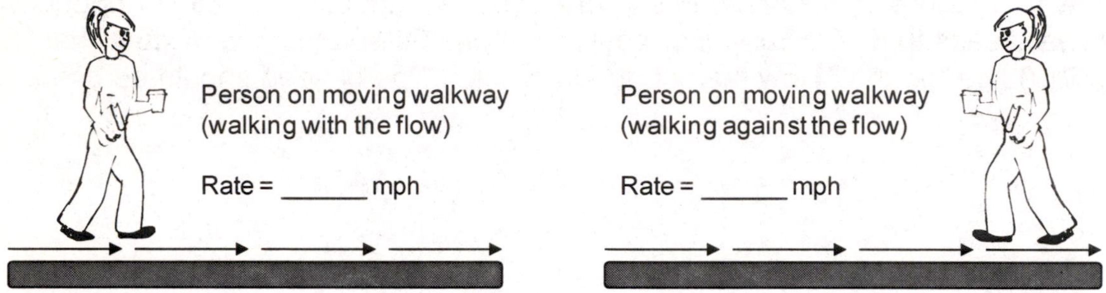

2. Often, you don't know one of the rates in a problem. In these cases, you have to express the rate as an algebraic expression. Assume the person is jogging at a rate of 4.7 mph and the rate of the moving walkway is x mph. Fill in the missing rates.

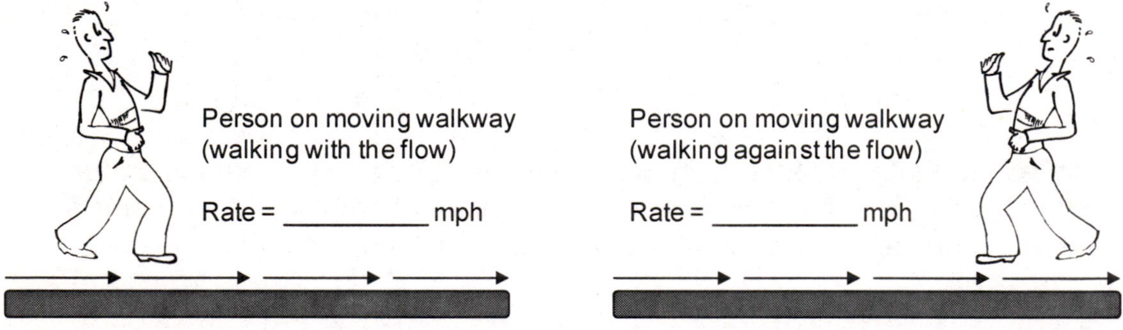

3. Now the rate of the person is R mph and the rate of the walkway is 2.3 mph. Fill in the missing rates.

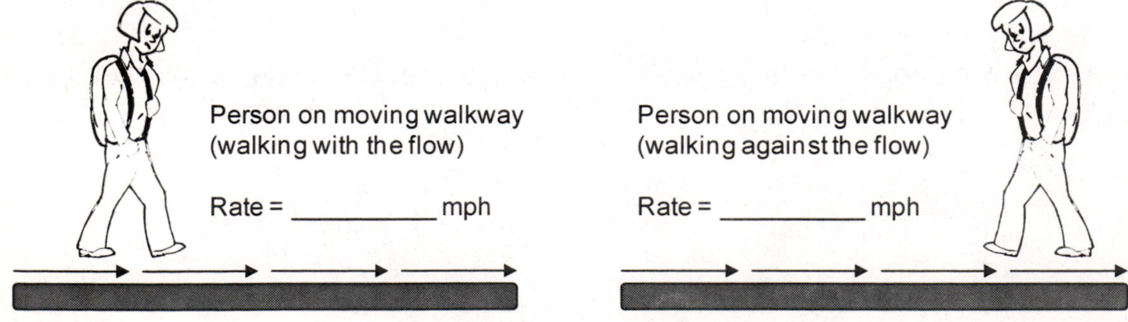

4. At the Montparnasse station in Paris, there is a high-speed moving walkway that was designed to help commuters get through the station faster. This moving walkway, which is 0.18 km long, moves at a speed of 9 km/hr (originally the walkway was designed to go 11 km/hr, but there were too many accidents). David walks at his normal pace in the direction of flow on the walkway – it takes 54 seconds to go the full distance of the walkway. How fast was David walking?

5. While on a layover at the Detroit International Airport, a bored math teacher decides to calculate the speed of one of the many moving walkways in the airport. He knows (from years of treadmill walking) that her normal walking speed is 2.5 mph (or 2.2 ft/sec). Walking with the flow of the walkway, it takes 43 seconds to go the distance of the walkway. Walking against the flow of the walkway, it takes 2 minutes and 47 seconds. What is the approximate rate of the moving walkway in feet per second and what is the length of the walkway in feet?

	Rate	Time	Distance
With Flow of Walkway			
Against Flow of Walkway			

6. The Central-Mid-Levels Escalator in Hong Kong is the world's longest covered escalator, stretching 800m (2,625 ft). A trip on the escalator, from top to bottom, takes about 20 minutes when a person stands still. What is the approximate rate of the escalator in feet per second? How long would the trip take if the person walked downward at a rate of 2 ft/sec?

Student Activity

Tic-tac-toe on Inequalities

Directions for Tic-tac-toe #1: If the ordered pair in the square **IS** a solution of the system of inequalities shown below, then circle the ordered pair (thus putting an **O** on the square). If it **IS NOT** a solution, then put an **X** over the ordered pair.

System of Inequalities: $\begin{cases} y < 4x + 2 \\ 3x - 3y \geq 6 \end{cases}$

$(-2,-6)$	$(1,1)$	$(4,1)$
$(6,2)$	$(8,6)$	$(4,4)$
$(2,0)$	$(0,2)$	$(0,-2)$

Directions for Tic-tac-toe #2: If the ordered pair in the square **IS** a solution of the graphed system of inequalities, then circle the ordered pair (thus putting an **O** on the square). If it **IS NOT** a solution, then put an **X** over the ordered pair.

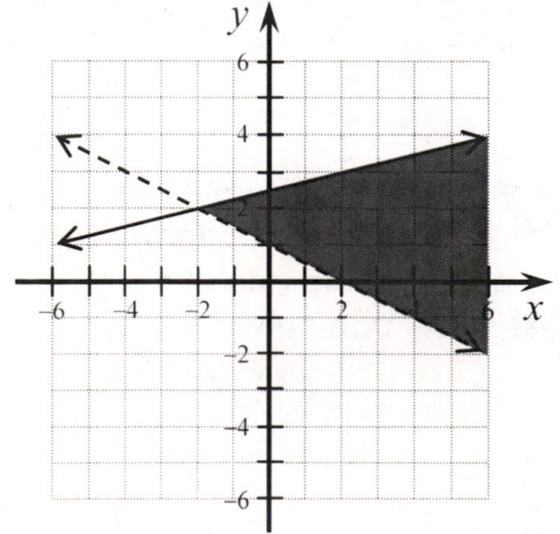

$(-2,2)$	$(2,0)$	$(2,3)$
$(8,3)$	$(3,4)$	$(0,4)$
$(6,4)$	$(6,-2)$	$(4,0)$

Student Activity

Testing Test Points

Directions: In each problem below you are given a system of inequalities, and the graphed inequality boundary lines. In each region separated by the boundary lines, a test point is provided. Test this point in both inequalities to determine if it is true (T) or false (F). Shade the region of the graph where the ordered pair **is** a solution to both inequalities. The first test point has been done for you.

1.

	$y<\frac{1}{2}x$	$y>2x-8$
$(0,3)$	F	T
$(2,-2)$		
$(-5,0)$		
$(-4,-4)$		

2.

	$y\le-\frac{1}{3}x+2$	$y\ge-\frac{3}{2}x-\frac{3}{2}$
$(-6,5)$		
$(-6,1)$		
$(6,4)$		
$(6,-2)$		

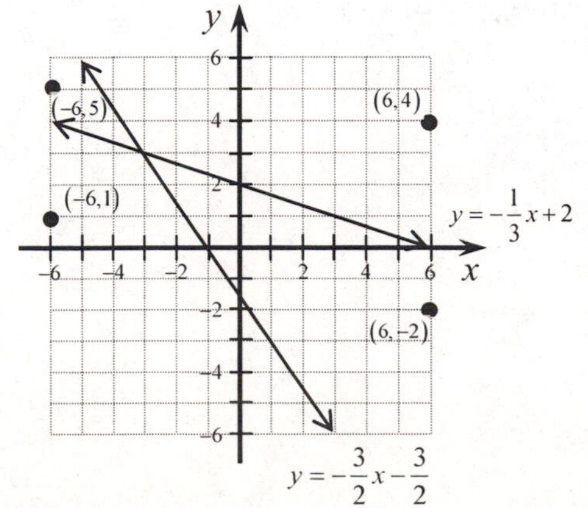

3.

	$y<\frac{2}{3}x+4$	$y>\frac{2}{3}x$
$(-6,3)$		
$(-3,1)$		
$(3,-2)$		

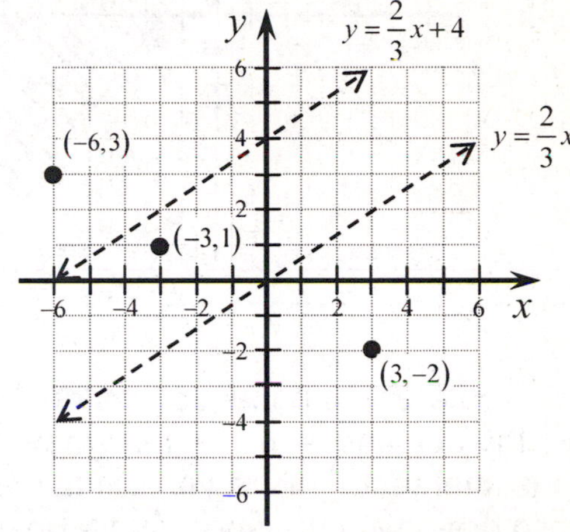

4.

	$y\ge-\frac{5}{4}x+\frac{7}{2}$	$y\le-\frac{5}{4}x-\frac{5}{2}$
$(-2,-3)$		
$(0,0)$		
$(2,4)$		

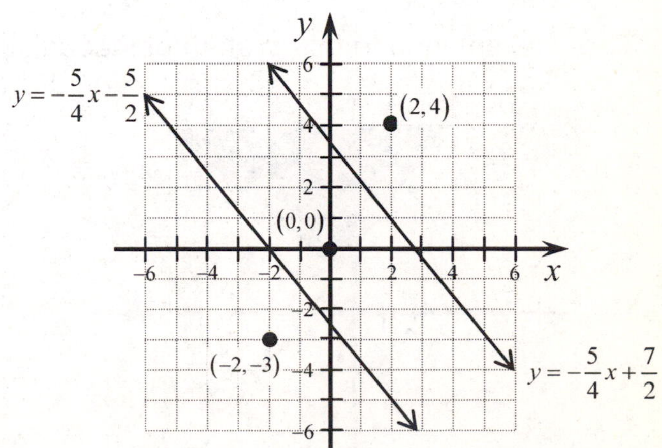

Guided Learning Activity

Graphing Systems of Linear Inequalities

Example: Consider the system of inequalities given by $y \ge -\frac{4}{3}x + 2$ and $y > \frac{5}{6}x + 1$.

What does the solution of a system of inequalities look like? Separately, the graphs of our inequalities look like this:

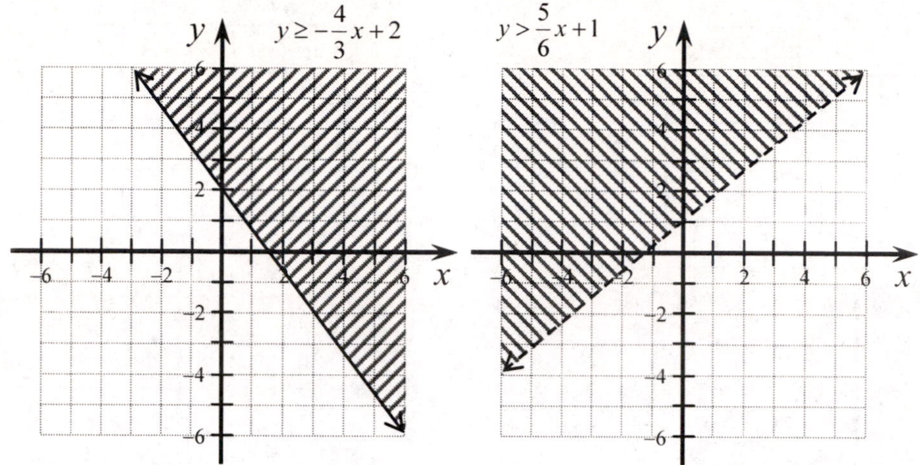

When you graph the inequalities on the same set of axes, the solution is the region where the two graphs intersect. This region of intersection (shown in solid gray) with the appropriate boundary lines is the solution to the system of inequalities.

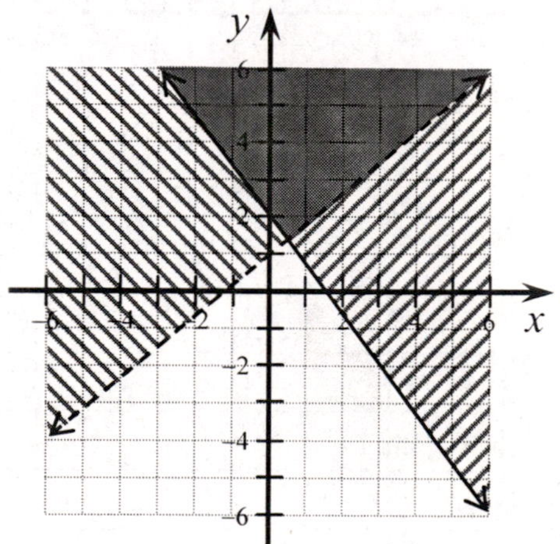

The final solution to this system of inequalities is shown below.

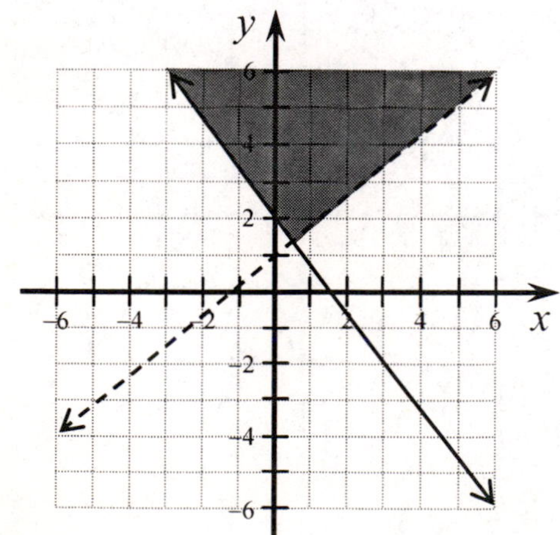

Now try your hand at graphing the solutions to these systems of inequalities.

1. $\begin{cases} y < -2x+3 \\ y < 2x-3 \end{cases}$

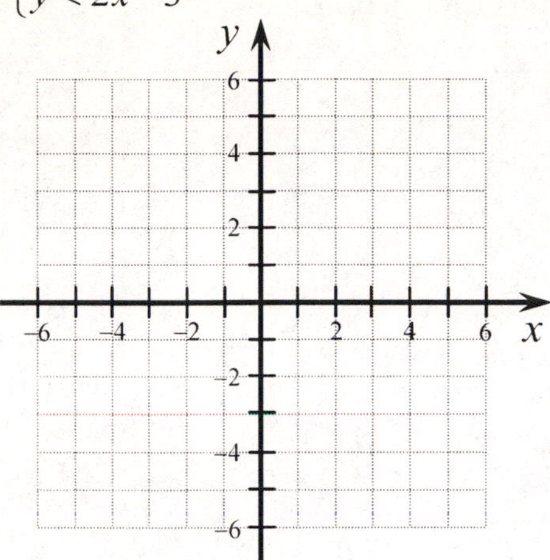

2. $\begin{cases} 2-y \le 0 \\ y-\frac{1}{2}x \le 4 \end{cases}$

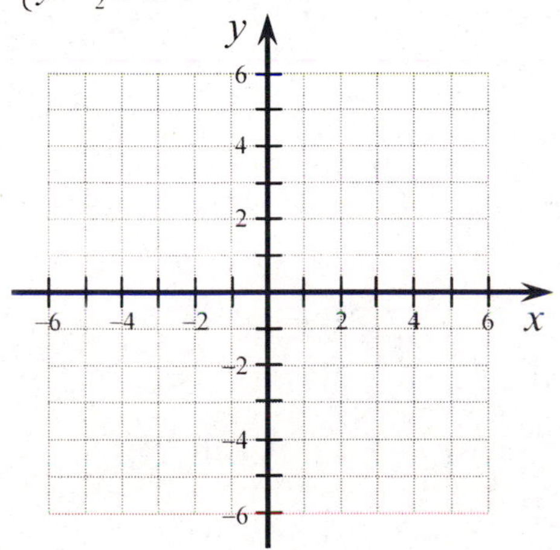

3. $\begin{cases} y > \frac{1}{3}x+1 \\ y < \frac{1}{3}x-3 \end{cases}$

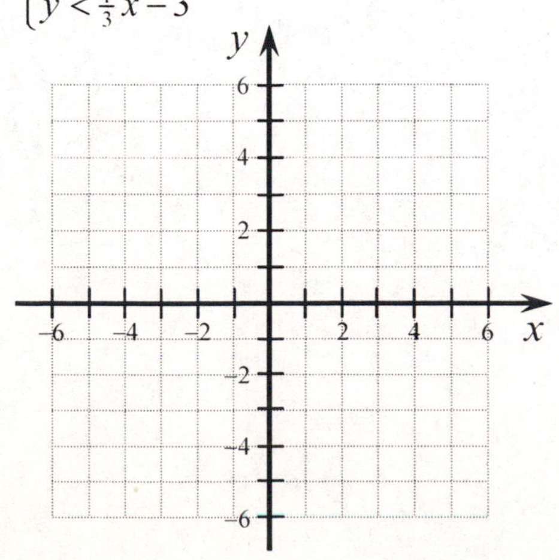

4. $\begin{cases} 3x+y \ge -5 \\ -3x-y \ge -5 \end{cases}$

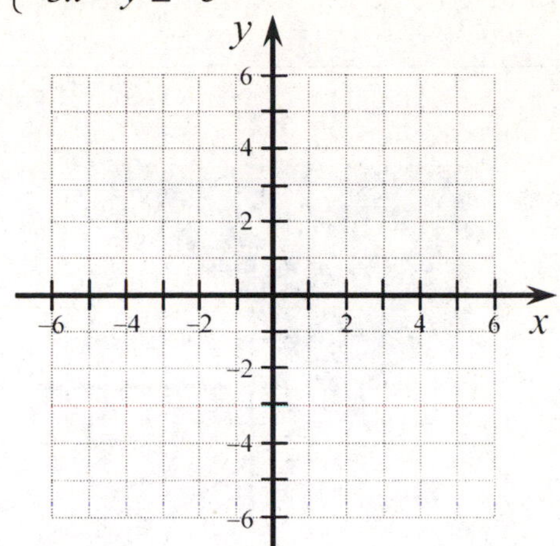

5. $\begin{cases} y \le 0 \\ x \le 0 \end{cases}$

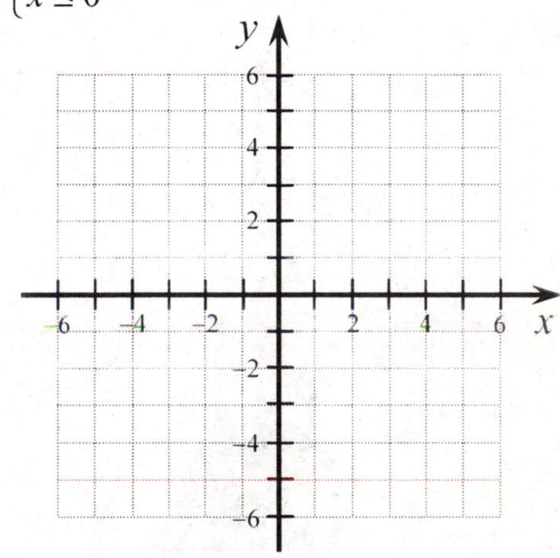

6. $\begin{cases} y \le -\frac{1}{2}x+2 \\ y > 2x+2 \end{cases}$

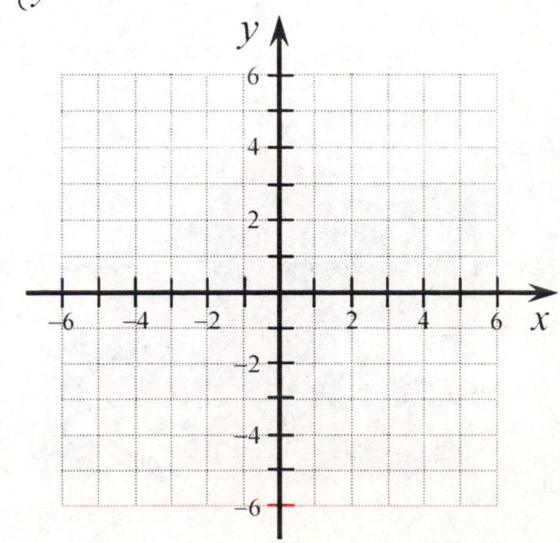

Student Activity

Following the Clues Back to the System of Inequalities

Directions: In each "crime-scene" below, you are shown the graph of a system of inequalities. Use your mathematical powers of reasoning (and detective skills) to determine what the inequality must have been to result in this graph.

1.

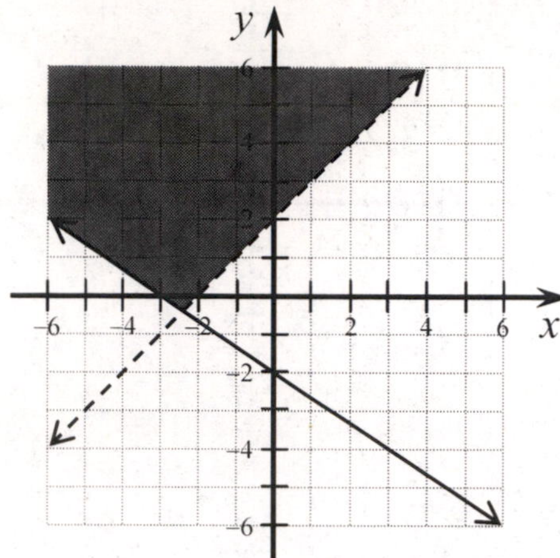

2.

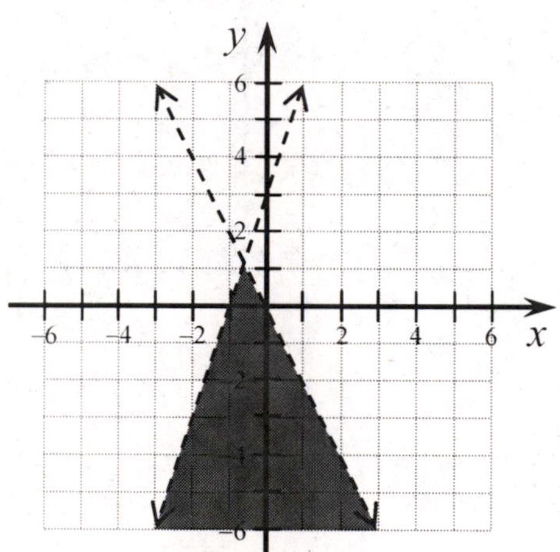

3.

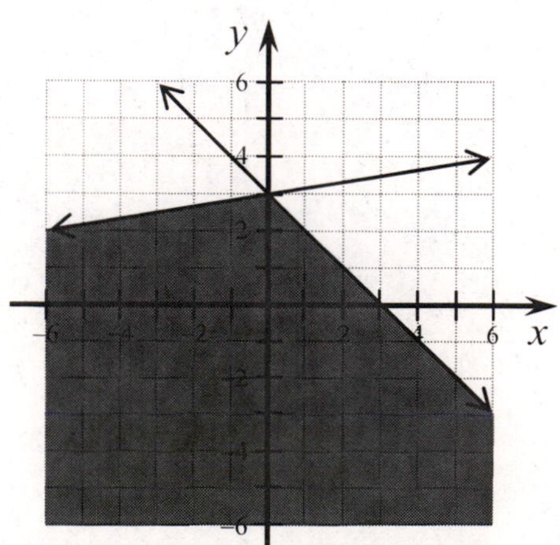

Assess Your Understanding

Solving Systems

For each of the following, describe the strategies or key steps that will help you **start** the problem. You do **not** have to complete the problems.

		What will help you to start this problem?
1.	Use substitution to solve the system: $\begin{cases} x-3y=-1 \\ 3x-4y=7 \end{cases}$	
2.	Is the system below consistent or inconsistent? $\begin{cases} x-4y=6 \\ 2x=8y+10 \end{cases}$	
3.	Is $(1,-2)$ a solution to the system? $\begin{cases} 3x+y=1 \\ x-y=-1 \end{cases}$	
4.	Use elimination to solve the system: $\begin{cases} x+3y=-2 \\ x+4y=0 \end{cases}$	
5.	Solve the system. $\begin{cases} x=y-7 \\ x-2y=14 \end{cases}$	
6.	Is the pair of equations dependent or independent? $\begin{cases} x-4y=6 \\ 2x=8y+10 \end{cases}$	
7.	Solve the system by graphing: $\begin{cases} y=-2x+4 \\ y=2x \end{cases}$	
8.	Solve the system using the method of your choice: $\begin{cases} 3x-4y=7 \\ 5x-8y=8 \end{cases}$	

		What will help you to start this problem?
9.	Is $(1,-3)$ a solution to the system? $\begin{cases}3x+y<1\\x-y>-1\end{cases}$	
10.	Rafi paddles his canoe upstream 24 miles in 4 hours. If he can make the downstream trip in 3 hours, find the rate that Rafi can paddle and the rate of the current.	
11.	Solve the system by the method of your choice: $\begin{cases}y-3x=2\\4x-y=-1\end{cases}$	
12.	Two angles are supplementary and one angle measures 30° more than the other angle. Find the measure of each angle.	
13.	Graph the solution to the system: $\begin{cases}x\le 3\\y\le \frac{1}{2}x-1\end{cases}$	
14.	Is the pair of equations dependent or independent? $\begin{cases}3x+y=8\\6x+2y=8\end{cases}$	
15.	Graph the solution to the system: $\begin{cases}3x+y<4\\2y-x>-2\end{cases}$	

Metacognitive Skills

Solving Systems

Metacognitive skills refer to the ability to judge how well you have learned something and to effectively direct your own learning and studying. This is a self-evaluation tool designed to help you focus your studying and to improve your metacognitive skills with regards to this math class.

Fill the 1st column out **before** you begin studying. Fill the 2nd column out after you study for your test.

Go back to this assessment after your test and circle any of the ratings that you would change – this identifies the "disconnects" between what you **thought** you knew well and what you **actually** knew well.

Use the scale below to assign a number to each topic.

5 *I am confident I can do any problems in this category correctly.*
4 *I am confident I can do most of the problems in this category correctly.*
3 *I understand how to do the problems in this category, but I still make a lot of mistakes.*
2 *I feel unsure about how to do these problems.*
1 *I know I don't understand how to do these problems.*

Topic or Skill	**Before Studying**	**After Studying**
Checking whether an ordered pair is a solution of a system of equations.		
Classifying a system of equations as consistent or inconsistent.		
Classifying a system of equations as dependent or independent.		
Graphing a system of equations to find the solution.		
Understanding what the "solution" to a system of equations is.		
Finding or identifying a substitution equation for a system of equations.		
Solving a system of equations by substitution.		
Clearing the fractions from an equation or inequality.		
Identifying a variable to eliminate and performing the steps to set up that elimination in a system of equations.		
Solving a system of equations by elimination.		
Deciding whether substitution or elimination would be an easier method for solving a system of equations.		
Clearing the decimals from an equation or inequality.		
Declaring the variables for an application involving a system of equations.		
Writing the system of equations for a problem involving $d = rt$.		
Writing the system of equations for a value-mixture problem or a percent-solution problem.		
Writing the system of equations for an investment problem.		
Solving a system of equations from an application problem.		
Writing the conclusion to an application problem involving a system of equations.		
Checking whether an ordered pair is a solution of a system of inequalities.		
Determining whether the boundary lines for a system of inequalities are dashed or solid.		
Graphing the boundary lines for a system of inequalities.		
Determining the proper region to shade in a system of inequalities.		

EXP: Exponent Rules

Student Activity

Exponent Duos

Directions: In each of the "duos" below, place two equivalent exponential expressions of the following format:

Expanded Expression using multiplication
Compact exponential expression

The first two duos have been done for you.

$(3x)(3x)(3x)$	$3\cdot x\cdot x\cdot x$	$(-5x)(-5x)$	$-(5x)(5x)$
$(3x)^3$	$3x^3$		

		$(x+1)(x+1)$	
$(2a^2)^3$	$2a^5$		$(a+b)^2$

$\left(\frac{1}{2}\right)\left(\frac{1}{2}\right)\left(\frac{1}{2}\right)\left(\frac{1}{2}\right)$		$(10ab)(10ab)$	$6\cdot x\cdot x\cdot y\cdot y$
	$\left(-\frac{2}{3}\right)^2$		

$7\cdot m\cdot m\cdot m\cdot p$		$\left(\frac{1}{z}\right)\left(\frac{1}{z}\right)$	$(a-2)(a-2)$
	$\left(\frac{a}{b}\right)^3$		

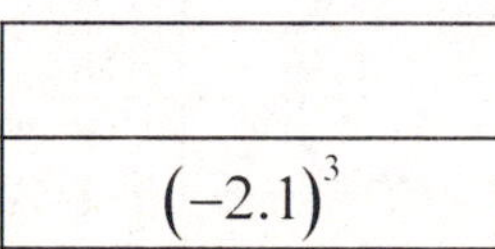

$-(4x)(4x)(4x)$		$\pi\cdot r\cdot r$	
	$(-2.1)^3$		$\frac{4}{3}\pi r^3$

Student Activity

Match Up on Basic Exponent Rules

Match-up: Match each of the expressions in the squares of the grid below with an equivalent simplified expression from the top. If an equivalent expression is not found among the choices A through D, then choose E (none of these).

A x^5

B x^6

C $9x^4$

D $12x^6$

E None of these

$(6x^2)(2x^4)$	$4x^4-(-5x^4)$	$x^3\cdot x^2$	$(-3x^3)(-3x)$	$x^5\cdot x$
$6x^3+6x^3$	$-4x^5+5x^5$	$12(x^2)^3$	$(-3x^3)(-4x^3)$	x^5+x
$(x^2)(x^2)(x)$	$(27x^3)\left(\frac{1}{3}x\right)$	$(x^3)^2$	$(12x^3)^2$	$(-3x^2)^2$
$(3x^2)^2$	$3(-2x^3)^2$	$3x^3+6x$	$(5x^2)\left(\frac{1}{5}x^3\right)$	$8x^6+4x^6$

Guided Learning Activity

Zero and Negative Exponents

To develop the concepts of zero and negative exponents, we will first look at several patterns of values.

1. Fill in the blanks to complete each sequence below:

$81, 27, 9,$ ___, ___, ___, ___ How do you get the next term? ________________

$16, 8, 4,$ ___, ___, ___, ___ How do you get the next term? ________________

$x^4, x^3, x^2,$ ___, ___, ___ How do you get the next term? ________________

$4^4, 4^3, 4^2,$ ___, ___, ___ How do you get the next term? ________________

$\frac{1}{125}, \frac{1}{25}, \frac{1}{5},$ ___, ___, ___ How do you get the next term? ________________

2. Now complete each table below by finishing the patterns:

81	3^4
27	3^3
9	3^2

16	2^4
8	2^3
4	2^2

$\frac{1}{125}$	
$\frac{1}{25}$	
$\frac{1}{5}$	
	5^2
	5^3

64	4^3
	4^2
	4^1

3. Based on your observations in the patterns above, finish the exponent rules below:

Rule for Zero Exponents:

$x^0 =$ ____ (for $x \neq 0$)

Rule for Negative Exponents:

$x^{-1} =$ ____ (for $x \neq 0$)

$x^{-2} =$ ____

$x^{-3} =$ ____

$x^{-n} =$ ____

4. Apply the rules you have just developed to write each of these exponential expressions without zero or negative exponents.

$6^{-2} =$ $10^0 =$ $y^{-4} =$ $2^{-5} =$ $z^{-1} =$ $(4a)^0$

Student Activity

Match Up on Trickier Exponent Rules

Match-up: Match each of the expressions in the squares of the grid below with an equivalent simplified expression from the top. If an equivalent expression is not found among the choices A through D, then choose E (none of these).

Mark my words! You harness that negative power of yours, and you can make it to the top just like me!

A 1 **B** $\dfrac{4}{x^2}$

C $9x^2y^3$ **D** $\dfrac{-9x^4}{y^3}$

E None of these

$\left(4x^{-2}\right)^0$	$(4x)^{-2}$	$4x^{-2}$	$4x^0$	$(2x)^{-2}$
$\left(\dfrac{x}{2}\right)^{-2}$	$\dfrac{(-3xy)^2}{y^{-1}}$	$\dfrac{3^{-2}y^{-3}}{x^{-4}}$	$\dfrac{-\left(3x^2y\right)^2}{y^5}$	$\left(\dfrac{100x^{27}y^{35}}{a^4b^5}\right)^0$
$\left(\dfrac{x}{2}\right)^2$	$\left(9x^2y^3\right)^0$	$\left(-\dfrac{y}{a^4b^4}\right)\left(\dfrac{3xab}{y}\right)^4$	$y^7\left(\dfrac{y^2}{3x}\right)^{-2}$	$(2yz)^2(xyz)^{-2}$
$4\left(\dfrac{1}{x^2}\right)^0$	$8x^2\left(\dfrac{x^{-2}}{8}\right)$	$3\left(x^2y^2\right)\left(3x^2y^2\right)^{-1}$	$\dfrac{\left(2x^{-1}z^2\right)^2}{z^4}$	$\dfrac{-12x^4}{5}\left(\dfrac{5}{-12x^4}\right)$

Student Activity

Double the Fun on Exponent Rules

Directions: For many problems involving the simplification of exponents, there are at least two ways to tackle the problem. For each of the problems below, try the problem both ways – your answers should be the same for both methods. If they are not, you'll have to go back and look for a mistake. The first one has been done for you.

Simplify this expression	Apply exponent rules directly	First move factors to avoid working with negative exponents
$\dfrac{x^{-2}y^{5}}{x^{-6}y^{-1}}$	$\dfrac{x^{-2}y^{5}}{x^{-6}y^{-1}} = x^{-2-(-6)}y^{5-(-1)}$ $= x^{-2+6}y^{5+1} = x^{4}y^{6}$	$\dfrac{x^{-2}y^{5}}{x^{-6}y^{-1}} = \dfrac{x^{6}y^{5}y^{1}}{x^{2}} = x^{6-2}y^{5+1} = x^{4}y^{6}$
$\dfrac{16a^{3}b^{-2}}{8a^{-1}b^{5}}$		
$3x^{-2}x^{3}x^{-1}$		
$(2x)^{-2}(-6x^{3})$		
$\left(\dfrac{x^{-2}y^{5}}{x^{-1}y^{6}}\right)^{3}$		
$\dfrac{(xy^{3})^{-1}}{2x^{4}y^{-1}}$		

Student Activity

Mathematical Heteronyms

Directions: In writing, there are words that are spelled the same but have different pronunciations and different definitions; these are called heteronyms. Many mathematical expressions look similar but are really very different (almost like mathematical heteronyms). In each set of expressions below, pay close attention to the use of parentheses and the mathematical operations and notation.

1.

$3-1$	$3(-1)$	3^{-1}	$3-(-1)$	-3^{-1}	$3\div(-1)$

2.

$2-5$	$2(-5)$	2^{-5}	$2^{-1}-5$	$2^{-1}-5^{-1}$	$2-5^{-1}$

3.

x^2x^{-3}	$x^2\left(-x^3\right)$	x^2x^3	$x^{-2}x^3$	x^2-x^3	$\left(x^2\right)^{-3}$

4.

$2a^{-3}$	$(-2a)^3$	$2^3(-a)$	$2-a^3$	$(2a)^{-3}$	$(-2a)\cdot 3$

5.

$\left(\frac{3}{4}\right)^{-2}$	$\frac{3^{-2}}{4}$	$\left(\frac{3}{4}\right)(-2)$	$\frac{3}{4}-2$	$\frac{3}{4^{-2}}$	$\left(-\frac{3}{4}\right)^2$

Student Activity

Exponents Using a Calculator

When you input exponents into a calculator, you must be careful to tell the calculator which part is the base and which part is the exponent. Each calculator requires a specific set of keystrokes to evaluate exponential expressions and this can become quite complicated when the expression also involves negatives. We will fill out the table below, first performing each calculation by hand, then finding the proper keystrokes to do the evaluation on your calculator.

First you need to locate your exponent button. It may look like $\boxed{\wedge}$ or $\boxed{x^y}$.
Reminders: There is a difference between the minus key and the negative key on your calculator. Fractions need to be placed inside parentheses for proper evaluation.

	Expression	Evaluate by hand	Calculator Keystrokes to get equivalent result
a.	2^3		
b.	2^{-1}		
c.	5^{-2}		
d.	$\left(\frac{1}{2}\right)^3$		
e.	$\left(\frac{2}{3}\right)^{-2}$		
f.	-2^4		
g.	$(-2)^4$		
h.	-4^{-2}		
i.	$(-4)^{-2}$		
j.	$2^{-1}+3^{-1}$		
k.	$(2+3)^{-1}$		
l.	$5-2^{-1}$		

Guided Learning Activity

Charting Scientific Notation

Scientific notation is based on using powers of 10. Fill in the Powers of 10 with their decimal notation in the table below. 10^3 has been done for you.

Power of 10	10^0	10^1	10^2	10^3	10^4	10^5	10^6
Value				1,000			

Power of 10	10^{-6}	10^{-5}	10^{-4}	10^{-3}	10^{-2}	10^{-1}	10^0
Value							

Scientific notation: $N \times 10^n$ where $1 \le N < 10$, and n is an integer.
With your class, fill out the table below. The first one has been done for you.

Number in standard notation (decimal notation)	**Put together the $N \times 10^n$ part in pieces**			
	N	**2nd factor in decimal notation**	**2nd factor as a power of 10**	$N \times 10^n$
0.00495	4.95	0.001	10^{-3}	4.95×10^{-3}
8,000,000,000				
0.00000034				
				9.2×10^5
	6.843	0.0000001		
				8.04×10^{-7}
	2	10,000,000		

Rather than remembering whether the decimal place moves left or right, it is easier to remember that 10 raised to negative powers means you are looking at a number that is small (between zero and 1) and 10 raised to positive powers means you are looking at a number that is large (>1). Simply move the decimal point the correct number of digits in the direction that creates the appropriate bigger or smaller number!

Decimal Value	**Is it a *small* or *large* number?**	**Is the power on the 10 *positive* or *negative?***	**Scientific Notation**
0.0000725			
			1.2×10^{7}
			4.534×10^{-6}
84,000,000,000			
48			
			9×10^{-4}
674,000			
			2.99×10^{8}
			6.67×10^{-11}
0.000000056704			

Student Activity

Escape the Matrix with Ten Power

Directions: Begin at the box marked START. By shading in pairs of adjacent squares that represent equal numbers, you will eventually find the path to "escape" this matrix of boxes. The first "step" in the path and two of the middle steps in the path have been shaded for you.

I've got TEN power!

START 0.00043	4.3×10^{-4}	5×10^{6}	5,000,000	7.6×10^{3}
	4.3×10^{-3}	500,000	5.0×10^{7}	50×10^{6}
	789×10^{-2}	7.89	0.023	2.3×10^{-2}
39×10^{8}	390,000,000	3.90×10^{7}	2.3×10^{2}	23,000
16×10^{4}	3.90×10^{8}	1,000,000	10×10^{5}	5.555×10^{-3}
1.6×10^{2}	0.00016	1×10^{5}	10×10^{3}	0.005555
32,300	32.30×10^{5}	3.230×10^{6}	0.003230×10^{6}	3230
3.230×10^{4}	6.89×10^{4}	689	689×10^{-1}	323×10^{0}
323.0×10^{-2}	3.230	32.30	3.230×10^{1}	**ESCAPE** the Matrix

Student Activity
Scientific Notation Using a Calculator

There is a special button on most calculators that is used for scientific notation. Sometimes it is denoted with $\boxed{\text{EE}}$ or $\boxed{\text{EXP}}$. This button stands for "$\times 10$ to the," so, for example we could enter 4×10^3 as $\boxed{4}$ $\boxed{\text{EE}}$ $\boxed{3}$. When using the scientific notation button, you **do not need to** multiply by 10 too.

1. First we need to find the scientific notation button on your calculator. Try entering the expression: 4×10^3 then press $\boxed{\text{ENTER}}$ or $\boxed{=}$. You should see either 4000 or 4e3 or $4\ ^{03}$ or something like this. Ask your instructor if you are unsure about whether you have correctly located the scientific notation button.

Draw the series of keystrokes you used to enter 4×10^3 here:

2. Try another one: 2.5×10^{12} and press $\boxed{\text{ENTER}}$ or $\boxed{=}$. This number is too big to display on most calculator screens, so you should see the calculator give the number to you in it's version of scientific notation, like 2.5e12, 2.5E12, or $2.5\ ^{12}$.

Draw how you entered 2.5×10^{12} **and** the way the number appears on the screen:

3. Now we try a number with a negative exponent. Key in the number 3.4×10^{-3} followed by $\boxed{\text{ENTER}}$ or $\boxed{=}$. This should display as 0.0034, 3.4e-3, or $3.4\ ^{-03}$. Remember to use the negative key and not the minus key to enter the negative exponent.

Draw the series of keystrokes you used to enter 3.4×10^{-3} here:

4. Try another one: 4.62×10^{-14} and press $\boxed{\text{ENTER}}$ or $\boxed{=}$. This number has too many zeros to display on most calculator screens, so you should see the calculator give the number to you in it's version of scientific notation, like 4.62e-14, 4.62E-14, or $4.62\ ^{-14}$.

Draw how you entered 4.62×10^{-14} **and** the way it appears on the screen:

The real usefulness of the scientific notation button becomes obvious when we begin to perform calculations that involve scientific notation.

For example, if I wanted to calculate $\dfrac{2.4\times10^{-3}}{8\times10^{-12}}$ **without** the scientific notation button, here's what I'd have to enter on *my* calculator:

[(] 2.4 [×] 10 [^] [(−)] 3 [)] [÷] [(] 8 [×] 10 [^] [(−)] 12 [)]

Using the scientific notation button, this is much easier:

2.4 [EE] [(−)] 3 [÷] 8 [EE] [(−)] 12

5. Try the calculation $\dfrac{2.4\times10^{-3}}{8\times10^{-12}}$ on your calculator. (the answer is 3×10^{8})

Draw how you entered $\dfrac{2.4\times10^{-3}}{8\times10^{-12}}$ here:

For practice, try these calculations and write down the answers using the proper scientific notation. In other words, don't write an answer like $8e3$; write the answer as 8×10^{3} instead.

6. $\left(8.2\times10^{9}\right)\left(2\times10^{4}\right)$

7. $\dfrac{1.5\times10^{-4}}{2.5\times10^{3}}$

8. $\left(1.1\times10^{-5}\right)^{2}$

9. $\dfrac{\left(5.4\times10^{-6}\right)\left(2\times10^{3}\right)}{9\times10^{8}}$

10. $\dfrac{3.6\times10^{-4}}{\left(1.2\times10^{3}\right)\left(1.5\times10^{-10}\right)}$

Assess Your Understanding

Exponent Rules

For each of the following, describe the strategies or key steps that will help you **start** the problem. You do **not** have to complete the problems.

		What will help you to start this problem?
1.	Evaluate: 3^{-2}	
2.	Simplify: $\dfrac{3a^2b^5}{a^3b}$	
3.	Simplify: $\left(-4x^3\right)^2$	
4.	Write 0.000027 in scientific notation.	
5.	Simplify: $5x^0$	
6.	Multiply: $\left(3.1\times10^5\right)\left(2\times10^{-3}\right)$	
7.	Simplify $\dfrac{6x^{-3}}{-2x^2}$ and write without negative exponents.	
8.	Write 3.14×10^5 in standard notation.	
9.	Simplify: $\left(-4x^2y\right)\left(2x^3y^4\right)$	
10.	Simplify: $\left(3x^{-2}\right)^3\left(2x^4\right)$	

Metacognitive Skills

Exponent Rules

Metacognitive skills refer to the ability to judge how well you have learned something and to effectively direct your own learning and studying. This is a self-evaluation tool designed to help you focus your studying and to improve your metacognitive skills with regards to this math class.

Fill the 1st column out **before** you begin studying. Fill the 2nd column out after you study for your test.

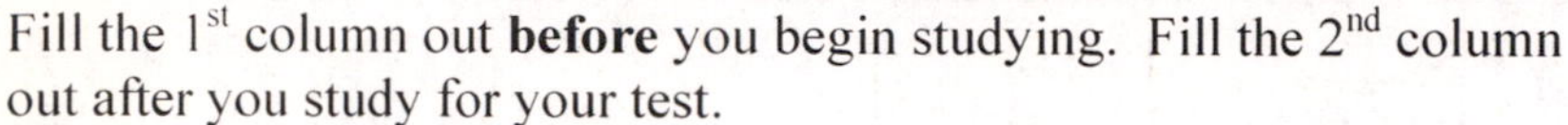

Go back to this assessment after your test and circle any of the ratings that you would change – this identifies the "disconnects" between what you **thought** you knew well and what you **actually** knew well.

Use the scale below to assign a number to each topic.

5 *I am confident I can do any problems in this category correctly.*
4 *I am confident I can do most of the problems in this category correctly.*
3 *I understand how to do the problems in this category, but I still make a lot of mistakes.*
2 *I feel unsure about how to do these problems.*
1 *I know I don't understand how to do these problems.*

Topic or Skill	Before Studying	After Studying
Understanding what an exponent represents (what does 2^5 mean?).		
Understanding how expressions like $(-3)^2$ and -3^2 OR $2x^2$ and $(2x)^2$ are different.		
Knowing the product, quotient and power rules for exponents.		
Correctly applying the product, quotient, and power rules for exponents.		
Knowing the exponent rules for zero and negative exponents.		
Rewriting an expression to eliminate the negative exponents.		
Simplifying exponential expressions involving negative exponents.		
Simplifying exponential expressions that have zero exponents.		
Simplifying expressions involving many exponent rules all mixed up (terms in parentheses, fractions, negative exponents, zero exponents, etc.).		
Converting back and forth between standard notation and scientific notation.		
Performing operations (like multiplication and division) using scientific notation		
Using the scientific notation button on a calculator to perform calculations involving scientific notation.		

POLY: Simplifying Polynomials

Guided Learning Activity

Language of Polynomials

A **polynomial** is a single term or a sum of terms in which all variables have whole-number exponents and no variable appears in a denominator. Recall that the **terms** of an algebraic expression are separated by addition (remember that subtraction can be rewritten as addition of a negative term). A polynomial with exactly one term is called a **monomial**; exactly two terms, a **binomial**; and exactly three terms, a **trinomial**. A polynomial can be written with one or more variables.

With your class, fill out the table below. The first one has been done for you.

Expression (rewrite with addition symbols if necessary)	**How many variables?**	**Classification**			
		Monomial	**Binomial**	**Trinomial**	**Polynomial**
$5x^2-2x+7$ $5x^2+(-2x)+7$	1			X	X
$y-2.5$					
$-5x^2y$					
$4x^2-\frac{2}{x^3}$					
$b^3+\frac{1}{3}b^2-\frac{1}{2}b+12$					
$x^2-3xy-10y^2$					

Polynomials are often written in **descending powers** of the variable (the variable exponents decrease from left to right). When a polynomial is written in descending powers, the first term is called the **lead term**. The coefficient of the lead term is called the **lead coefficient**. Recall that a term that consists of a single number is called the **constant term**.

Expression	Rewrite the expression in descending powers of x	Lead term	Lead coefficient	Constant term (if there is one)
$4x-x^2$	$-x^2+4x$	$-x^2$	-1	None
$3x+5-2x^2$				
$x^3+200x+300x^2$				
$7x^3-2-3x^4$				

The **degree of a term** of a polynomial in one variable is the value of the exponent on the variable. Thus, the degree of $7x^4$ is equal to 4. If the polynomial has more than one variable, the degree of a term is the *sum* of the exponents on the variables in that term. Thus, the degree of $7x^2y^3$ is 5. For a constant term, we can imagine an unwritten variable with a zero power, consider that 7 could be written as $7x^0$. Thus, the **degree of a nonzero constant** is zero. We can also discuss the **degree of a polynomial** which is the same as the degree of the highest degree term of the polynomial.

Expression	Degree of...				Degree of polynomial
	1st term	2nd term	3rd term	4th term	
$5x^2-3x+7$	2	1	0	None	2
$y-2.5$					
$-5x^4y+2x^2y^2$					
$b^3+\frac{1}{3}b^2-\frac{1}{2}b+12$					
$x^2-3xy-10y^2$					
$x^3+300x^2+200x-1000$					

Student Activity

Match Up on Polynomial Evaluation

Remember that it may be helpful to first create a parentheses skeleton for each polynomial before you substitute the designated values.

For example, $x^2 - 3x$ would be first rewritten as $(\quad)^2 - 3(\quad)$.

Match-up: Match each of the answers in the squares of the grid below with its result in choices A through D. If you do not see the result in choices A through D, then choose choice E (none of these).

A 9 **B** 1 **C** −2 **D** 0 **E** None of these

Evaluate $5x^2 + 10x$ for $x = -2$.	Evaluate $5x^2 - 5x + \frac{9}{4}$ for $x = \frac{1}{2}$.	Simplify: $3x^2 + 5 - 2x^2 - 7 - x^2$
Evaluate $x^2 y$ for $x = -1$ and $y = -2$.	Evaluate $x^2 + 9x + \frac{23}{9}$ for $x = \frac{2}{3}$.	Evaluate $x^3 + x^2 + x - 2$ for $x = -2$
Simplify: $\frac{x^5}{2} - 2x^5 + \frac{3x^5}{2}$	Let $y = 3x^2 - 2x - 7$; find y when $x = 2$.	Let $y = 2.25x^3 + 7.75x - 1$; find y when $x = 1$.

Student Activity

Caught in the Net

Directions: Which fish get "trapped" between the two nets?

Graph $y = \frac{1}{9}x^3 + 1$ and $y = -\frac{3}{4}x^2 + 10$ (the "nets") to find out.

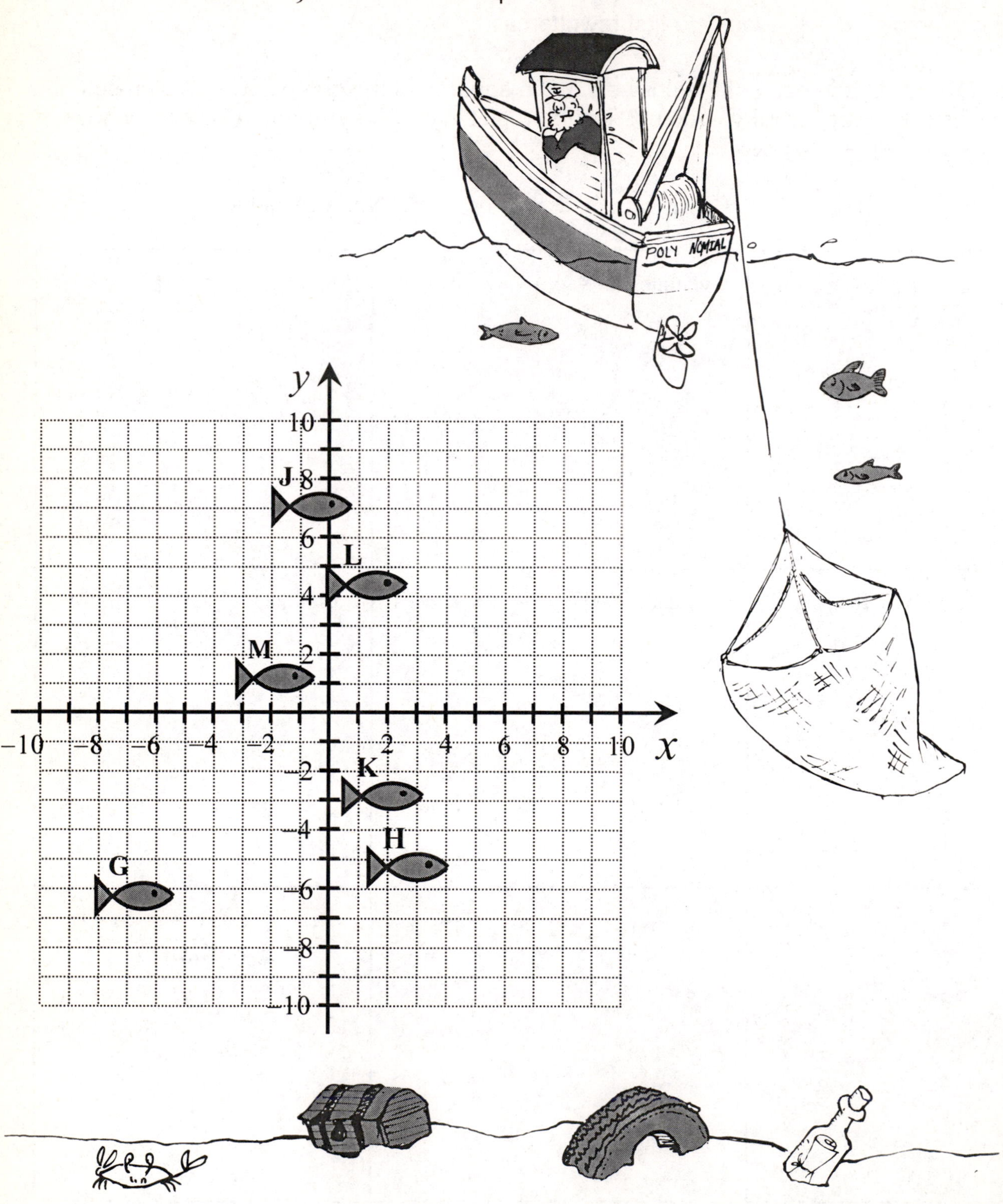

Guided Learning Activity

Vertical Form of Polynomial Addition and Subtraction

The expressions $342+605$ and $(3x^2+4x+2)+(6x^2+5)$ are simplified in almost the same way: For both expressions, we line up the like terms vertically; then add.

Example 1:

$$\begin{array}{rrrr} & 3 & 4 & 2 \\ + & 6 & 0 & 5 \\ \hline & 9 & 4 & 7 \end{array} \qquad \begin{array}{rlll} & \text{3 hundreds} & \text{4 tens} & \text{2 ones} \\ + & \text{6 hundreds} & \text{0 tens} & \text{5 ones} \\ \hline & \text{9 hundreds} & \text{4 tens} & \text{7 ones} \end{array} \qquad \begin{array}{rrrr} & 3x^2 & +4x & +2 \\ + & 6x^2 & +0x & +5 \\ \hline & 9x^2 & +4x & +7 \end{array}$$

One of the differences between the vertical addition of the expressions is that we cannot "carry" coefficients greater than 9 to the next column like we carry numbers greater than 9. Another difference is that we can have negative coefficients if that is part of the original expression.

Example 2: $(4x^2+5x+9)+(-2x^2+5x-6)$ becomes

$$\begin{array}{rrrr} & 4x^2 & +5x & +9 \\ + & -2x^2 & +5x & -6 \\ \hline & 2x^2 & +10x & +3 \end{array}$$

Now try these. Make sure to **line up the like terms** in the same columns and keep the signed coefficients with the terms in the columns.

a. $(y^3+3y^2-4y+2)+(9y^2+4y-7)$ **b.** $(a^2+4ab-2b^2)+(a^2+8ab-3b^2)$

In subtraction, we have to remember that the second row of each column is being subtracted from the first row **for every entry**. Below we look at the similar expressions $749-615$ and $(7x^2+4x+9)-(6x^2+x+5)$.

Example 3:

$$\begin{array}{rrrr} & 7 & 4 & 9 \\ - & 6 & 1 & 5 \\ \hline & 1 & 3 & 4 \end{array} \qquad \begin{array}{rlll} & \text{7 hundreds} & \text{4 tens} & \text{9 ones} \\ - & \text{6 hundreds} & \text{1 ten} & \text{5 ones} \\ \hline & \text{1 hundred} & \text{3 tens} & \text{4 ones} \end{array} \qquad \begin{array}{rrrr} & 7x^2 & +4x & +9 \\ - & 6x^2 & +1x & +5 \\ \hline & 1x^2 & +3x & +4 \end{array}$$

We've learned on many occasions that subtraction of grouped terms can be tricky as the -1 must be distributed to the grouped terms. Because some of the terms in the second row have a + in front of the coefficients or constants, it is confusing to think of this as a subtraction problem. To be clear, we should write the vertical format of subtraction like this:

$$\begin{array}{rrrr} & 7x^2 & +4x & +9 \\ - & (6x^2 & +x & +5) \\ \hline \end{array}$$

And since polynomial subtraction can be rewritten without the grouping as polynomial addition (after the distribution), we can further alter the problem:

$$\begin{array}{rrrr} & 7x^2 & +4x & +9 \\ - & (6x^2 & +x & +5) \\ \hline \end{array} \quad \text{becomes} \quad \begin{array}{rrrr} & 7x^2 & +4x & +9 \\ + & -6x^2 & -x & -5 \\ \hline & 1x^2 & +3x & +4 \end{array}$$

Example 4: Let's go through this process with $(4x^2+3x-6)-(3x^2-2)$

First we line up the like terms, still using the grouping symbols for the subtraction:

$$\begin{array}{rrrr} & 4x^2 & +3x & -6 \\ - & (3x^2 & +0x & -2) \\ \hline \end{array} \quad \text{becomes} \quad \begin{array}{rrrr} & 4x^2 & +3x & -6 \\ + & -3x^2 & -0x & +2 \\ \hline & 1x^2 & +3x & -4 \end{array}$$

Now try these. Write the polynomial subtraction in the vertical format, first with grouping symbols as subtraction, then performing the distribution by -1 and rewriting as addition.

c. $(y^3+9y^2-4y+2)-(6y^2+2y-7)$

d. $(a^2+4ab-2b^2)-(a^2-8ab+3b^2)$

e. $(6x^3-5x-4)-(2x^2-8x+3)$

Student Activity

When Does Order Matter?

Directions: Add or subtract in each of the expressions below and then combine the like terms. Write the simplified expression in descending powers of x.

1. Add: $(x^2+3x-7)+(4x^2+5x-2)$

2. Add: $(4x^2+5x-2)+(x^2+3x-7)$

3. What lesson can we learn from problems **1** and **2**?

---

4. Subtract: $(x^2+3x-7)-(4x^2+5x-2)$

5. Subtract: $(4x^2+5x-2)-(x^2+3x-7)$

6. What lesson can we learn from problems **4** and **5**?

---

7. Subtract: $(x^2+3x-7)-(4x^2+5x-2)$

8. Simplify: $x^2+3x-7-4x^2+5x-2$

9. What lesson can we learn from problems **7** and **8**?

---

10. Add: $(x^2+3x-7)+(4x^2+5x-2)$

11. Simplify: $x^2+3x-7+4x^2+5x-2$

12. What lesson can we learn from problems **10** and **11**?

Student Activity

Monomial Addition and Multiplication Tables

Here are simple addition and multiplication tables with monomial inputs.

Addition:

+	x	$2x$	x^2	$2x^2$
x	$2x$	$3x$		
$2x$	$3x$	$4x$		
x^2			$2x^2$	$3x^2$
$2x^2$			$3x^2$	$4x^2$

Multiplication:

•	x	$2x$	x^2	$2x^2$
x	x^2	$2x^2$	x^3	$2x^3$
$2x$	$2x^2$	$4x^2$	$2x^3$	$4x^3$
x^2	x^3	$2x^3$	x^4	$2x^4$
$2x^2$	$2x^3$	$4x^3$	$2x^4$	$4x^4$

Notice that there are shaded spaces left in the addition table where unlike terms can not be combined.

Directions: The following tables are addition or multiplication tables involving monomial inputs. Fill in the missing squares with the appropriate monomials. Write NL (for "not like") or shade the grid spaces where the terms *cannot* be combined.

+	$-3x^2$	$-x^2$	$-3x$	$-x$	0	x	$3x$	x^2	$3x^2$
$-3x^2$									
$-x^2$									
$-3x$									
$-x$									
0									
x									
$3x$									
x^2									
$3x^2$									

•	$-3x^2$	$-x^2$	$-3x$	$-x$	0	x	$3x$	x^2	$3x^2$
$-3x^2$									
$-x^2$									
$-3x$									
$-x$									
0									
x									
$3x$									
x^2									
$3x^2$									

Directions: The following tables are addition or multiplication tables with missing information. Fill in the missing squares with the appropriate monomials. Write NL for "not like" in the squares where unlike terms cannot be added.

+	$7x$				x^2
$5x$				$-2x$	
	$7x$	$-2x^2$	0		x^2
		$-6x^2$	$-4x^2$		$-3x^2$
$-4x$				$-11x$	
			$6x^2$		$7x^2$

•		x^6		
x^4	$-5x^6$	x^{10}		
	$-15x^7$	$3x^{11}$		$24x^8$
$-7x$	$35x^3$			
		x^6		$8x^3$
		x^7	$2x^2$	

Guided Learning Activity

Vertical Form of Polynomial Multiplication

The expressions $304 \cdot 21$ and $(3x^2+4)+(2x+1)$ are simplified in almost the same way: For both expressions, we line up the like terms vertically; then multiply.

Example 1:

	3	0	4
×		2	1
	3	0	4
6	0	8	
6	3	8	4

	3 hundreds	0 tens	4 ones
×		2 tens	1 one
	3 hundreds	0 tens	4 ones
6 thousands	0 hundreds	8 tens	
6 thousands	3 hundreds	8 tens	4 ones

	$3x^2$	$+0x$	$+4$
×		$+2x$	$+1$
	$+3x^2$	$+0x$	$+4$
$6x^3$	$+0x^2$	$+8x$	
$6x^3$	$+3x^2$	$+8x$	$+4$

One of the differences between the vertical multiplications of the expressions is that we cannot "carry" coefficients greater than 9 to the next column like we carry numbers greater than 9. Another difference is that we can have negative coefficients if that is part of the original expression.

Example 2: $(4x+5)(2x-6)$ becomes

	$4x$	$+5$
×	$2x$	-6
	$-24x$	-30
$8x^2$	$+10x$	
$8x^2$	$-14x$	-30

Now try these. Make sure to **line up the like terms** in the same columns and keep the signed coefficients with the terms in the columns.

a. $(7y-6)(y+5)$

b. $(x^2-3x+6)(x-9)$

c. $(a^2+8a+6)(a^2+5)$

d. $(2x^2+5x-4)(x^2-6x+7)$

Student Activity

Not All Multiples are the Same

Directions: All the expressions below are written in sets of triples. First multiply and simplify each expression. Then circle the pairs from each problem that are really equivalent (if there are any).

1. $(x+3)(x-5)$ $(x-5)(x+3)$ $(x-3)(x+5)$

2. $(x-2)(x+6)$ $(x+2)(x-6)$ $(x-6)(x+2)$

3. $(x-2)(x-2)$ $(2-x)(2-x)$ $(x-2)(2-x)$

4. $(x+3)(x+4)$ $(3+x)(4+x)$ $(x-3)(x-4)$

5. $(x+2)(x^2-2x+4)$ $(x-2)(x^2+2x-4)$ $(x-2)(x-2)(x-2)$

6. $(x+5)(x-5)$ $(x-5)^2$ $(5+x)(5-x)$

Student Activity
Multiplying Binomials with FOIL

When we multiply two **binomials**, we can use a shortcut method called the FOIL method. FOIL stands for **F**irst **O**uter **I**nner **L**ast. The first one has been done for you.

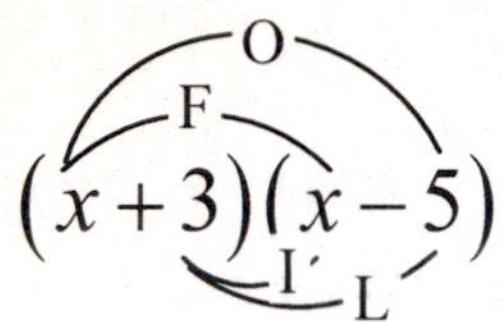

Trent began looking for new ways to use his algebra vocabulary.

Expression	Find these terms of the multiplication…				Combine like terms and simplify
	First	Outer	Inner	Last	
$(x+3)(x-5)$	x^2	$-5x$	$3x$	-15	$x^2-2x-15$
$(u-4)(u-8)$					
$(y+3)(y+3)$					
$(x+7)(x-7)$					
$\left(a+\frac{1}{3}\right)\left(a+\frac{2}{3}\right)$					
$(2x+5)(3x-4)$					
$(4x+3)(6x+7)$					
$(x^2+4)(x^2+6)$					
$\left(2w+\frac{1}{2}\right)\left(w-\frac{3}{2}\right)$					
$(s^3-2)(s^3-8)$					
$(3.6x+1)(2x-4.5)$					

Student Activity

POLY-13

Charting Squared Binomials

When we **square a binomial**, we are really just multiplying two binomials and can use the FOIL method to do it. For example, $(x-6)^2$ is really the same as $(x-6)(x-6)$. Fill in the empty boxes in the table below. The first one has been done for you.

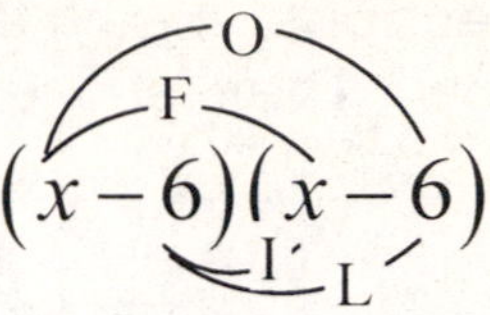

Expression	Find these terms of the multiplication...				Combine like terms and simplify
	First	Outer	Inner	Last	
$(x-6)^2$ $(x-6)(x-6)$	x^2	$-6x$	$-6x$	36	$x^2-12x+36$
$(u+4)^2$ $(u+4)(u+4)$					
$(a-5)^2$ $(a-5)(a-5)$					
$(x+\frac{1}{2})^2$ $(x+\frac{1}{2})(x+\frac{1}{2})$					
$(A+B)^2$ $(A+B)(A+B)$					
$(A-B)^2$ $(A-B)(A-B)$					

Notice that in the first row, $-6x-6x=-12x$ is the same as $2(-6x)=-12x$. If you look back through the rows in the table above, you should see a similar pattern in every row.

We can thus use the formula below to determine the product of a binomial squared.

$$(A+B)^2 = A^2+2\cdot A\cdot B+B^2$$

The middle term is found by taking twice the product of A and B. Make sure that you use parentheses when you are squaring the A-term and the B-term! For example, the term like $2x^2$ is not the same as $(2x)^2$.

Directions: Using the binomial squared formulas, fill in the empty boxes in the table below. The first one has been done for you.

Expression	**Find these terms of binomial squared.** The square of the first term: A^2	Twice the product of the first and second terms: $2 \cdot A \cdot B$	The square of the last term: B^2	**Final expression** $A^2 + 2AB + B^2$
$(x-6)^2$	$(x)^2$	$2(x)(-6)$	$(-6)^2$	$x^2 - 12x + 36$
$(u+4)^2$				
$(a-5)^2$				
$\left(x+\frac{1}{2}\right)^2$				
$(2y+5)^2$				
$(3a-1)^2$				
$(5x+4y)^2$				
$(9-u)^2$				
$(v^3+2)^2$				
$\left(6z-\frac{4}{3}\right)^2$				
$(x^4-y^3)^2$				

Student Activity

Charting a Sum-Difference

There's one more special product to look at, but again, we'll start by going back to the FOIL method to figure it out. Fill in the empty boxes in the table below. The first one has been done for you.

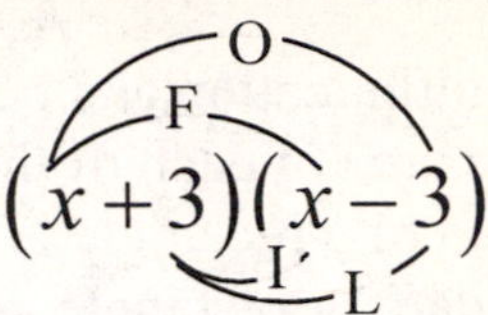

Expression	Find these terms of the multiplication...				Combine like terms and simplify
	First	Outer	Inner	Last	
$(x+3)(x-3)$	x^2	$-3x$	$3x$	-9	x^2-9
$(x-3)(x+3)$					
$(x-3)(x-3)$					
$(x+3)(x+3)$					
$(a-5)(a+5)$					
$(a+5)(a-5)$					
$(a-5)(a-5)$					
$(a+5)(a+5)$					
$(x+\frac{1}{2})(x-\frac{1}{2})$					
$(A+B)(A-B)$					

Circle all the rows where the inner and outer terms were additive inverses (where their sum was zero). Did it happen in every row? ____

What is special about the beginning expressions where the middle terms **were** additive inverses?

Student Activity

Visual Representation for Multiplying Binomials

Multiplication can be represented using a diagram called an area model. Find an expression for the area of each of the squares and rectangles below. Remember that the formula for finding the area of a rectangle is Area = (length)(width). Write the expression for the area inside each square or rectangle, and then write it inside the same square or rectangle in the composite figure below. **Thus each area will be written in two places.**

Problem 1:

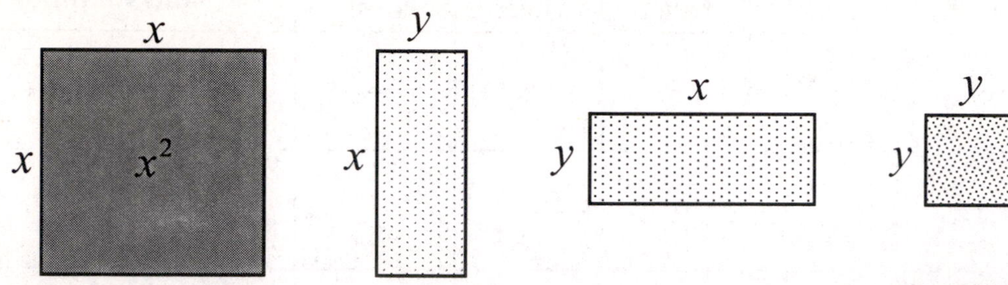

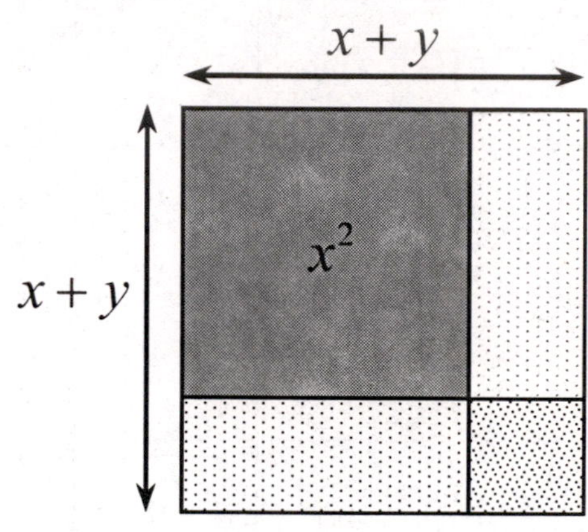

To find the area of the composite rectangle, we could either add the areas of all the smaller pieces, which would give this area:

____

Or we could find the area by using the area formula, which would give Area $= (x+y)(x+y) = (x+y)^2$

This tells us that $(x+y)^2 =$ ____________________ .

Problem 2: Do this one the same way.

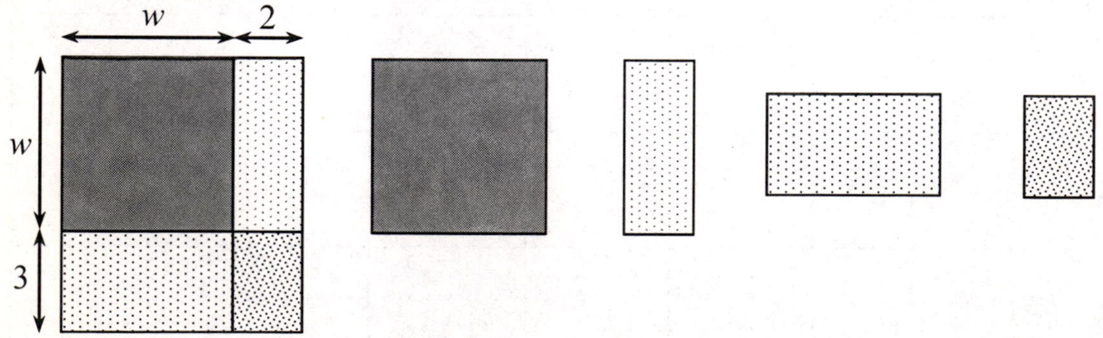

What mathematical equation does this set of figures tell us?

____________________ = ______________________________

Student Activity

Exponential and Polynomial Heteronyms

Directions: In writing, there are words that are spelled the same but have different pronunciations and different definitions; these are called heteronyms. Many polynomial and exponential expressions look similar but are really very different (almost like mathematical heteronyms). In each set of expressions below, pay close attention to the use of parentheses and the mathematical operations and notation.

1.

$(3x)^2$	$2(3x)$	$2(x+3)$	$(x+3)^2$	$(2x)^3$

2.

$(x-4)^2$	$2(x-4)$	$2(-4x)$	$(-2x)^4$	$(4x)^2$

3.

$(x+3)(x-2)$	$(3x)(-2x)$	$3x(x-2)$	$-2x(x+3)$	$x^3(-2x)$

4.

$(x+5)(x-5)$	$(5x)(x-5)$	$5x(-5x)$	$x^5(x-5)$	$x^5 \cdot x^5$

5.

$5x^2(4x^3)$	$(5+2x)(4x+3)$	$5x^2(4x+3)$	$(5x)^2(4x^3)$	$(5x)^2(4x+3)$

Guided Learning Activity

Polynomial Long Division

The expressions $168 \div 14$ and $(1x^2+6x+8)\div(1x+4)$ are simplified in almost the same way using long division. For both expressions, we line up the like terms vertically; then divide.

Example 1:

$$\begin{array}{r}
12 \\
14\overline{)168} \\
-14 \\
\hline
28 \\
-28 \\
\hline
0
\end{array}
\qquad\qquad
\begin{array}{r}
1x\quad +2 \\
1x\;+4\overline{)1x^2\;\;+6x\;\;+8} \\
-(1x^2\;\;+4x) \\
\hline
2x\;\;+8 \\
-(2x\;\;+8) \\
\hline
0
\end{array}$$

Notice that in both long division problems, the like terms line up in the same columns.

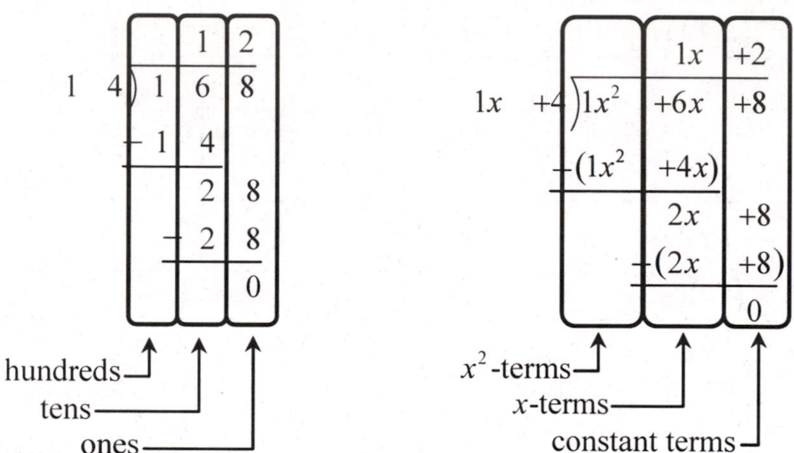

As you perform the operations in polynomial long division, you will want to make sure that you line up the like terms as well.

If you are missing a column of terms, you will need to insert a placeholder. For example, $12\overline{)1008}$ would not give the same quotient as $12\overline{)18}$, the two zeros in 1008 act as placeholders for the tens place and the hundred place. Also, we would not change the order of the digits in the columns, for example $12\overline{)1008}$ is not the same as $12\overline{)8010}$.

In a similar fashion, we would not write $x+2\overline{)x^3+8}$ because the x-term and the x^2-term columns are missing. Instead we write $x+2\overline{)x^3+0x^2+0x+8}$ to hold the places of the necessary columns.

In polynomial long division, the terms of the dividend and the divisor must be written in descending order with no missing terms.

1. Rewrite these long division problems so that all the columns are accounted for and the terms are in the proper order. Write in any unwritten ones. For example, write x^2 as $1x^2$.

a. $x-2\overline{)x^3-4x+5}$ **b.** $x-1\overline{)4-5x+x^2}$ **c.** $2+a\overline{)a^2-4}$

Here is a polynomial long division example shown step by step. Your instructor will walk you through the steps. You should add your own notes so that you will remember what has happened on each step. At the very least, highlight the new part of each step in color so that it is easier to see later on when you go back through your notes.

Example 2:

$$\begin{array}{r} x \\ x+8\overline{)x^2+5x-24} \end{array}$$

$$\begin{array}{r} x \\ x+8\overline{)x^2+5x-24} \\ \underline{x^2+8x} \end{array}$$

$$\begin{array}{r} x \\ x+8\overline{)x^2+5x-24} \\ \underline{-(x^2+8x)} \end{array}$$

$$\begin{array}{r} x \\ x+8\overline{)x^2+5x-24} \\ \underline{-x^2-8x} \end{array}$$

$$\begin{array}{r} x \\ x+8\overline{)x^2+5x-24} \\ \underline{-x^2-8x} \\ -3x \end{array}$$

$$\begin{array}{r} x\ -3 \\ x+8\overline{)x^2+5x-24} \\ \underline{-x^2-8x} \\ -3x-24 \end{array}$$

$$\begin{array}{r} x\ -3 \\ x+8\overline{)x^2+5x-24} \\ \underline{-x^2-8x} \\ -3x-24 \\ \underline{-3x-24} \end{array}$$

$$\begin{array}{r} x\ -3 \\ x+8\overline{)x^2+5x-24} \\ \underline{-x^2-8x} \\ -3x-24 \\ \underline{-(-3x-24)} \end{array}$$

$$\begin{array}{r} x\ -3 \\ x+8\overline{)x^2+5x-24} \\ \underline{-x^2-8x} \\ -3x-24 \\ \underline{+3x+24} \end{array}$$

$$\begin{array}{r} x\ -3 \\ x+8\overline{)x^2+5x-24} \\ \underline{-x^2-8x} \\ -3x-24 \\ \underline{+3x+24} \\ 0 \end{array}$$

Usually, in practice, we do not write this step:

$$\begin{array}{r} x \\ x+8\overline{\smash{)}\,x^2+5x-24} \\ \underline{-\left(x^2+8x\right)} \end{array}$$

We skip directly from

$$\begin{array}{r} x \\ x+8\overline{\smash{)}\,x^2+5x-24} \\ \underline{x^2+8x} \end{array}$$

to

$$\begin{array}{r} x \\ x+8\overline{\smash{)}\,x^2+5x-24} \\ \underline{-x^2-8x} \\ -3x \end{array}$$

by changing the signs.

Change the signs using a colored pencil, so that you can see that you have done this step *after* writing the terms for the original row.

Directions: Try each of the problems below. Stop after each problem and wait for your instructor to go over the answer. If you make mistakes, correct them using a colored pencil so that you will see the corrections (and original mistakes) when you go back to study later.

2. $\left(2x^2-5x-12\right)\div(x-4)$

3. $\left(7x^2+x^3-28+5x\right)\div(x+4)$

If there is a remainder after the long division is done, we write this as a ratio with the divisor. Consider the following example to see why it is done this way.

Example 3:

$\frac{25}{4}=6\frac{1}{4}$ or we could write it like this $4\overline{)\,25}$ with quotient 6, -24, remainder 1, which gives the result $6\text{ R }1$.

$$\begin{array}{r} 6 \\ 4\overline{)\,25} \\ \underline{-24} \\ 1 \end{array}$$

If we want to write $6\text{R}1$ mathematically, it should be $6\frac{1}{4}$, which is the same as $6+\frac{1}{4}$.

Example 4:

Likewise, when we do the following division, we get a remainder:

$$\begin{array}{r} x\ +3 \\ x-5\overline{)\,x^2-2x-16} \\ \underline{-x^2+5x} \\ 3x-16 \\ \underline{-3x+15} \\ -1 \end{array}$$

So we write the answer: $x+3+\frac{-1}{x-5}$ $\begin{array}{l}\leftarrow \text{remainder} \\ \leftarrow \text{divisor}\end{array}$

Directions: Try each of the problems below. Stop after each problem and wait for your instructor to go over the answer. If you make mistakes, correct them using a colored pencil so that you will see the corrections (and original mistakes) when you go back to study later.

4. $(x^2+4x-8)\div(x+6)$

5. $(x^3+4x^2-5x-22)\div(x+4)$

Student Activity

Match Up on Polynomial Division

Directions: Match each of the expressions in the squares of the grid below with an equivalent simplified expression from the top. If an equivalent expression is not found among the choices A through D, then choose E (none of these). Be careful – you may have to simplify rational expression terms.

A $x+4$ **B** $2x-3$ **C** $x+4+\dfrac{3}{x-1}$ **D** $2x-3-\dfrac{4}{x+3}$ **E** None of these

$x-1\overline{)x^2+3x-1}$	$x+7\overline{)x^2+11x+28}$	$x+3\overline{)x^2+7x+8}$
$x+3\overline{)2x^2+3x-9}$	$\left(4x^2+6x-26\right)\div(2x+6)$	$\left(x^2+7x+16\right)\div(x+3)$
$\left(x^2-16\right)\div(x-4)$	$\dfrac{6x^2y+2x^3y}{2x^2y}$	$\dfrac{36a^2b^2cx-54a^2b^2c}{18a^2b^2c}$

Student Activity

Pay Attention!

Directions: When you're simplifying polynomials, many problems look similar. You **must** pay attention to the operation and directions for the problems. Try the problems below, but please, **pay attention** to what you're doing!

1. Multiply: $(x^2+5x-14)(x+7)$

2. Add: $(x^2+5x-14)+(x+7)$

3. Divide using long division: $(x^2+5x-14)\div(x+7)$

4. Subtract: $(x^2+5x-14)-(x+7)$

5. Add: $(27x^3+12x^2)+(3x^2)$

6. Divide: $(27x^3+12x^2)\div(3x^2)$

7. Multiply: $(27x^3+12x^2)(3x^2)$

8. Subtract: $(27x^3+12x^2)-(3x^2)$

Student Activity

Language of Polynomial Operations

Directions: Now that we've looked at adding, subtracting, multiplying, and dividing polynomials, let's practice with the language of polynomials. You **do not** have to carry out the mathematics in each problem; simply translate the sentence into mathematical notation. The first one has been done for you.

	Directions in words	**Write the expression with mathematical notation, using the proper grouping terms, fraction bars, or long division notation where necessary.**
1.	Find the sum of $x+4$ and $x-1$.	$(x+4)+(x-1)$
2.	Divide x^2-3x+2 by $x-1$.	
3.	Find the difference of x^2-2x-7 and x^2-4.	
4.	Find the quotient of $-6x^2y^3$ and $2x^2y$.	
5.	Find the product of x^2+4x+3 and $x+5$.	
6.	Find the product of $x+6$ and $x-9$.	
7.	Find the total of $x+3$, $x+4$, and $x-2$.	
8.	Find twice the sum of x^2 and x^2-4.	
9.	Find the quotient of $10a^3b-25a^4$ and $5a^2$.	
10.	What is $x^2-10x+16$ less $2x+4$?	

Assess Your Understanding

Simplifying Polynomials

For each of the following, describe the strategies or key steps that will help you **start** the problem. You do **not** have to complete the problems.

		What will help you to start this problem?
1.	Write $6-5x^2+3x$ in descending powers of x.	
2.	What is the leading coefficient of x^2+5x?	
3.	Expand the binomial: $(3x-5)^2$	
4.	Divide: $(4x^2-9)\div(2x+3)$	
5.	Multiply: $(x-5)(x-6)$	
6.	Subtract: $(2x^2-4x)-(3x-7)$	
7.	Multiply: $(12+a)(12-a)$	
8.	Divide: $(14x^3-21x^2+28x)\div(7x)$	
9.	Evaluate $y=x^3-4x^2+12$ for $x=-3$.	
10.	Multiply: $(x+3)(x^2-6x+8)$	

Metacognitive Skills

Simplifying Polynomials

Metacognitive skills refer to the ability to judge how well you have learned something and to effectively direct your own learning and studying. This is a self-evaluation tool designed to help you focus your studying and to improve your metacognitive skills with regards to this math class.

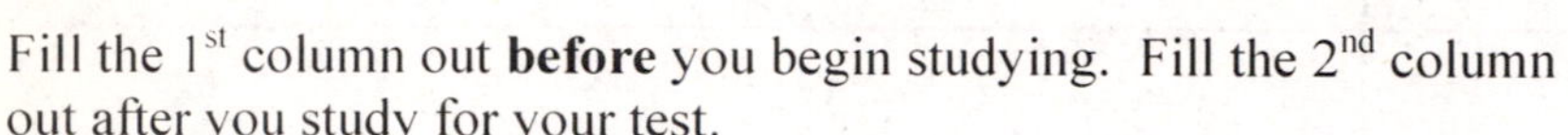

Fill the 1st column out **before** you begin studying. Fill the 2nd column out after you study for your test.

Go back to this assessment after your test and circle any of the ratings that you would change – this identifies the "disconnects" between what you **thought** you knew well and what you **actually** knew well.

Use the scale below to assign a number to each topic.

5 *I am confident I can do any problems in this category correctly.*
4 *I am confident I can do most of the problems in this category correctly.*
3 *I understand how to do the problems in this category, but I still make a lot of mistakes.*
2 *I feel unsure about how to do these problems.*
1 *I know I don't understand how to do these problems.*

Topic or Skill	Before Studying	After Studying
Identifying the number of terms in a polynomial.		
Writing a polynomial in descending order.		
Identifying the degree or coefficient of a term or of a polynomial.		
Evaluating a polynomial for a given value (especially negative values).		
Graphing an equation that involves x^2 or x^3 by making a table of values and plotting points.		
Adding or subtracting polynomials.		
Multiplying a monomial by a polynomial. Example: $3x^2(x^2-2x+4)$.		
Multiplying two binomials using FOIL. Example: $(x+3)(x+7)$		
Multiplying any two polynomials. Example: $(x+3)(x^2-4x+7)$		
Multiplying a special sum-difference product. Example: $(a+7)(a-7)$		
Squaring a binomial. Example: $(x-4)^2$ (be careful, the answer's **not** x^2-16)		
Distinguishing between addition and multiplication, because the rules are not the same! For example, $(x+5)^2$ vs. $(5x)^2$ OR $(x+2)(x+3)$ vs. $(x+2)+(x+3)$.		
Dividing a polynomial by a monomial. Like $(6x^3+3x^2-9x)\div 3x$		
Using polynomial long division, including what to do with a remainder! Example: $(x^2-6x+8)\div(x-2)$		
Knowing how to make adjustments in polynomial long division when there's a term missing.		

FACT: Factoring and More

Guided Learning Activity

FACT-1

Charting Factors and GCFs

Directions Part I: For each expression, decide if each of the terms in the top row is a factor of the given expression. Place an **X** in any column for which the term **would** be a factor, then use those decisions to help you find the GCF.

		2	3	4	6	8	12	x	x^2	x^3	x^4	y	y^2	y^3	y^4	GCF
a.	$6x^3-3x^2$		X					X	X							$3x^2$
b.	$18y^2+24y$															
c.	$19x^4y-3xy^2$															
d.	$16+40x^3$															
e.	$24x^4y^4+36x^2y^2$															
f.	x^4+y^4															
g.	$36x+2x^2$															
h.	$9y^4+3xy^4$															
i.	$x^3y^2w+x^2y^3z$															

Directions Part II: Given each expression, find the numerical GCF, the x GCF, and the y GCF. Then use the product to find the overall GCF for the expression. The first one has been done for you.

	Expression	Constant GCF	x GCF	y GCF	Overall GCF	Factored Expression
a.	$15x^4-25x^3$	5	x^3		$5x^3$	$5x^3(3x-5)$
b.	$7y^3+49y^2$					
c.	$16xy^2+32y^4$					
d.	$24x^8-46x^7$					
e.	$4x^6+4y^6$					
f.	$12x^2+144$					
g.	$33xy-44x^2y^2$					
h.	$9x^{10}y^5+3x^7y^4$					

Student Activity

Hatch the Missing Factors

Directions: Fill in the blank in each equation to make each statement true. All of the solutions are provided inside the "egg". When you have correctly placed all of the factors, you have "hatched" the egg.

$8x^2 \cdot \underline{\qquad\qquad} = 24x^3$

$-4x \cdot \underline{\qquad\qquad} = 36x^4$

$9y^2 \cdot \underline{\qquad\quad} = 54y^6$

$\underline{\qquad\quad} \cdot 6x^3 = -36x^3$

$2x^2y \cdot \underline{\qquad\quad} = -8x^2y^3$

$5x^2 \cdot \underline{\qquad\quad} = 5x^2$

$5x^2 \cdot \underline{\qquad\quad} = -5x^2$

$\underline{\quad\;} \cdot (-3z) = 3z^2$

$7w^3 \cdot \underline{\qquad\quad} = -49w^5$

$5uv \cdot \underline{\quad\;} = 5u^4v$

$16ab^2 \cdot \underline{\quad\;} = 32a^2b^2$

$9r^2 \cdot \underline{\quad\;} = 36r^3s^2$

$t^7u \cdot \underline{\quad\;} = 16t^8u^2v$

$\underline{\quad\;} \cdot (-7x) = 14x$

$c \cdot \underline{\qquad\quad} = mc^2$

$12x^4 \cdot \underline{\quad\;} = -144x^{12}$

$18x^4 \cdot \underline{\quad\;} = -36x^5y^3$

$-10x^3y^2z \cdot \underline{\qquad\quad} = -120x^5y^3z^4$

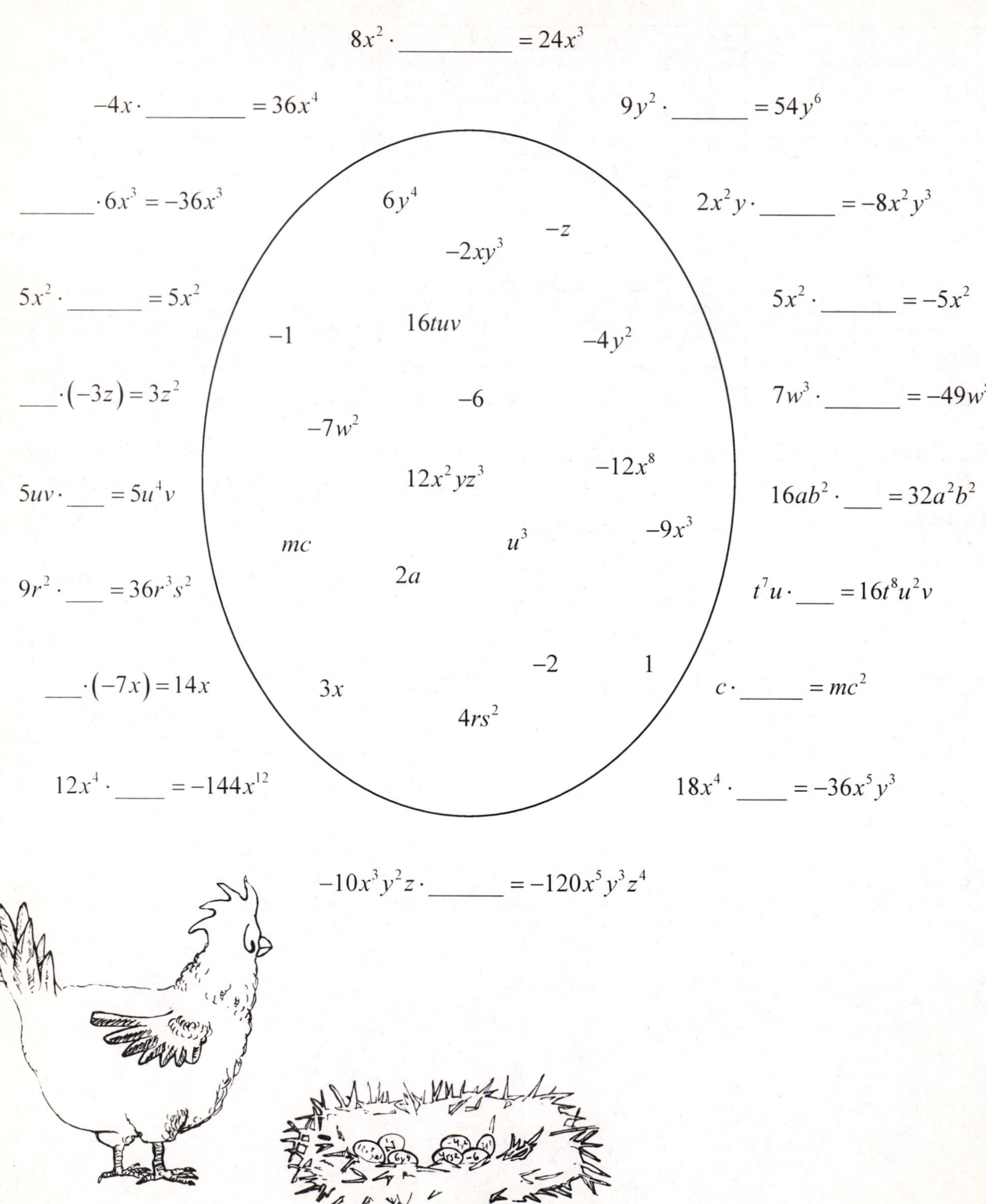

Student Activity

The First Factoring Matchup

Directions (READ them): In each box of the grid, you will find an expression that needs to be factored. Once you have factored each expression, look to see whether the factor appears in the list at the top. If it does, list that letter, if none of the factors are listed, then choose F (none of these are factors). The first one has been done for you. Some boxes may have more than one answer.

A $x+3$ **B** $x-2$ **C** $x+2y$ **D** $y-5$ **E** $x+4$ **F** None of these are factors

$x^2-3x+2xy-6y$ $x(x-3)+2y(x-3)$ Ans: $(x+2y)(x-3)$ C	$3x^2y^2+9xy^2$	$4xy^4+8y^4$	$xy+3y-5x-15$
$x^2+4x+2xy+8y$	$3x^2y^2z^3-6xy^2z^3$	$xy+4y-5x-20$	$7xy-14y-3x+6$
$6xy^2+10y^2+3xy+5y$	$x^2+3x+2xy+6y$	$3a^2b^3x+12a^2b^3$	$10x^2y-50x^2+9xy-45x$
$xy+3y-2x-6$	$x^2+2xy+2xy+4y^2$	$xy+2y^2-5x-10y$	$xy-2y-5x+10$

Student Activity

Revenge of Factor Pairings

Directions: In each diagram, there is a number in the top box and exactly enough spaces beneath it to write all the possible factor-pairs involving **integers**. The number −12 has been done for you. See if you can find all the missing factor-pairs.

−12	
1(−12)	−1(12)
2(−6)	−2(6)
3(−4)	−3(4)

−32	

−45	

50	

63	

−18	

28	

−30	

100	

−36	
	X

56	

−42	

Directions: In the tables below, you are given a factor pair. If you **ADD** the two factors instead of multiplying them, what is the result? Complete the tables to find the sum of each factor pair. The first one has been started for you.

−12	Sum
−1(12)	11
1(−12)	−11
−2(6)	
2(−6)	
−3(4)	
3(−4)	

32	Sum
1(32)	
−1(−32)	
2(16)	
−2(−16)	
4(8)	
−4(−8)	

−64	Sum
−1(64)	
1(−64)	
−2(32)	
2(−32)	
−4(16)	
4(−16)	
−8(8)	

30	Sum
1(30)	
−1(−30)	
2(15)	
−2(−15)	
3(10)	
−3(−10)	
5(6)	
−5(−6)	

Student Activity

Find the Prime Trinomial

Directions: In each row there are three trinomials; two are factorable and one is not. If the trinomial is factorable, factor it. If it is not factorable, write "PRIME."

1.	x^2-5x-6	x^2-x+6	x^2-5x+6
2.	$x^2-5x+24$	$x^2-5x-24$	$x^2+5x-24$
3.	$x^2-8x+12$	x^2-x-12	$x^2+4x+12$
4.	$x^2-11x-12$	$x^2+11+12$	$x^2+13x+12$
5.	x^2+3x+2	$x^2+3x+10$	$x^2+3x-28$
6.	$x^2-19x+60$	$x^2-19x-60$	$x^2+19x+60$
7.	$x^2-4x-60$	$x^2+4x-60$	$x^2-4x+60$
8.	$x^2-16x+60$	$x^2+16x+60$	$x^2-16x-60$

Student Activity

Match Up on Factoring Trinomials

Directions (READ them): In each box of the grid, you will find an expression that needs to be factored. Once you have factored each expression, look to see whether the factor appears in the list at the top. If it does, list that letter, if none of the factors are listed, then choose F (none of these are factors). The first one has been done for you. Some boxes may have more than one answer.

A $x+4$ **B** $x-2$ **C** $x-1$ **D** $x+5$ **E** $x-7$ **F** None of these are factors

x^2-3x+2 $(x-2)(x-1)$ Ans: B, C	$x^2+12x+35$	$x^2-2x-35$	$x^2+8x+16$
x^2-3x-4	x^2-4x+4	x^2+x-2	x^2-x-42
$x^2+13x+42$	x^2-6x-7	x^2+2x-8	x^2+x-20
$3x^2y-18xy+24y$	$5x^2+10x-15$	$5x^2+20x+15$	$10x^2y^2-30xy^2-280y^2$

Student Activity

Life After the GCF, Part I

Directions: For each expression below, find and factor out the GCF. Then categorize the resulting expression in the table as one of the following:

- GCF only: if no other factoring can be done within the parentheses
- GCF and form $x^2 + bx + c$: expression within parentheses is of the form $x^2 + bx + c$
- GCF and form $ax^2 + bx + c$ expression within parentheses is of the form $ax^2 + bx + c$

The first one has been done for you.

	Expression (and factor out the GCF)	**GCF only**	**GCF and form** $x^2 + bx + c$	**GCF and form** $ax^2 + bx + c$
1.	$3x^2 - 12x - 15$ $3(x^2 - 4x - 5)$		**X**	
2.	$8x^3 + 16x^2 - 24x$			
3.	$20x^2yz + 15x^2$			
4.	$9x^3y + 15x^2y + 6xy$			
5.	$60t^2x^2 + 78t^2x + 24t^2$			
6.	$27x^2 + 27x + 6$			
7.	$150a^2b^2c^2 - 125ab^2c$			
8.	$25x^5 + 75x^4 + 50x^3$			
9.	$26y^2 + 13y$			
10.	$36x^2 + 99x + 54$			

Student Activity
Is it Completely Factored?

Is $(8x-2)(x+6)$ the completed factoring for the expression $8x^2+46x-12$? Let's see.

1. Multiply $(8x-2)(x+6)$: ____________________

2. Is the result of the multiplication equivalent to $8x^2+46x-12$? _____

3. From the expression $8x^2+46x-12$, first factor out the 2: ________________

4. Then continue to factor by examining the trinomial within the parentheses:

5. Was the original expression, $(8x-2)(x+6)$, completely factored? _____

6. How can you make $(8x-2)(x+6)$ a completely factored expression?

7. Tic-tac-toe Directions: If the expression in the square **IS** completely factored, then circle it (thus putting an **O** on the square). If the expression **IS NOT** completely factored, then put an **X** on the square and finish the factoring.

$(x-2)(3x-12)$	$(7y-35)(2y+3)$	$(9x-28)(x+2)$
$(4x-9)(2x-5)$	$(3x+4)(4x-9)$	$(2a+8)(3a+12)$
$(2x+6)(2x-6)$	$(3w-16)(6w+21)$	$(5x-9)(6x+25)$

Student Activity

Spotting Perfect Squares and Cubes

Directions: For each expression below, determine whether it might be a perfect square or a perfect cube.

Then write the term in its squared $(\quad)^2$ or cubed $(\quad)^3$ form.

Be careful, a couple of these terms are both squares and cubes.

The first two have been done for you.

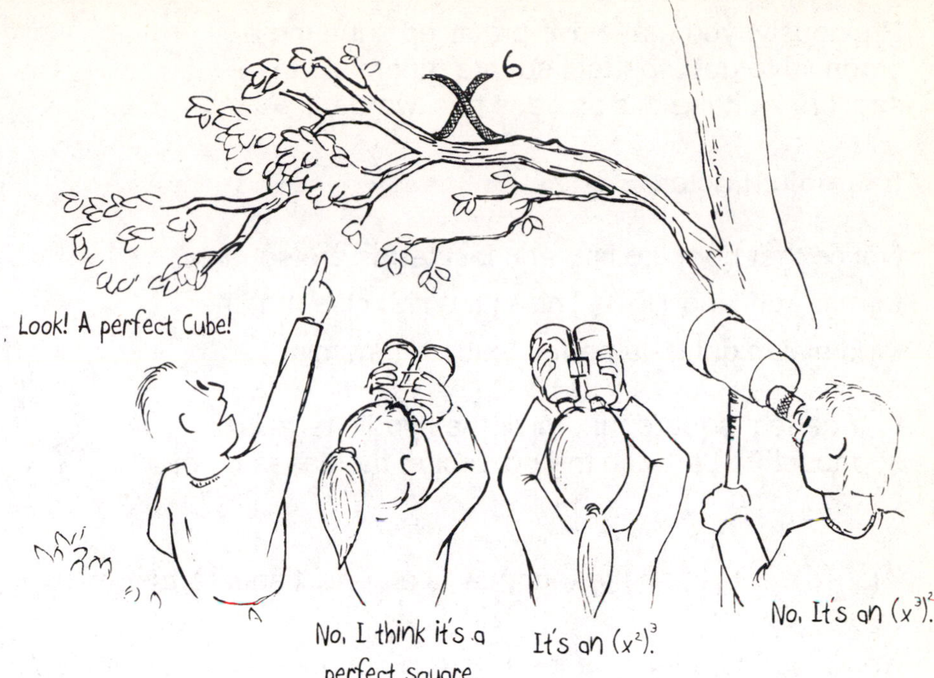

-27 $(-3)^3$	125	9	512	$216x^3y^3$
$16b^2$ $(4b)^2$ or $(-4b)^2$	-64	1	100	$1000x^3$
729	$49a^4b^4$	-1	$8x^3$	$81h^6$
144	$9x^2$	$25z^2$	$-125z^3$	64
$27a^3$	0	169	$216x^3y^9$	$1000a^3$

Student Activity
Skeleton of a Perfect Square Trinomial

Previously, you may have practiced squaring a binomial to get a perfect square trinomial. In this section, we take this process backwards.

Example: Factor: $x^2 - 12x + 36$

Notice that both the first and last terms are squared terms: $(x)^2$ and $(6)^2$. This is the first clue that the trinomial might be a perfect square trinomial.

In a perfect square trinomial, the middle term is supposed to be twice the product of the bases (in this case, x and 6).

$2(x)(6)$ is $12x$, so the trinomial is a perfect square trinomial.

Remember the formula for a perfect square trinomial is the sum (or difference) of the bases, the quantity squared:

$$A^2 + 2AB + B^2 = (A+B)^2 \qquad A^2 - 2AB + B^2 = (A-B)^2$$

Write the correct factored form: $(x)^2 - 2(x)(6) + (6)^2 = (x-6)^2$.

Now you try these! For each problem, fill in the skeleton for the sum or difference of squares, then factor appropriately. If the trinomial is NOT a perfect square trinomial, say so!

1. Factor: $x^2 - 10x + 25$ $\quad = (\quad)^2 - 2(\quad)(\quad) + (\quad)^2 = (\qquad)^2$

2. Factor: $y^2 + 2y + 1$ $\quad = (\quad)^2 + 2(\quad)(\quad) + (\quad)^2 = (\qquad)^2$

3. Factor: $25x^2 - 60x + 36$ $\quad = (\quad)^2 + 2(\quad)(\quad) + (\quad)^2 = (\qquad)^2$

4. Factor: $x^2 + 7x + 49$ $\quad = (\quad)^2 + 2(\quad)(\quad) + (\quad)^2 = (\qquad)^2$

5. Factor: $v^4 + 6v^2 + 9$ $\quad = (\quad)^2 + 2(\quad)(\quad) + (\quad)^2 = (\qquad)^2$

6. Factor: $16 - 8x + x^2$ $\quad = (\quad)^2 + 2(\quad)(\quad) + (\quad)^2 = (\qquad)^2$

7. Factor: $w^4 - 4w + 4$ $\quad = (\quad)^2 + 2(\quad)(\quad) + (\quad)^2 = (\qquad)^2$

8. Factor: $a^2b^2 + 14abc + 49c^2$ $\quad = (\quad)^2 + 2(\quad)(\quad) + (\quad)^2 = (\qquad)^2$

Student Activity

Skeleton of the Difference or Sum of Two Squares

The best way to learn the sum and difference of squares formulas is to learn a procedure for constructing the formulas rather than memorizing them. Here are three examples, with the description of each step:

Example 1: Factor: $x^2 - 25$

Construct a skeleton of the difference of squares: $(\quad)^2 - (\quad)^2$

Fill in the skeleton with the appropriate bases: $(x)^2 - (5)^2$

Construct the skeleton of the difference of squares formula: $(\quad + \quad)(\quad - \quad)$

Insert the bases into both factors: $(x+5)(x-5)$

Example 2: Factor: $64y^2 - 81z^2$

Construct a skeleton of the difference of squares: $(\quad)^2 - (\quad)^2$

Fill in the skeleton with the appropriate bases: $(8y)^2 - (9z)^2$

Construct the skeleton of the difference of squares formula: $(\quad + \quad)(\quad - \quad)$

Insert the bases into both factors: $(8y+9z)(8y-9z)$

Example 3: Factor: $4x^2 + 9$

Construct a skeleton of the sum of squares: $(\quad)^2 + (\quad)^2$

Fill in the skeleton with the appropriate bases: $(2x)^2 + (3)^2$

A sum of squares is prime. This binomial does not factor.

Now you try these! For each problem, fill in the skeleton for the sum or difference of squares, then factor appropriately.

1. Factor: $36 - a^2 \quad = (\quad)^2 + (\quad)^2 =$ ______________________

2. Factor: $49y^2 + x^2 = (\quad)^2 + (\quad)^2 =$ ______________________

3. Factor: $a^2b^2 - 4 \quad = (\quad)^2 - (\quad)^2 =$ ______________________

4. Factor: $w^4 - 49v^2 = (\quad)^2 - (\quad)^2 =$ ______________________

5. Factor: $81 + u^2 \quad = (\quad)^2 + (\quad)^2 =$ ______________________

6. Factor: $9w^2 - z^6 \quad = (\quad)^2 - (\quad)^2 =$ ______________________

Guided Learning Activity

Learning the Cubes Procedure

The best way to learn the sum and difference of cubes formulas is to learn a **procedure** for constructing the formulas rather than memorizing the formulas.

Here are two examples, with a description of each step.

After you learn the procedure, try your own hand at using the cubes procedures by doing the four problems at the end.

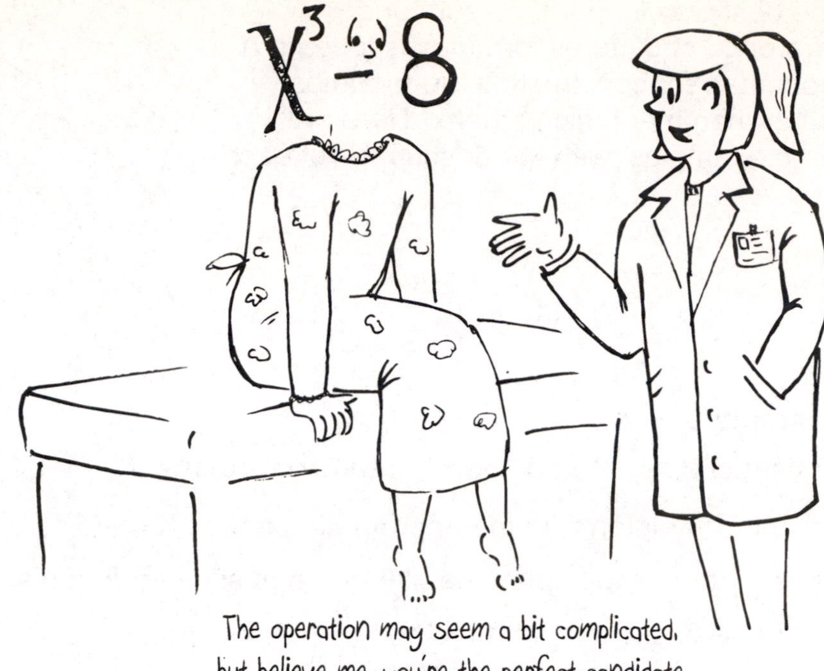

Example 1: Factor: $x^3 - 125$

Step 1: Construct a skeleton of the difference of cubes: $(\quad)^3 - (\quad)^3$

Step 2: Fill in the skeleton with the appropriate bases: $(x)^3 - (5)^3$

Step 3: Construct the skeleton of the cubes formula: $(\quad - \quad)(\quad + \quad + \quad)$

- same sign as original expression (the – sign)
- opposite of first sign (the first + sign)
- always positive (the second + sign)

Step 4: Insert the bases: $(x-5)(\quad + \quad + \quad)$

Step 5: Insert the squares of those bases: $(x-5)(x^2 + \quad + 25)$

x^2 ← $(x)^2$, 25 ← $(5)^2$

Step 6: Insert the product of the bases: $(x-5)(x^2 + 5x + 25)$

$5x$ ← $(x)(5)$

Example 2: Factor: $8w^3 + 27x^6$

Step 1: Construct a skeleton of the sum of cubes: $(\quad)^3 + (\quad)^3$

Step 2: Fill in the skeleton with the appropriate bases: $(2w)^3 + (3x^2)^3$

Step 3: Construct the skeleton of the cubes formula: $(\quad + \quad)(\quad - \quad + \quad)$

same sign as original expression (the + in the first factor)

opposite of first sign (the − in the second factor)

always positive (the last + in the second factor)

Step 4: Insert the bases: $(2w + 3x^2)(\quad - \quad + \quad)$

Step 5: Insert the squares of those bases: $(2w - 3x^2)(4w^2 + \quad + 9x^4)$

$\uparrow$ $(2w)^2$ $\qquad$ $\uparrow$ $(3x^2)^2$

Step 6: Insert the product of the bases: $(2w - 3x^2)(4w^2 + 6x^2w + 9x^4)$

$\uparrow$ $(2w)(3x^2)$

Now you try these!

1. Factor: $x^3 - 8 = (\quad)^3 - (\quad)^3$

$= (\qquad)(\qquad\qquad)$

$\uparrow \qquad \uparrow \qquad \uparrow$

$(\quad)^2 \quad (\quad)(\quad) \quad (\quad)^2$

2. Factor: $64y^3 + 125 = (\quad)^3 + (\quad)^3$

$= (\qquad)(\qquad\qquad)$

$\uparrow \qquad \uparrow \qquad \uparrow$

$(\quad)^2 \quad (\quad)(\quad) \quad (\quad)^2$

3. Factor: $a^9 + 8b^6 = (\quad)^3 + (\quad)^3$

$= (\qquad)(\qquad\qquad)$

$\uparrow \qquad \uparrow \qquad \uparrow$

$(\quad)^2 \quad (\quad)(\quad) \quad (\quad)^2$

Student Activity

Match Up on Factoring Binomials

Match-up: In each box of the grid, you will find an expression. Decide how to classify the expression and look to see whether the classification appears in the list at the top. If it does, list that letter, if the classification is not listed, then choose E (none of these). The first one has been done for you. **Some boxes may have more than one answer.**

Also, draw the appropriate skeleton and insert the appropriate bases.

A Sum of Squares: $(\ )^2+(\ )^2$

B Difference of Squares: $(\ )^2-(\ )^2$

C Sum of Cubes: $(\ )^3+(\ )^3$

D Difference of Cubes: $(\ )^3-(\ )^3$

E None of these

x^2-16	$4z^4-9$	$9x^2+27$	y^2+8
w^4+4z^2	$8x^3-27$	x^6-64	$8x^9+125$
$729-x^6$	$4a^2+144b^6$	$169x^4-121y^2$	$343x^3+64$
x^2-3	x^3-1000	w^6-27x^3	w^6+1

Student Activity

Life After the GCF, Part II

Directions: For each expression below, find and factor out the GCF. Then categorize the resulting expression in the table by examining the expression that is within the parentheses. Some expressions may have more than one categorization. The first one has been done for you.

	Expression (factor out the GCF)	GCF only	GCF and form x^2+bx+c	GCF and form ax^2+bx+c	GCF and perfect-square trinomial	GCF and sum or difference of squares	GCF and sum or difference of cubes
1.	$12x^3+12x^2-72x$ $12x\left(x^2+x-6\right)$		**X**				
2.	$125x^5-625x^3$						
3.	$12x^3y^3+42x^2y^3+18xy^3$						
4.	$3t^2x^3+24t^2$						
5.	$40x^2-250$						
6.	$2x^2y^2+4xy^2+2y^2$						
7.	$24y^3+27$						
8.	$27x^4y^6-8x^4$						
9.	$4x^4-16x^3-16x^2$						
10.	$32t^2x^3-2t^2x$						

Student Activity

Leftovers

Directions: Factor each of the binomials below and then mark the factors that they contain. Make sure to look for GCFs first! The first one has been done for you.

Factor these binomials.	Mark the factors that appear in the binomials.						
	$x+2$	$x-3$	$x+5$	$x-4$	$2x-3$	x^2+4	Leftovers
1. x^3+125 $(x+5)(x^2-5x+25)$			X				$(x^2-5x+25)$
2. $3x^2-27$							
3. $4x^3-9x$							
4. x^3-64							
5. $2x^4-32$							
6. $4x^3+32$							
7. $2x^3-54$							
8. $8x^3-27$							
9. x^6+64							
10. $128x-8x^5$							

Student Activity

FACT-17

Skeletons of Tricky Binomials

Directions: Categorize each of the expressions using the choices below, then write the appropriate skeleton for the problem, and factor the expression. If the expression cannot be factored, say so. The first one has been done for you.

A Sum of Squares: $(\ )^2 + (\ )^2$

B Difference of Squares: $(\ )^2 - (\ )^2$

C Sum of Cubes: $(\ )^3 + (\ )^3$

D Difference of Cubes: $(\ )^3 - (\ )^3$

E None of these

Polynomial	Category	Skeleton	Factored Form
1. $x^4 - (a+b)^2$	B	$(x^2)^2 - (a+b)^2$	$(x^2 + (a+b))(x^2 - (a+b))$ or $(x^2 + a + b)(x^2 - a - b)$
1. $(a-b)^2 - x^4$			
2. $1 + (x+y)^3$			
3. $(a-b)^2 + x^2$			
4. $\frac{8}{27} - x^6$			
5. $x^3 - 36$			
6. $x^2 + 2xy + y^2 - 100$			
7. $(x+y)^2 + 27$			
8. $(a+b)^3 - (c-d)^3$			

Guided Learning Activity

GCF to the Rescue

It's tempting to start a factoring problem by immediately setting up a factor table or two sets of parentheses. However, you really should look for a GCF first. Yes, it's true that you can still factor a GCF out at the end of the problem, but it's so much easier if you find the GCF first!

Greatest Common Factor comes to the rescue!

Directions (READ them): In each box of the grid, you will find an expression that needs to be factored. Once you have completely factored the expression, look to see whether any of the factors appear in the list at the top. If none of the factors are listed, then choose E (none of these are factors). Some boxes may have more than one answer. The first one has been done for you.

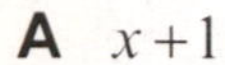

A $x+1$ **B** $2x+3$ **C** $4x-1$ **D** $x-5$ **E** None of these are factors

$40x^2+100x+60$ $20(2x^2+5x+3)$ $20(2x^2+2x+3x+3)$ $20[2x(x+1)+3(x+1)]$ $20(2x+3)(x+1)$ **A, B**	$120x^2+90x-30$	$70-70x^2$
$-60x^2+315x-75$	$40x^2+80x+40$	$8x^2+10x-3$
$60x^2-260x-200$	$20x^2-40x+20$	$40x^2-140x-300$

Student Activity

Finding Factors

Directions: Factor each polynomial and mark which factors it contained. The first one has been done for you.

Factor these polynomials.	Mark the factors that appear in the polynomials.							
	x	$x+2$	$x-3$	$x+5$	$2x+1$	$3x-2$	x^2+4	x^2-2
1. $x^2+7x+10$ $(x+2)(x+5)$		X		X				
2. x^3-2x								
3. $x^2+2x-15$								
4. $2x^2-5x-3$								
5. x^3-3x^2-2x+6								
6. $6x^2-x-2$								
7. $3x^3-2x^2+12x-8$								
8. $2x^2+5x+2$								
9. x^4+2x^2-8								
10. $3x^3-11x^2+6x$								

Student Activity

Factoring Strategizing

For each type of factoring problem, describe the classification of the factoring problem and the strategies and key steps (like the parentheses skeleton) to remember while doing the problem. You do not have to complete the problems.

Choose from these classifications (more than one may apply):

Factor out a GCF	Trinomial ax^2+bx+c	Sum of Cubes
Factor by Grouping	Sum of Squares	Difference of Cubes
Trinomial x^2+bx+c	Difference of Squares	Perfect Square Trinomial

Expression to be factored	Classification	Strategies and Key Steps
1. $9x^2-4$		
2. $4u^4+32u$		
3. $t^2-12t+35$		
4. x^2+49		
5. $x^3+5x^2-3x-15$		
6. $15y^2-36y+12$		
7. $x^2+14x+49$		
8. $64z^3-27$		
9. $a^2+15ab+54b^2$		
10. $4x^4+4x^3-80x^2$		

Student Activity

Factoring Variations on a Number

Directions: In each set there are six expressions to be factored, all ending with the same constant (positive or negative). If the expression is factorable, factor it. If it is not factorable, write "PRIME."

Variations on 1			
	$x^2 - 1$	$x^3 - 1$	$x^2 - 2x + 1$
	$x^2 + 1$	$x^3 + 1$	$x^2 + 2x + 1$

Variations on 64			
	$x^2 - 64$	$x^2 + 64$	$x^2 - 16x + 64$
	$x^3 - 64$	$x^3 + 64$	$x^2 + 16x + 64$

Variations on 36			
	$x^2 - 36$	$9x^2 + 36$	$x^2 + 12x + 36$
	$4x^2 - 36$	$9x^2 - 36$	$x^2 - 12x + 36$

Student Activity

Factors in Hiding

Directions: Factor each polynomial and then find each factor in the grid at the bottom of the page. You should find every factor in the grid.

1. Factor: $(a+b)^2+5(a+b)+6$ HINT: Treat it like a trinomial.

2. Factor: $(a-b)^2+9(a-b)-22$

3. Factor: $(a+b)^2-4(a+b)+4$

4. Factor: $4(a-b)+11(a-b)-3$

5. Factor: $(a-b)(x+y)+2(x+y)$ HINT: Is there a GCF?

6. Factor: $(a+b)^2+x(a+b)$

7. Factor: $a^2b^2+5a^2b+3a^2+10b^2+50b+30$ HINT: Factor by grouping.

8. Factor: $a^2b+4ab+2b-3a^2-12a-6$

$a+b-2$	$a+b-2$	$a-b-2$	$a+b$
$a+b+2$	b^2+5b+3	$b-3$	$x+y$
a^2+10	$a-b+11$	$a-b+2$	$a-b+3$
$a+b+3$	$4a-4b-1$	a^2+4a+2	$a+b+x$

Student Activity

Checking the Factoring with a Calculator

Directions: One way you can check your factoring is to evaluate both the original expression and the factored expressions for the same value of the variable. It is not wise to use the values of 0, 1, or 2 for these types of checks since these three numbers have some quirky properties:

$0+0=0$ and $0\cdot 0=0$ $\quad$ $1\cdot 1=1$ $\quad$ $2+2=4$ and $2\cdot 2=4$

If a check by evaluation results in different answers, then you should reexamine the factoring. Note that getting the same result from evaluation does not **ensure** that your factors are correct, but it is likely to find a mistake in your work if there is one.

In each table below, check the student's factoring by evaluating for the given values with a calculator. It will help to write out the skeleton with the value substituted first. Then input the numerical expression into your calculator to get the result. If you find a factoring mistake, correct it! The first one has been started for you.

1.

	Evaluate for $x=3$.	**Evaluate for $x=2.4$.**
$3x^2+7x+2$	$3(3)^2+7(3)+2=50$	
$(x+2)(3x+1)$	$(3+2)(3(3)+1)=50$	

2.

	Evaluate for $x=4$.	**Evaluate for $x=1/3$.**
$x^2+4x-21$		
$(x-7)(x+3)$		

3.

	Evaluate for $x=1/2$.	**Evaluate for $x=-3$.**
$4x^2-25$		
$(2x+5)(2x-5)$		

4.

	Evaluate for $x=3.5$.	**Evaluate for $x=-5/4$.**
$4x^2+20x+25$		
$(4x+5)^2$		

5.

	Evaluate for $x=5$.	**Evaluate for $x=3$.**
x^3-8		
$(x-2)(x^2+4x+4)$		

Student Activity

Escape the Matrix by Solving Quadratic Equations

Directions: In each box of the grid, you will find a quadratic equation that needs to be solved. Once you have solved the equations, use pairs of matching solution numbers to navigate your way out of the matrix. The first one has been done for you. For example, the repeated solution of -3 leads you to the box on the right, with solutions of -3 and 4. To move the next step on the escape route, you need to find an adjacent box with a solution of 4 and something else.

START HERE Solve: $(x+3)^2=0$ Solutions: -3 and -3	Solve: $x^2-x-12=0$ $(x-4)(x+3)=0$ Solutions 4 and -3	Solve: $x^2-8x-48=0$	Solve: $(x+12)^2=0$
Solve: $x^2+2x-8=0$	Solve: $(x-4)(x-\frac{1}{2})=0$	Solve: $x^2-\frac{5}{6}x+\frac{1}{6}=0$	Solve: $3x^2-19x+6=0$
Solve: $x^2-8x+15=0$	Solve: $x^2-x-42=0$	Solve: $x^2+x=2$	Solve: $(x-6)(x+2)=0$
ESCAPE the Matrix Solve: $x^2+8x+16=0$	Solve: $x^2-16=0$	Solve: $x^2-5x+4=0$	Solve: $x^2-11x=-24$

Student Activity

FACT-25

Multiply, Factor, or Solve

Directions: The procedures for multiplying polynomials, factoring polynomials, and solving quadratic equations using factoring are most likely stored very near to each other in your memory because the procedures are entwined. Here is some practice to make sure that the procedures are clearly separated in your mind.

1. Factor: $y^2 - 49$

2. Solve: $21x^2 - 25x - 4 = 0$

3. Multiply: $(x-2)^2$

4. Solve: $x^2 - 16 = 0$

5. Factor: $2x^5 - 2y^2$

6. Solve: $x^2 = 8x - 16$

7. Multiply: $x(x-2)(x+3)$

8. Factor: $x^2 - 3xy - 4y^2$

9. Multiply: $(x-5)(x^2 + 5x + 25)$

10. Factor: $2x^4 + 16x^3 - 40x^2$

Describe what each of the directions mean in your own words.

Factor:

Multiply:

Solve:

Student Activity

Parentheses, Brackets, and Braces, Oh My!

Some mathematical situations require very specific notation with parentheses, brackets, and braces. In these cases, the notation is **not** interchangeable.

- Ordered pairs are written with parentheses. For example, $(-2,5)$ is an ordered pair with x-coordinate -2 and y-coordinate 5.
- Interval solutions use parentheses and brackets. For example $(-2,5]$ is an interval in which 5 is included in the solution.
- Solution sets use braces. For example $\{-2,5\}$ shows two solution values, -2 and 5.

1. Match the solution with the appropriate graph:

___ **a.** $(-2,5)$ ___ **b.** $\{-2,5\}$ ___ **c.** $(-2,5]$

I

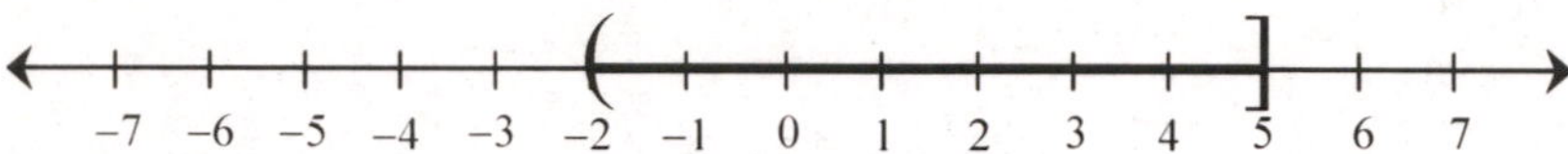

II

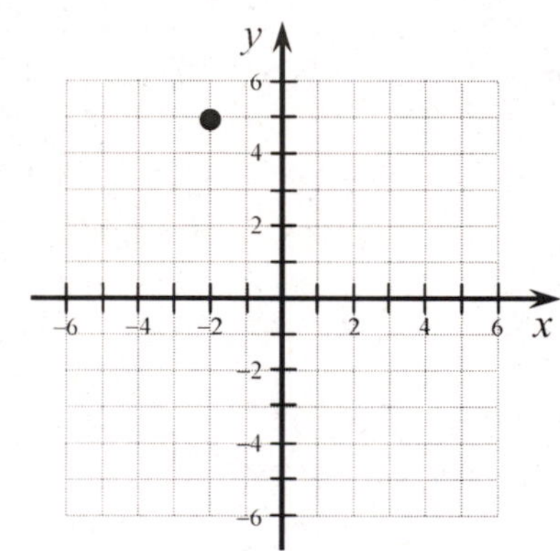

III

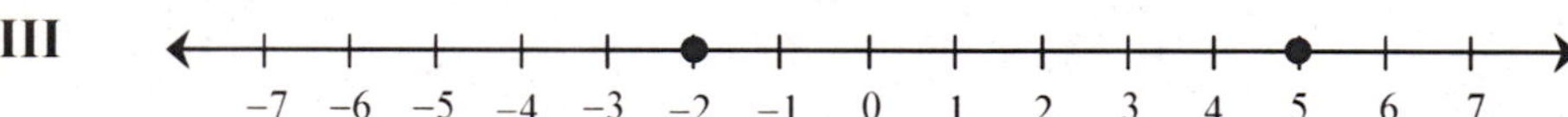

2. Solve $3-2x \le 9$ and write the solution in interval notation.

3. Find the ordered pair solution for $3x - y = 4$ if $x = 3$.

4. Solve $x^2 - 5x - 36 = 0$ and write the answer as a solution set.

5. Solve $-2 \le x + 4 \le 7$ and write the solution in interval notation.

6. Solve $5(x-2) + 3x = 4x - 8$ and write the solution as a solution set.

7. Is $(-1, 3)$ a solution of $4x + y = 1$?

8. Write the solution that is represented by the graph below.

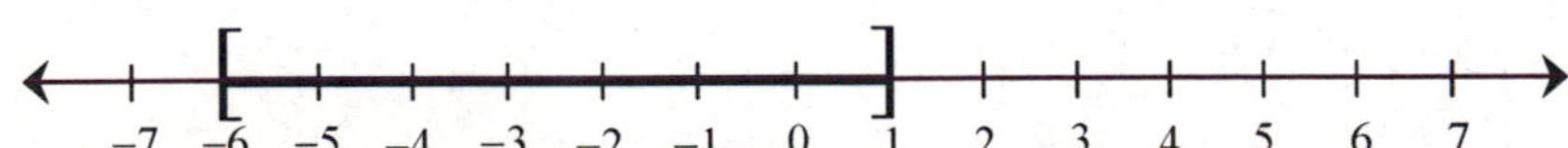

9. Write the ordered pair that is represented by the graph below.

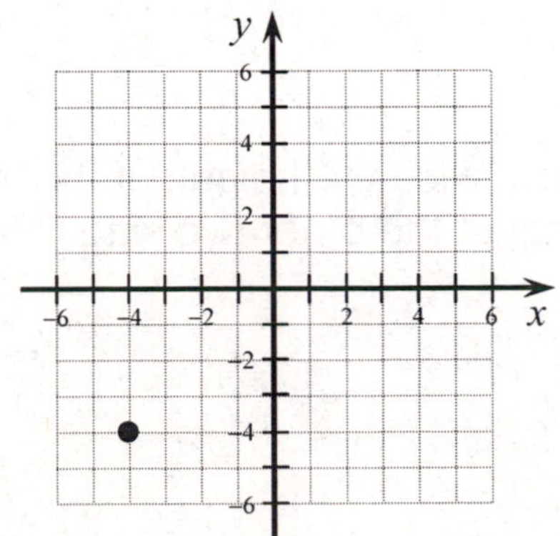

10. Write the solution that is represented by the graph below.

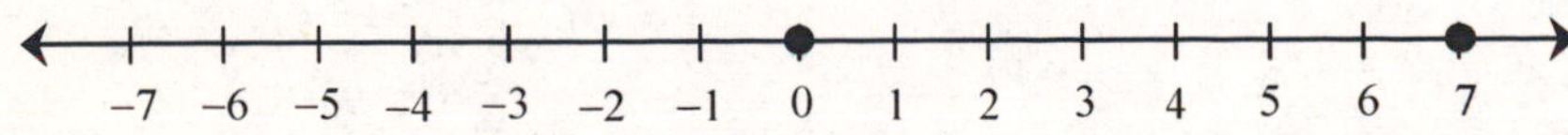

Guided Learning Activity

Using the Problem Solving Strategy with Factoring

1. **Analyze.** Draw diagrams, tables, and charts to help make sense of the problem.
2. **Declare.** Declare the variables and the quantitative givens in the problem.
3. **Formulate.** Write equations to represent the problem.
4. **Solve.** Solve the equation or system of equations.
5. **Validate.** Check the result. Does it make sense? Is it reasonable?
6. **Conclude.** State the conclusion. Don't forget units.

Problem 1: The area of a vegetable garden is 32 m^2. If the length is 4 meters longer than the width, what are the dimensions (length and width) of the garden?

Problem 2: The area of a right triangle is 10 ft^2. If the base of the triangle is 1 foot longer than the height, what are the dimensions of the triangle?

Problem 3: The base of a right triangle is one less than twice the height. If the length of the hypotenuse is 17 cm, what are the measurements of the other two sides?

Student Activity

FACT-29

Proof of the Pythagorean Theorem

Materials required: Paper for tracing, scissors, and tape.

This is a geometric proof of the Pythagorean Theorem (by dissection) given by the Arabian Mathematician Thâbit ibn Kurrah, who lived from 836 to 901 A.D. in the Middle East.

Sources: http://mathworld.wolfram.com/PythagoreanTheorem.html, http://www.britannica.com/eb/article-9071897/Thabit-ibn-Qurra

Step 1: Trace the figure below onto a **new sheet of paper**.
Cut out the square with side *a* and the square with side *b*.

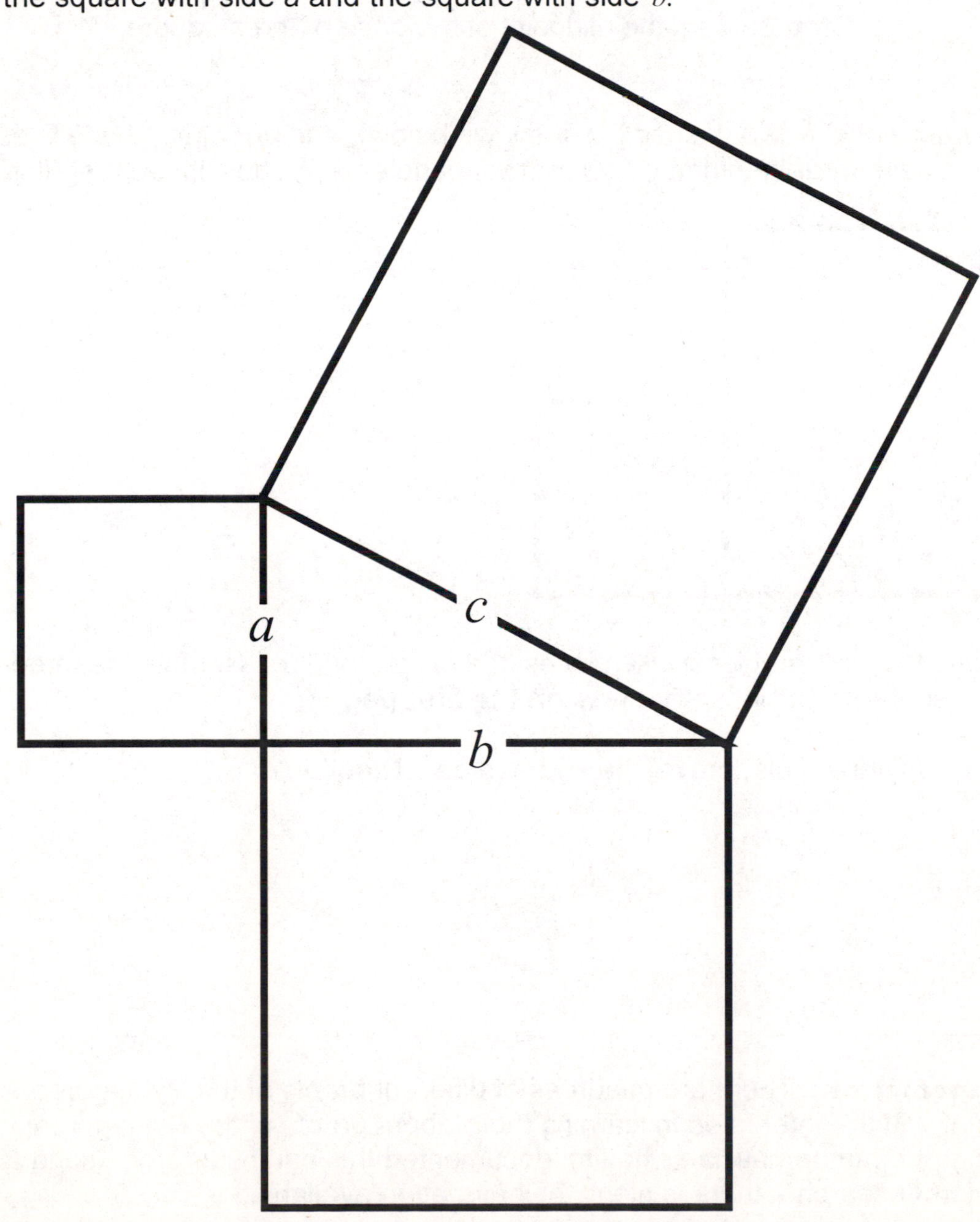

Step 2: Securely tape these two squares together like so.

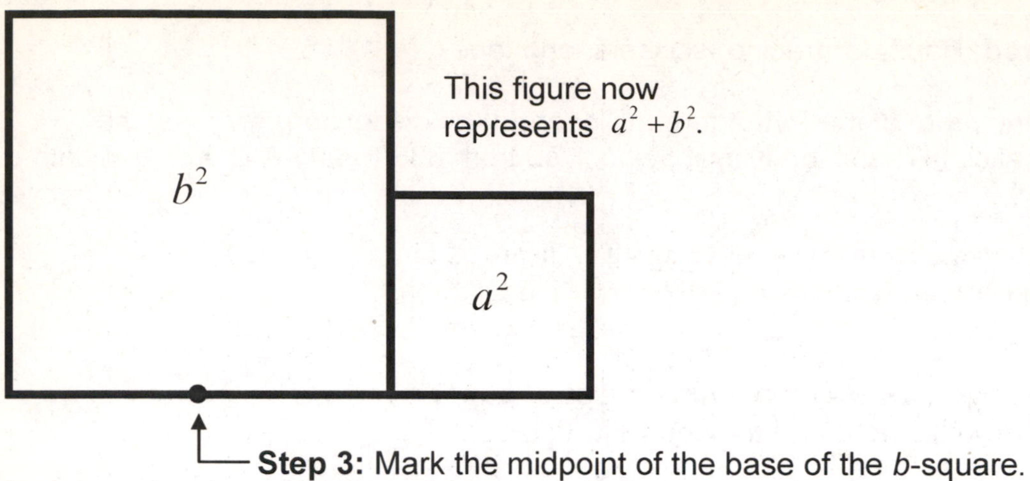

Step 4: Draw in these two dashed lines (shown below), and **cut** along these lines. You should now have three geometric figures (two triangles and one with an irregular shape

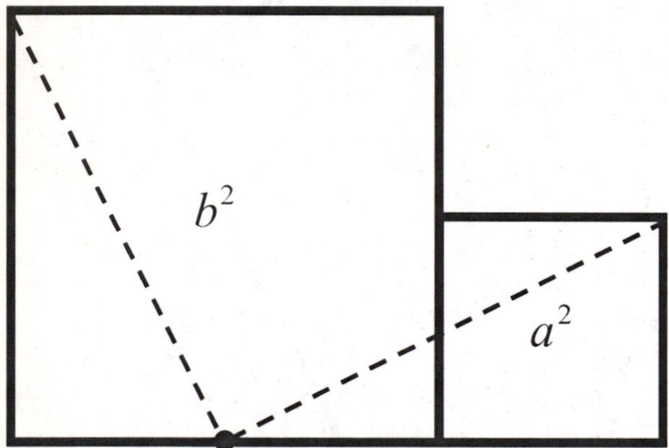

Step 5: Use these three pieces like pieces of a puzzle and reassemble the three pieces to fit in the square with side *c* **that was on the first page**.

Question: Why does this "prove" the Pythagorean Theorem?

More Information: There are hundreds of different proofs of the Pythagorean Theorem. If you are interested in learning more about proofs of the Pythagorean Theorem or the mathematician who first documented this particular proof, conduct a simple Internet search – there is plenty of information available.

Student Activity

Pumpkin Launch

Ballistics is the study of the flight characteristics of projectiles. For ballistic flights, Littlewood's Law can be used to *approximate* the peak height of a projectile by using the total flight time.

Littlewood's Law: $h = \frac{g}{8} \cdot t^2$ where h is the peak height in meters, t is the total flight time in seconds, and g is the gravity constant, which is $g = 9.8 \text{ m/s}^2$ near the surface of the earth.

The "Pumpkin Rocket II"
(source: http://www.et.byu.edu/~wheeler/benchtop/pumpkin2.php)

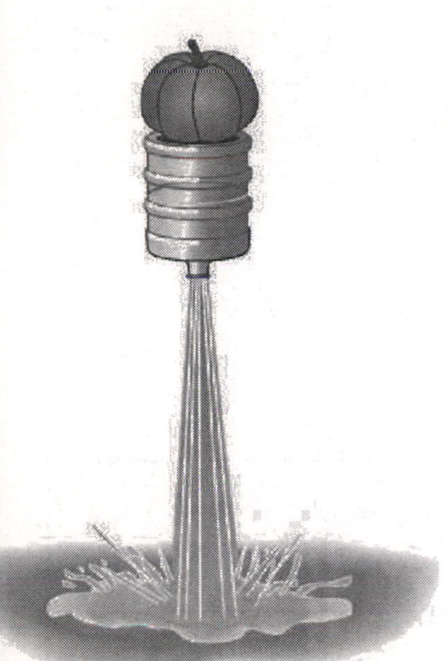

The "Pumpkin Rocket II" was launched using a hydrogen-power method to pressurize the launch vehicle (a large water bottle).

Question 1: The pumpkin and bottle reached a peak height of 18 meters. Try substituting different values of t into Littlewood's Law to find how long the flight must have been to reach this peak height.

t (seconds)	h (meters)
1	
2	
3	
4	
5	
6	

Question 2: Assuming we could build a pumpkin rocket that launch a pumpkin to a height of 78.4 meters, what would the total flight time be for this rocket? **This time, simplify the equation and write it in standard quadratic form, then use factoring to solve the problem.**

Assess Your Understanding

Factoring and More

For each of the following, describe the strategies or key steps that will help you **start** the problem. You do **not** have to complete the problems.

		What will help you to start this problem?
1.	Factor: $y^2+7y-18$	
2.	Factor: $3x^2-12$	
3.	Factor: $3x^2-14x-24$	
4.	Factor: $xy^2-9x+4y^2-36$	
5.	Is -2 a solution of $x^2-5x=14$?	
6.	Factor: $4x^2+11x-20$	
7.	Factor: $5x^2-80$	
8.	The area of a rectangle is $18\,\text{cm}^2$. If the length is twice the width, find the dimensions of the rectangle.	
9.	Solve: $x^2-8x+12=3(x-4)$	
10.	Factor: $x^3y^3-64z^6$	

Metacognitive Skills

Factoring and More

Metacognitive skills refer to the ability to judge how well you have learned something and to effectively direct your own learning and studying. This is a self-evaluation tool designed to help you focus your studying and to improve your metacognitive skills with regards to this math class.

Fill the 1st column out **before** you begin studying. Fill the 2nd column out after you study for your test.

Go back to this assessment after your test and circle any of the ratings that you would change – this identifies the "disconnects" between what you **thought** you knew well and what you **actually** knew well.

Use the scale below to assign a number to each topic.

5 *I am confident I can do any problems in this category correctly.*
4 *I am confident I can do most of the problems in this category correctly.*
3 *I understand how to do the problems in this category, but I still make a lot of mistakes.*
2 *I feel unsure about how to do these problems.*
1 *I know I don't understand how to do these problems.*

Topic or Skill	Before Studying	After Studying
Finding the GCF for several terms.		
Factoring out the GCF.		
Factoring out a -1. Why is it sometimes necessary to factor out a -1?		
Factoring by grouping and knowing when to employ this method.		
Distinguishing between trinomials of the form x^2+bx+c and the form ax^2+bx+c.		
Factoring a trinomial of the form x^2+bx+c.		
Factoring a trinomial of the form ax^2+bx+c.		
Knowing the numbers that are perfect squares and perfect cubes.		
Recognizing expressions that are perfect squares and perfect cubes.		
Recognizing binomials that are either a sum or difference of squares.		
Factoring a difference of squares.		
Knowing about the factorability of a sum of squares.		
Recognizing binomials that are either a sum or difference of cubes.		
Factoring a sum or difference of cubes.		
Recognizing a perfect-square trinomial and properly factoring it.		
Using the general factoring strategy. What should you always do first?		
How can the number of terms in a polynomial help you to make decisions about how to factor it?		
Understanding when an expression is not yet completely factored, and how to complete it. Ex: Is $x(2x+4)(x^2-9)$ completely factored?		

Continued on next page.

Use the scale below to assign a number to each topic.

5 *I am confident I can do any problems in this category correctly.*
4 *I am confident I can do most of the problems in this category correctly.*
3 *I understand how to do the problems in this category, but I still make a lot of mistakes.*
2 *I feel unsure about how to do these problems.*
1 *I know I don't understand how to do these problems.*

Topic or Skill	Before Studying	After Studying
Being able to check the answer to a factoring problem by multiplying out the answer.		
Understanding why you must first get "=0" on one side of a quadratic equation before solving by factoring.		
Rewriting a quadratic equation in standard quadratic form.		
Solving a quadratic equation by factoring.		
Knowing how to properly write a solution set.		
Checking the solution to a quadratic equation.		
Distinguishing between the instructions: multiply, factor, or solve.		

QUAD: Quadratics

Student Activity

QUAD-1

Paint by Double-Signs

Directions: Simplify each of the expressions below. Then shade in the equivalent expressions in the grid below. When you're done, you should see a picture!

1. -1 ± 6

2. -3 ± 5

3. $\dfrac{5 \pm 10\sqrt{2}}{5}$

4. $\dfrac{6 \pm 2\sqrt{2}}{2}$

5. $\dfrac{-8 \pm 24}{4}$

6. $\dfrac{12 \pm \sqrt{18}}{3}$

7. $\dfrac{-2 \pm \sqrt{25-9}}{2}$

".... and I thought I'd never need to use math if I became an artist! Ugh!"

$12-\sqrt{2}$	$12+\sqrt{2}$	$4+\sqrt{2}$	$4+\sqrt{18}$	$4-\sqrt{6}$
-2	$4-\sqrt{18}$	-7	7	$3-2\sqrt{2}$
2	$4-\sqrt{2}$	-3	-8	$3-\sqrt{2}$
$3+2\sqrt{2}$	$6+\sqrt{2}$	$3+\sqrt{2}$	-14	$6-\sqrt{2}$
$4+\sqrt{6}$	-6	$1+2\sqrt{2}$	$1+10\sqrt{2}$	3
8	-5	0	$1-10\sqrt{2}$	-4
-8	5	1	$1-2\sqrt{2}$	4

Guided Learning Activity

The Square Root Property

I am thinking of a number. If I square the number, I get 9. What is the number?

The easy choice here is 3, since $3^2 = 9$.

What **other** number could satisfy the condition? $(\_\_\_\_)^2 = 9$

You could also set up the equation $x^2 = 9$ to solve this problem. We have learned how to solve this equation **by factoring**. Go ahead and do that. *Hint: Remember to get =0 on one side first.*

Solve: $x^2 = 9$

We can also solve this equation (often in less steps) by **taking a square root** on both sides of the equation.

Solve:
$$x^2 = 9$$
$$\sqrt{x^2} = \sqrt{9}$$
$$x = 3$$

But there is a problem with the process we have written above; we are missing one of the possible solutions! Recall that 9 has two square roots, -3 and 3. However, the notation $\sqrt{9}$ only gives the principal (positive) square root. Thus, we have to solve $x^2 = 9$ by listing both square root solutions, as follows.

Solve:
$$x^2 = 9$$
$$x = \sqrt{9} \quad \text{or} \quad x = -\sqrt{9}$$
$$x = 3 \quad | \quad x = -3$$

We now have both of the correct solutions using the square root property $\{\pm 3\}$

The Square Root Property of Equations:
For any nonnegative real number c, if $x^2 = c$, then $x = \sqrt{c}$ or $x = -\sqrt{c}$.

To solve with the Square Root Property, we must **isolate the squared part** of the equation first. Try these together with your instructor.

Example 1: Solve $5x^2 - 10 = 70$.

Example 2: Solve $\dfrac{(x-5)^2}{2} = 32$.

Example 3: Solve $3x^2 - 33 = 0$.

Example 4: Solve $4 = (4x-3)^2 - 9$.

Example 5: Solve $x^2 + 25 = 0$.

Student Activity
Square Isolation Tic-Tac-Toe

The square root property can be used to solve a quadratic equation if

1) you can isolate the part of the equation that is written as a square.
2) the only variables in the equation are in the squared part.

Directions: If the equation **can** be solved easily with the square root property using the rules above, then isolate the squared part, and circle the equation (placing an O on the square). If the equation cannot be easily solved with the square root property, then place an X on the square.

$3(x+2)^2-25=50$	$x^2=5x-4$	$-5(3x-1)^2=-45$
$3x^2=75$	$24=(x+7)^2+8$	$x^2+8x-20=0$
$16x^2-24x=-9$	$(x-3)^2+2x=0$	$x^2+6x=2(3x+2)$

Student Activity

Double the Fun on Quadratic Equations

Directions: When you solve some quadratic equations, you can choose to solve by factoring or by using the Square Root Property. **If possible**, solve the following equations both ways. If it is not possible to solve the problem using one of the methods, say so. The first one has been started for you.

	Quadratic Equation	Solve by Factoring	Solve by Square Root Property
1.	$x^2 = 25$	$x^2 = 25$ $x^2 - 25 = 0$	$x^2 = 25$ $x = \sqrt{25}$ or $x = -\sqrt{25}$
2.	$4x^2 - 81 = 0$		
3.	$(x-3)^2 = 4$	Hint: Expand the binomial first.	
4.	$x^2 - 8 = 0$		
5.	$x^2 + 16x + 64 = 25$		Hint: Do you see a perfect square?

Student Activity

The Square Egg

Directions: Fill in the blanks in each equation to complete the square and write the expression as a perfect square. All of the solutions are provided inside the "egg". When you've crossed out all the solutions, you have "hatched" the egg. The first one has been done for you.

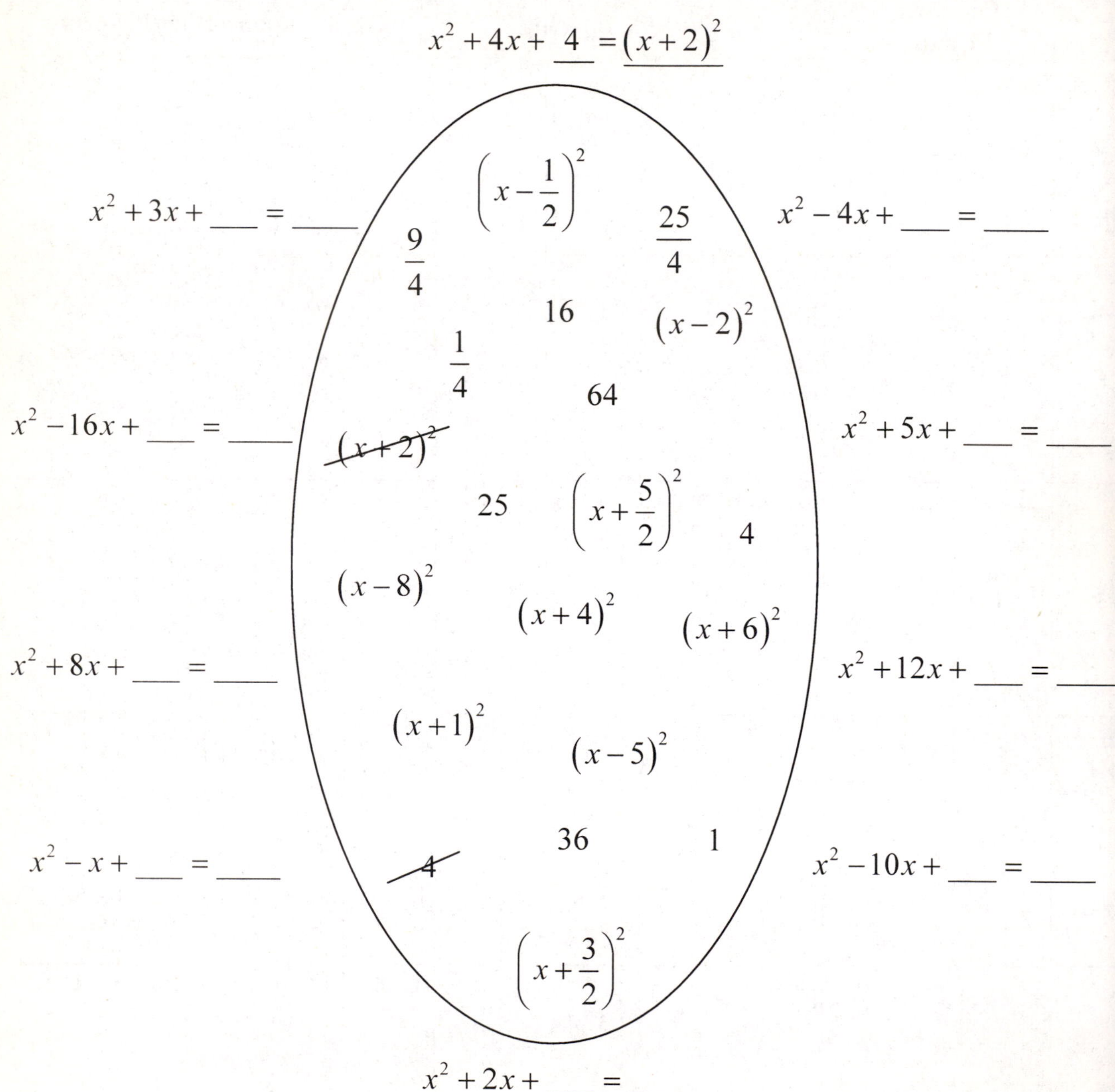

Guided Learning Activity

Learning to Complete

Recall that we can use the square root property to solve quadratic equations, if all of the variables appear within a quantity squared.

Example 1: Solve $(x+2)^2 = 9$

Now we will use the process of completing the square together with the square root property to help us solve quadratic equations.

Example 2: Solve $x^2 - 8x - 9 = 0$ by completing the square.

a) Get the constant term on the right side of the equation and leave a "blank" to complete the square in on the left side. (Note: Sometimes the equation is already set up this way.)

$$x^2 - 8x \underline{\qquad\qquad} = 9$$

b) Complete the square by adding the same constant to both sides.

$$x^2 - 8x \underline{+16} = 9 + 16$$

c) Write the left side as a perfect square, combine constant terms on the right side.

$$(x-4)^2 = 25$$

d) Now solve using the square root property.

e) And check your answers.

Example 3: Solve $x^2 - 10x + 5 = 0$ by completing the square.

Example 4: Solve $a^2 = 24(a-1)$ by completing the square. *Hint: Simplify first!*

Example 5: Solve $x^2 + 5x + \frac{1}{4} = 0$ by completing the square.

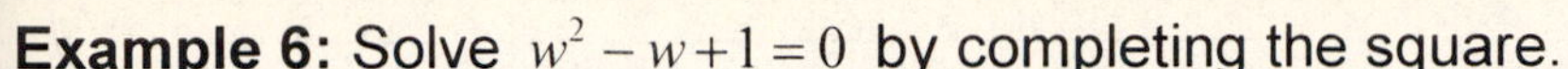

Example 6: Solve $w^2 - w + 1 = 0$ by completing the square.

What if the lead coefficient is not 1? We then begin by dividing both sides of the equation by the lead coefficient. Let's work through some of these problems together.

Example 7: Solve $4x^2 + 16x - 48 = 0$ by completing the square.

Example 8: Solve $3x^2 + 13x - 30 = 0$ by completing the square.

Student Activity

Recognizable Foe

Directions: Follow the directions for each step and fill in the boxes with the missing pieces as you go. Careful – as you move from line to line, the boxes might contain different expressions! At the end, you should get a formula that you might recognize!

Solve: $ax^2 + bx + c = 0$

Divide both sides by *a*: $x^2 + \square + \frac{c}{a} = \square$

Subtract $\square$ from both sides: $x^2 + \square = -\frac{c}{a}$

Complete the square: $\frac{1}{2}\left(\square\right) = \square$ and $\left(\square\right)^2 = \square$

Add $\square$ to both sides: $x^2 + \frac{b}{a}x + \square = -\frac{c}{a} + \square$

Write the left side as a perfect square: $\left(x + \square\right)^2 = -\frac{c}{a} + \square$

Find a common denominator on the right: $\left(x + \square\right)^2 = -\frac{\square}{4a^2} + \frac{b^2}{4a^2}$

Combine the terms on the right side: $\left(x + \square\right)^2 = \frac{\square - 4ac}{4a^2}$

Apply the square root property: $x + \square = \pm\sqrt{\frac{\square - 4ac}{4a^2}}$

Simplify the denominator on the right side: $x + \square = \pm\frac{\sqrt{\square - 4ac}}{\square}$

Subtract $\square$ from both sides: $x = \square \pm \frac{\sqrt{\square - 4ac}}{\square}$

Combine terms with like denominators: 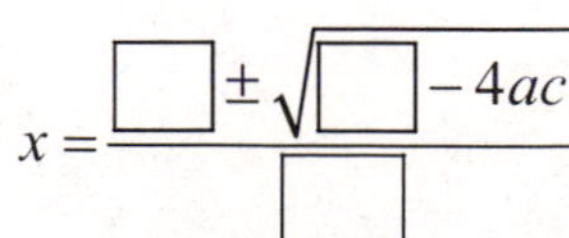

$x = \frac{\square \pm \sqrt{\square - 4ac}}{\square}$

If you're really good… try it on your own on a blank sheet of paper without peeking!

Student Activity

Quadratic Formula with a Calculator

It can be tricky to evaluate the quadratic formula using your calculator because of all the order of operations that are implied by the symbols.

For example, suppose we wanted to calculate the value of:

$$\frac{-3\pm\sqrt{(-3)^2-4\cdot1\cdot(-2)}}{2\cdot1}.$$

There are many extra parentheses we would have to add to do the calculation directly. First, the numerator and denominator need parentheses, since a fraction bar separates two distinct groups of terms:

$$\frac{\left(-3\pm\sqrt{(-3)^2-4\cdot1\cdot(-2)}\right)}{(2\cdot1)}$$

Then the radicand needs a set of parentheses, since the radicand must be simplified before the square root is taken:

$$\frac{\left(-3\pm\sqrt{\left((-3)^2-4\cdot1\cdot(-2)\right)}\right)}{(2\cdot1)}$$

Then we need to calculate two expressions because of the double-sign:

$$\frac{\left(-3-\sqrt{\left((-3)^2-4\cdot1\cdot(-2)\right)}\right)}{(2\cdot1)} \text{ or } \frac{\left(-3+\sqrt{\left((-3)^2-4\cdot1\cdot(-2)\right)}\right)}{(2\cdot1)}$$

Entering this into **my** calculator would look like this:

[(] [(−)] [3] [−] [2ND]√ [(] [(] [(−)] [3] [)] [^] [2] [−] [4] [×] [1] [×] [(−)] [2] [)] [)] [)] [÷] [(] [2] [×] [1] [)] [ENTER]

The point is, **you will want to simplify your quadratic formula** to a more manageable expression **before** you enter it into your calculator.

$$\frac{-3\pm\sqrt{(-3)^2-4\cdot1\cdot(-2)}}{2\cdot1}=\frac{-3\pm\sqrt{9+8}}{2}=\frac{-3\pm\sqrt{17}}{2}$$ ←This is much more manageable!

Now we calculate the value of the two expressions: (your keystrokes may vary)

$\frac{\left(-3-\sqrt{17}\right)}{2}$ [(] [(−)] [3] [−] [2ND]√ [1] [7] [)] [÷] [2] [ENTER] ≈ -3.56

$\frac{\left(-3+\sqrt{17}\right)}{2}$ [(] [(−)] [3] [+] [2ND]√ [1] [7] [)] [÷] [2] [ENTER] ≈ 0.56

NOTE: On some calculators, you can edit an old entry. For example, on most TI graphing calculators, use ENTRY to pull up the last entry. You can use this to change the [−] to a [+] and then press ENTER to get the second answer.

Now try finding these with your calculator. The first row has been done for you. Be careful not to round values until the **end** of the problem. Use a ≈ symbol if the calculated result is rounded. Round to two decimal places if necessary.

Quadratic Formula Expression	Simplified Expression	Calculated Result
1. $\dfrac{-(-4)\pm\sqrt{(-4)^2-4(3)(1)}}{2(3)}$		
2. $\dfrac{-5\pm\sqrt{5^2-4(2)(-3)}}{2\cdot 2}$		
3. $\dfrac{-11\pm\sqrt{11^2-4(-2)(-6)}}{2(-2)}$		
4. $\dfrac{-(-3,850)\pm\sqrt{(-3,850)^2-4(550)(-5,500)}}{2(550)}$		
5. $\dfrac{-(-4.18)\pm\sqrt{(-4.18)^2-4(1.32)(-1.54)}}{2(1.32)}$		

Scrambled Answers

0.5	≈ 0.61	≈ −0.33	−3	≈ 8.22
1	3.5	≈ 0.33	≈ −1.22	≈ 4.89

Student Activity

Triple the Fun on Quadratic Equations

Directions: For many quadratic equations, you have the choice of solving by using the factoring method, the square root method (completing the square if necessary), or the quadratic formula. Solve each of the quadratic equations by the indicated methods (if possible). If it is **not** possible to use one of the methods, say so. We've given you some hints on the first problem. For each problem, put a ☺ by the method that seemed easiest.

1. Solve: $x^2 - 3x - 40 = 0$

Solve by Factoring (if possible):

$$x^2 - 3x - 40 = 0$$

$$(\qquad)(\qquad) = 0$$

Solve by the Square Root Property:
(completing the square if necessary)

$$x^2 - 3x - 40 = 0$$

$$x^2 - 3x + \_\_\_\_ = 40 + \_\_\_\_\_$$

Solve using the Quadratic Formula:

$$x^2 - 3x - 40 = 0$$

$a =$ ____, $b =$ ____, $c =$ ____

2. Solve: $x^2 + 2x - 2 = 0$

Solve by Factoring (if possible):

$$x^2 + 2x - 2 = 0$$

Solve by the Square Root Property:
(completing the square if necessary)

$$x^2 + 2x - 2 = 0$$

Solve using the Quadratic Formula:

$$x^2 + 2x - 2 = 0$$

3. Solve: $4x^2 - 5x - 6 = 0$

Solve by Factoring (if possible):

$4x^2 - 5x - 6 = 0$

Solve by the Square Root Property:
(completing the square if necessary)

$4x^2 - 5x - 6 = 0$

Solve using the Quadratic Formula:

$4x^2 - 5x - 6 = 0$

4. Solve: $x^2 - 8 = 0$

Solve by Factoring (if possible):

$x^2 - 8 = 0$

Solve by the Square Root Property:
(completing the square if necessary)

$x^2 - 8 = 0$

Solve using the Quadratic Formula:

$x^2 - 8 = 0$

Student Activity

Match Up on the Discriminant

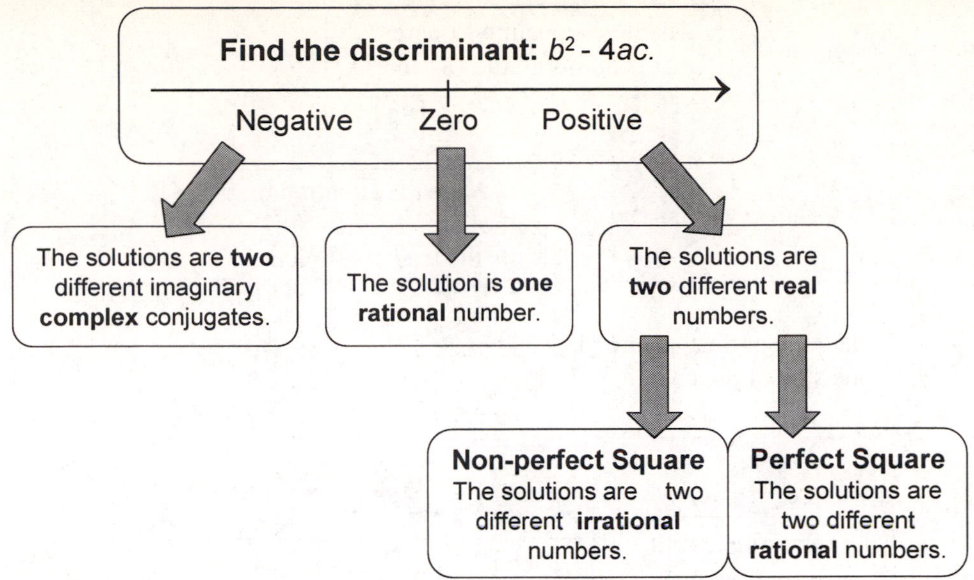

Directions: In each box, find the discriminant and then classify it.

A Two different rational numbers

B Two different irrational numbers

C One rational number

D Two different imaginary numbers

$25x^2 - 40x + 16 = 0$ $d = (-40)^2 - 4(25)(16) = 0$ C	$3x^2 + 14x - 24 = 0$	$x^2 - 2x + 2 = 0$
$9x^2 = 4$	$6x^2 - 3 = 0$	$x^2 + 3 = 5x$
$5(5x^2 + 2x) = -1$	$2x^2 + 3x = -3$	$3x^2 + 11x = 5$
$45x^2 + 24 = 67x$	$144x^2 - 264x + 121 = 0$	$4x^2 + 36x = -81$

Student Activity

Strategizing on Quadratic Equations

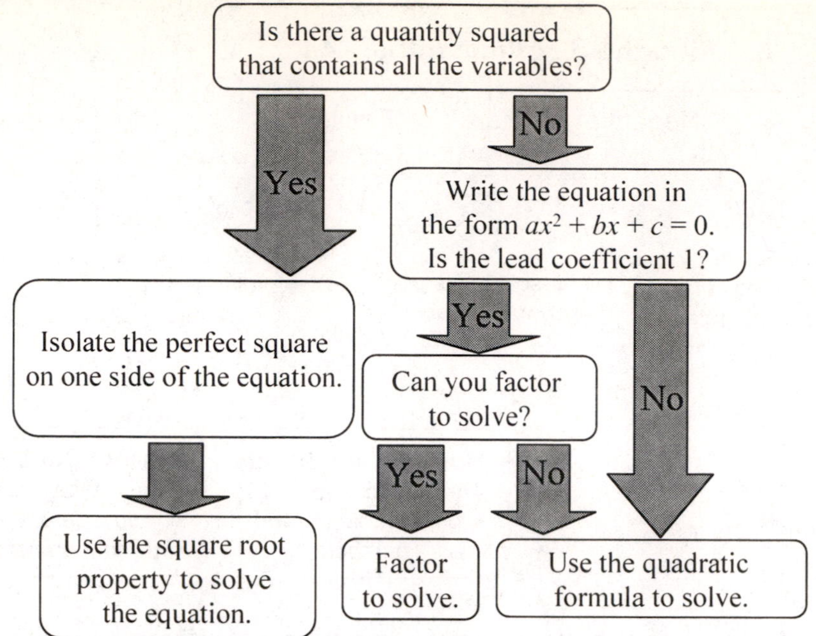

Directions: For each problem, use the flow chart above to help you make a decision about **how** to solve the quadratic equation (the square root property, factoring, or the quadratic formula). You do not need to solve the problems.

	Square root property	Factoring	Quadratic formula
1. Solve: $x^2 - 8x - 20 = 0$			
2. Solve: $4(x+3)^2 = 25$			
3. Solve: $4x^2 = 9$			
4. Solve: $x^2 + 5x - 3 = 0$			
5. Solve: $6x^2 + 7x = 20$			
6. Solve: $x^2 - 6x = 0$			
7. Solve: $x^2 - x + 6 = 0$			
8. Solve: $8x(x-2) = 2x + 5$			
9. Solve: $(x-2)^2 - 2x = 16$			
10. Solve: $(x-2)^2 = 16$			

Student Activity

Sink the Sub

Directions: For each problem, follow these steps:

- Find an appropriate y-substitution to make the equation quadratic in form.
- Solve the quadratic equation.
- Reverse the y-substitution to determine the solution set of the original equation.

The first problem is done for you. Two submarines holding the possible answers for $u = ?$ and $x = ?$ are provided to help you stay on track.

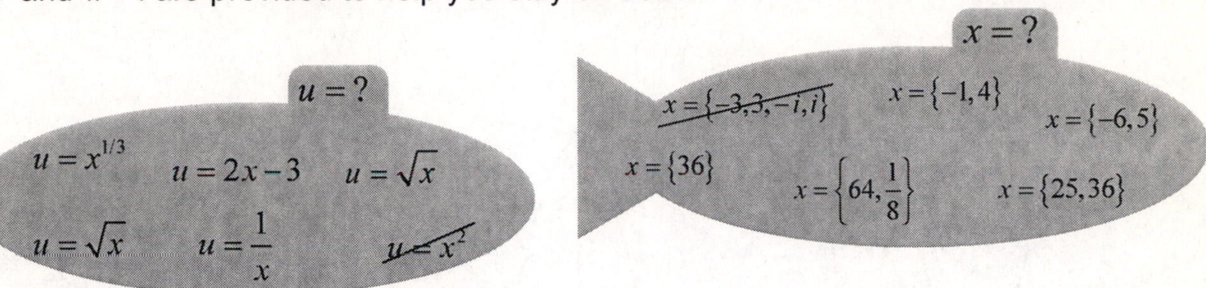

Solve.	$y = ?$	Substitute, then solve for y.	Reverse the substitution	Solution set
1. $x^4 - 8x^2 - 9 = 0$	$y = x^2$	$y^2 - 8y - 9 = 0$ $(y-9)(y+1) = 0$ $y = 9, -1$	$x^2 = 9$ $x^2 = -1$ $x = \pm 3$ $x = \pm i$	$x = \{-3, 3, -i, i\}$
2. $x - 11\sqrt{x} + 30 = 0$				
3. $x - \sqrt{x} - 30 = 0$				
4. $2x^{2/3} - 9x^{1/3} + 4 = 0$				
5. $(2x-3)^2 = 25$				
6. $\frac{30}{x^2} - \frac{1}{x} - 1 = 0$				

Guided Learning Activity

Everything's Coming Up Parabolas

The graph of $y = ax^2 + bx + c$, where $a \neq 0$, is a quadratic function, and its graph is called a parabola. If $a > 0$, then the parabola opens up. If $a < 0$, then the parabola opens down. The vertex is the maximum or minimum point of the parabola (depending on whether it opens up or down. Each parabola is symmetric about the axis of symmetry.

1. On each parabola below, label the **vertex**, the **axis of symmetry**, whether the parabola **opens up** or **opens down** and whether ***a* > 0** or ***a* < 0**.

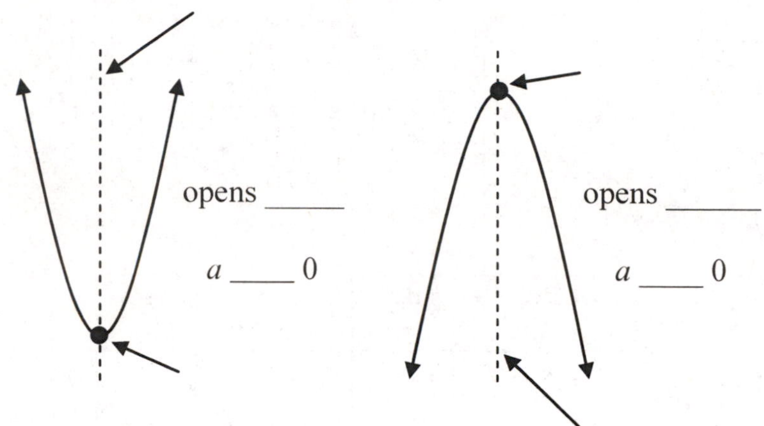

2. Draw the graph of $y = -x^2 + 2x + 3$ by plotting points. On the graph, clearly indicate the point that is the vertex and a dashed line for the axis of symmetry.

x	y
−3	
−2	
−1	
0	
1	
2	
3	

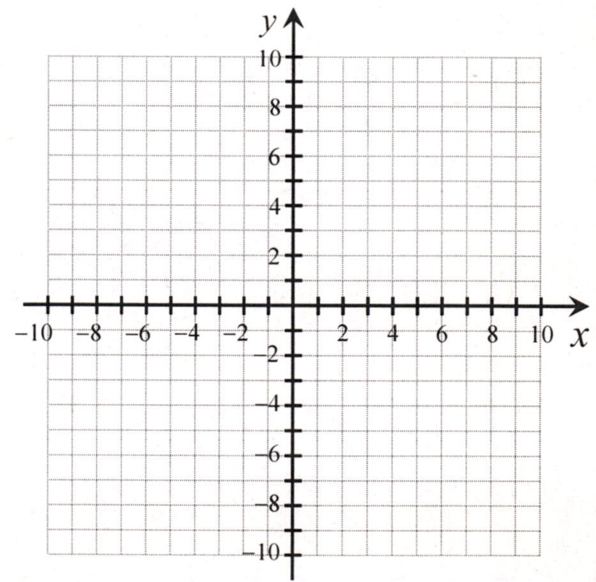

What is the vertex of the parabola? (___,___)

What is the equation for the axis of symmetry of the parabola? __________

What is the y-intercept of the parabola? (___,___)

What are the x-intercepts of the parabola? (___,___) and (___,___)

3.

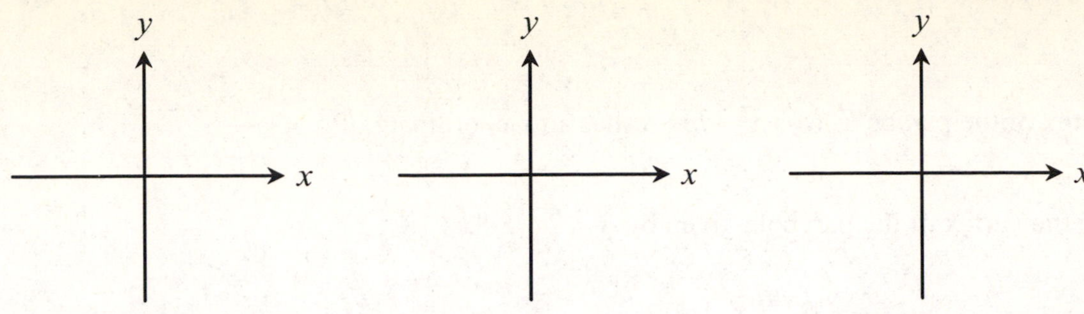

Draw a parabola with **two** x-intercepts.

Draw a parabola with **one** x-intercept.

Draw a parabola with **no** x-intercepts.

Will a parabola ever have **more** than one y-intercept? Why or why not?

Will a parabola ever have **less** than one y-intercept? Why or why not?

To find the y-intercept, we find y when $x = 0$. Thus, the ***y*-intercept** is at a point given by $(0,\_\_\_)$. To find x-intercepts, we need to find x when $y = 0$. Likewise, the ***x*-intercepts** are at point(s) given by $(\_\_\_,0)$.

4. Find the y- and x-intercepts of the graph of $y = x^2 + 4x - 12$.

5. Find the y- and x-intercepts of the graph of $y = 2x^2 + 3$.

The **vertex** of the parabola $y = ax^2 + bx + c$ has an x-coordinate of $x = -\dfrac{b}{2a}$.

6. Find the vertex of the parabola given by $y = 2x^2 - 8x + 13$.

Find the x-coordinate: $x = \dfrac{-b}{2a} = \underline{\qquad} =$

Substitute this value of x into the given function to find the y-coordinate:

$$y = 2x^2 - 8x + 13$$
$$y = 2(\quad)^2 - 8(\quad) + 13 = \underline{\qquad}$$

Vertex: (____,____)

7. Find the vertex of the parabola given by $y = -9x^2 - 6x - 8$.

8. Put it all together to graph $y = -x^2 + 2x + 8$.

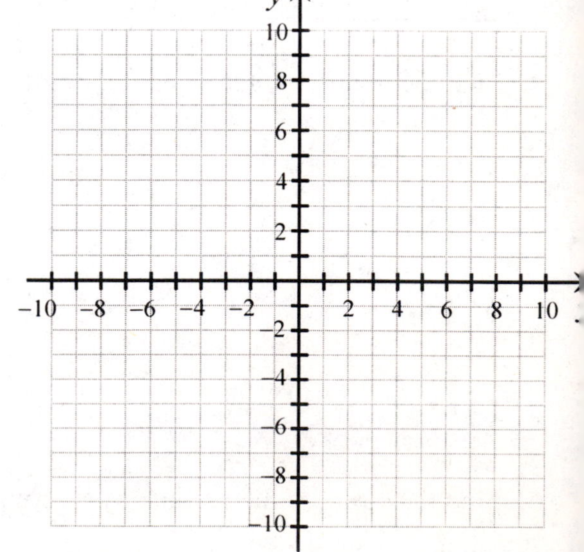

Does the parabola open up or down? _____

Find the y-intercept: (0,___)

Find the x-intercept(s), if there are any:

(____,0) (____,0)

Find the vertex: (____,____)

Write the equation for the axis of symmetry: ________

Student Activity

Finding the Vertex and Intercepts Using a Calculator

Example: Consider the quadratic equation $y = 8x^2 - 18x - 5$. First we will find all the characteristics of the graph using our **algebra** skills.

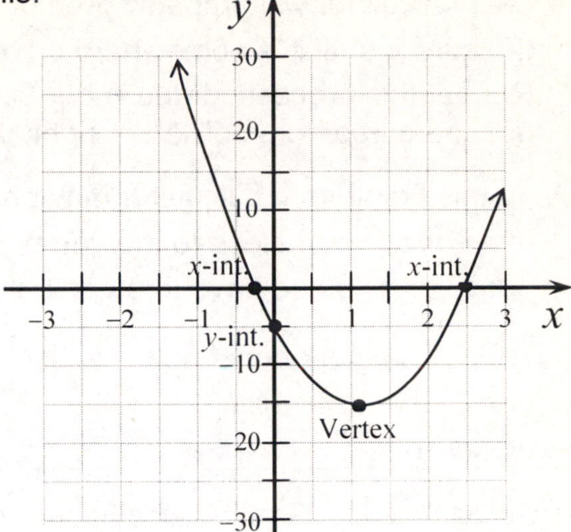

a. Find the vertex: (____,____)

b. Find the x-intercepts:

(____,0) and (____,0)

c. Find the y-intercept: (0,____)

Now we'll perform the same analysis using a **graphing calculator**.

☐ Using the $\boxed{y=}$ button or menu, graph the function $y = 8x^2 - 18x - 5$.

☐ Now use $\boxed{\text{WINDOW}}$ to find the window that best displays the parabola. Make sure you can see the vertex of the parabola in your graphing window.

Maximum or Minimum

Because the vertex of the parabola **is** the maximum or minimum point of the parabola, we can use this feature of the calculator to find the vertex.

☐ Find the graphing menu that has an option to find the MAX or MIN. You may find this in a MATH menu in the graphing panel, or under the CALC function. We will refer to this menu as the Graph Calculations Menu.

☐ Write down how you get to the Graph Calculations Menu on **your** calculator:

☐ Select the option to find the minimum of the graph (since in this case, the vertex is the minimum point).

☐ Your calculator will probably prompt you for the *Left Bound* or *Lower Bound*. Move the cursor until it is somewhere on the left of the vertex and then press ENTER. Repeat this process for the *Right Bound* or *Upper Bound*. Note: If you don't want to use the cursor to scroll left or right, you may just enter an *x*-value to be the bound.

☐ At this point, your calculator may prompt you to *Guess*. If this happens, you can move the cursor close to the vertex and press ENTER (note that you don't *have* to guess, but you do have to press enter).

☐ Your calculator should now display the coordinates of the vertex: (____,____).

y-intercept

☐ Go back to the Graph Calculations Menu. This time, select *Value* on the menu. Note: You can also use TRACE to do this.

☐ To find the *y*-intercept, we let $x = 0$. When prompted, enter 0 for the *x*-value.

☐ Your calculator should now display the coordinates of the y-intercept: $(0,$____$)$

x-intercept(s)

☐ Go back to the Graph Calculations Menu. This time, select *Zero* on the menu. It is called *Zero* because we let *y* be zero when we want to find the *x*-intercepts.

☐ Your calculator will probably prompt you for the *Left Bound* or *Lower Bound*. Move the cursor until it is somewhere on the left of the vertex and then press ENTER. Repeat this process for the *Right Bound* or *Upper Bound*.

☐ At this point, your calculator may prompt you to *Guess*. If this happens, you can move the cursor close to the *x*-intercept and press ENTER (note that you don't *have* to guess, but you do have to press enter).

☐ Your calculator should now display the coordinates of the x-intercept: (____,____).

☐ Repeat this process to find the coordinates of the second x-intercept: (____,____).

Possible Errors:

ERR: BOUND	You guessed a value that was not within the bounds you set.
ERR: INVALID	You tried to have the calculator find something with an *x*-value outside the viewing window.

Now use your calculator to find the vertex and the x – and y -intercepts of the following quadratic equations.

1. $y = 5x^2 + 15x + 10$

Vertex: (____,____)
y-intercept: (____,____)
x-intercept(s): (____,____) (____,____)

2. $y = -4x^2 + 20x - 24$

Vertex: (____,____)
y-intercept: (____,____)
x-intercept(s): (____,____) (____,____)

3. $y = x^2 - 4x + 14$

Vertex: (____,____)
y-intercept: (____,____)
x-intercept(s): (____,____) (____,____)

4. $y = \frac{1}{16}x^2$

Vertex: (____,____)
y-intercept: (____,____)
x-intercept(s): (____,____) (____,____)

Student Activity
Parabola Calisthenics

Directions: On each graph you are given the graph of the parabola $y = x^2$, the most basic quadratic function. You will be asked to finish a table of values to draw the graphs of other parabolas. If possible, use different colored pencils or pens to fill in each table and draw the corresponding graph.

1. Draw the graphs of y_1 and y_2 by filling in the missing values and plotting the points on the coordinate axes provided.

$y_1 = 2x^2$

x	y_1
−2	8
−1	
0	0
1	2
2	

$y_2 = 10x^2$

x	y_3
−1	
$-\frac{1}{2}$	2.5
0	
$\frac{1}{2}$	
1	10

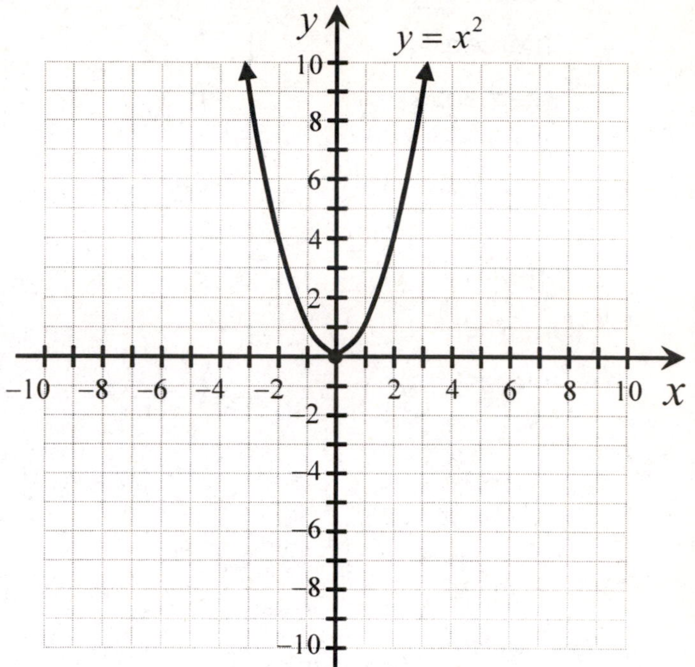

2. How do the graphs of y_1 and y_2 compare to the basic graph of $y = x^2$?

3. Circle the equations for the graphs that should fall between $y_1 = 2x^2$ and $y_2 = 10x^2$:

$y = 5x^2$

$y = 12x^2$

$y = 9x^2$

$y = 1.5x^2$

4. Draw the graphs of y_3 and y_4 by filling in the missing values and plotting the points on the coordinate axes provided.

$y_3 = \frac{1}{2}x^2$

x	y_2
–3	4.5
–2	
–1	
0	0
1	0.5
2	
3	

$y_4 = \frac{1}{10}x^2$

x	y_4
–5	2.5
–3	
–1	0.1
0	0
1	
3	0.9
5	

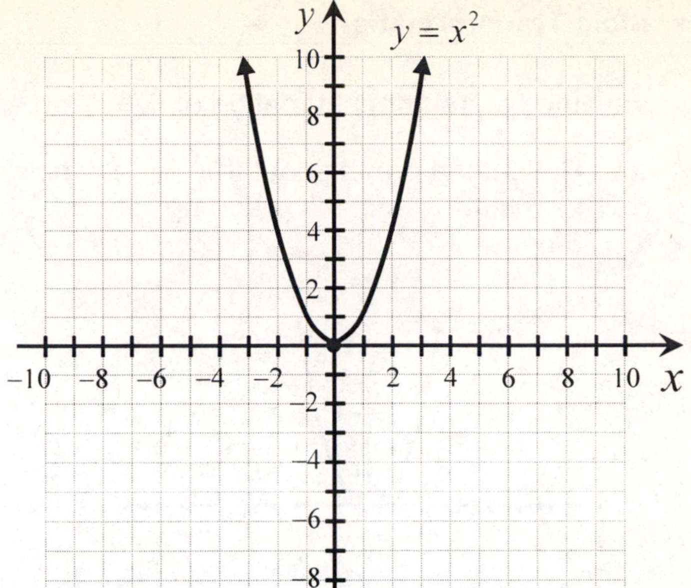

5. How do the graphs of y_4 and y_5 compare to the basic graph of $y = x^2$?

6. Circle the equations for the graphs that should fall between $y_3 = \frac{1}{2}x^2$ and $y_4 = \frac{1}{10}x^2$:

$y = \frac{3}{4}x^2$ $\qquad$ $y = \frac{1}{4}x^2$ $\qquad$ $y = \frac{1}{5}x^2$ $\qquad$ $y = \frac{1}{20}x^2$

7. To draw $y = a \cdot x^2$ if $a > 1$, how do you transform the graph of $y = x^2$?

8. What happens to the graph of $y = a \cdot x^2$ if a is negative?

Student Activity
Parabola Transformations

1. Draw the desired transformation of $y = x^2$ on each graphing grid below

   a. Reflect the graph about the x-axis, then shift the graph up 3 units.

   b. Shift the graph up 3 units, then reflect the graph about the x-axis.

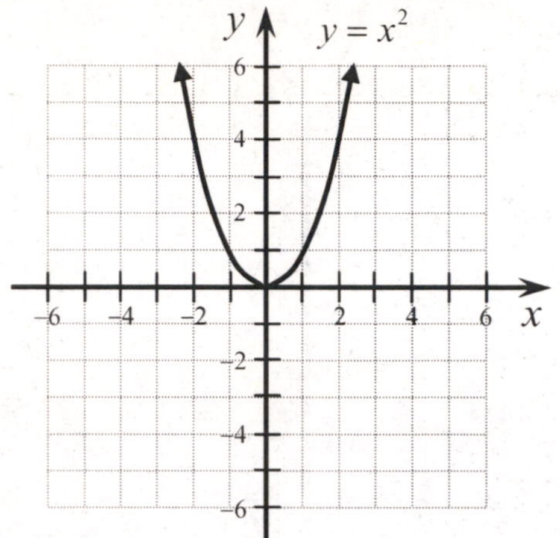

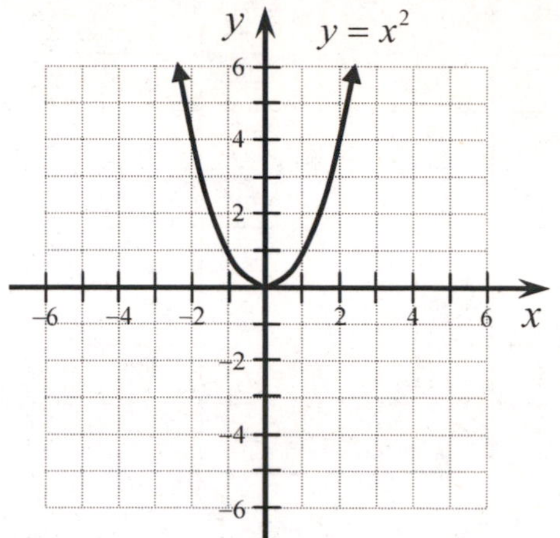

   c. Do you get the same result for both graphs? ____ Calculate a few points on the graph of $y = -x^2 + 3$ to see which graph (**a** or **b**) is the correct graph.

2. Let's examine some functions that are of the form $y = ax^2$.

   $y = 5x^2$ is narrower than $y = x^2$.

   $y = \frac{1}{3}x^2$ is wider than $y = x^2$.

   $y = -x^2$ is the graph of $y = x^2$ reflected about the x-axis.

   In all these cases, the stretch, compression, or reflection is caused by ________________ x^2 by a constant.

3. Let's examine some functions that include translations:

   Up 3 units: $y = x^2 + 3$ Down 3 units: $y = x^2 - 3$

   Right 3 units: $y = (x-3)^2$ Left 3 units: $y = (x+3)^2$

In all these cases, the translation is caused by ___________ or ___________ of a constant.

When we **graph**, the order of operations still places multiplication before addition or subtraction.

Graphing Order of Operations:
1. Draw the graph of $y = ax^2$.
2. Identify the translations of $y = ax^2$ (up or down, left or right) and draw the final graph.

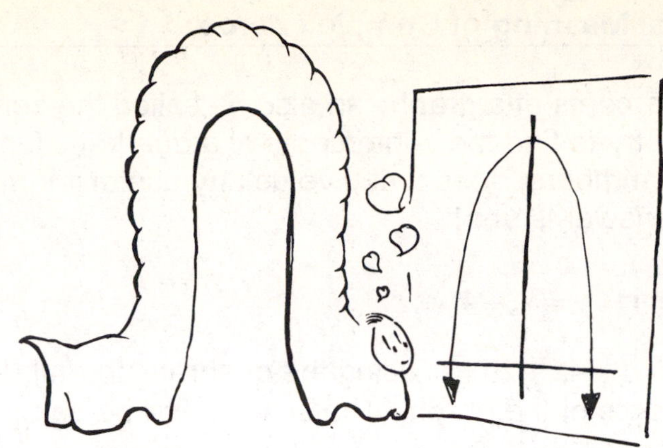

Yeah, she may look good, but can she function in a relationship?

4. Graph $y=-(x+1)^2+4$.

a. Draw the graph of $y=-x^2$ with a dashed line.

b. Identify any translations. Use the graph from part **a** to sketch the graph that includes these changes with a solid line.

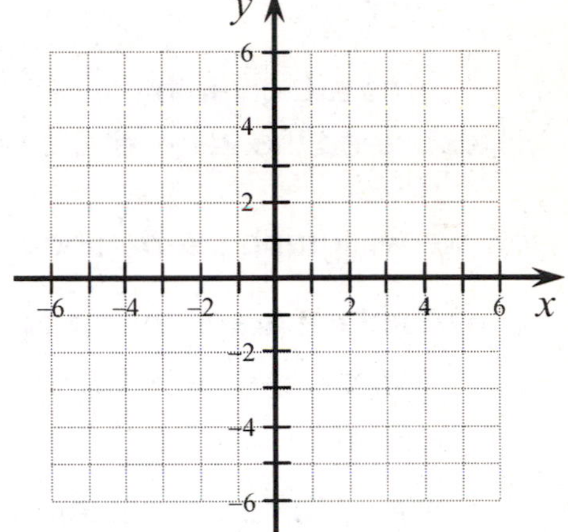

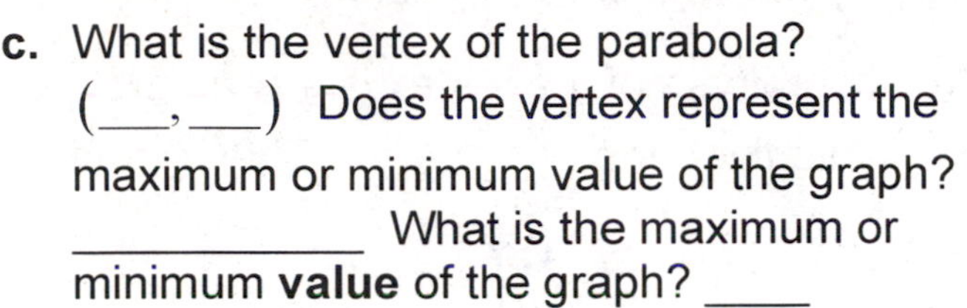

c. What is the vertex of the parabola? (___,___) Does the vertex represent the maximum or minimum value of the graph? ___________ What is the maximum or minimum **value** of the graph? ____

d. What is the axis of symmetry of the graph? _______

5. Graph $y=\frac{1}{2}(x+2)^2-3$.

a. Draw the graph of $y=\frac{1}{2}x^2$ with a dashed line.

b. Identify any translations. Use the graph from part **a** to sketch the graph that includes these changes with a solid line.

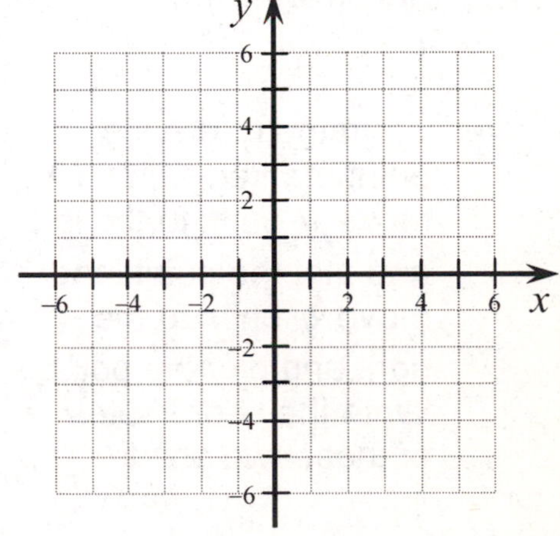

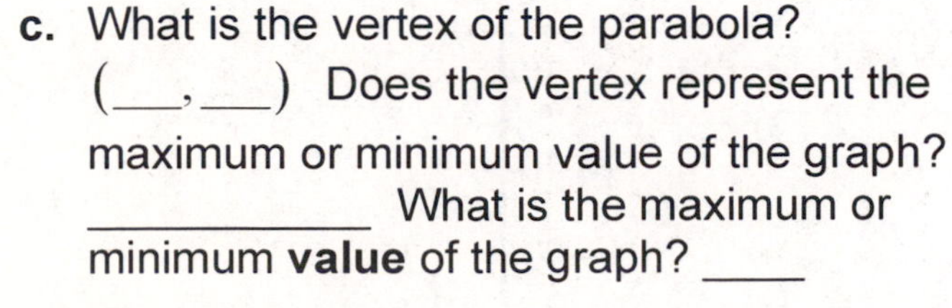

c. What is the vertex of the parabola? (___,___) Does the vertex represent the maximum or minimum value of the graph? ___________ What is the maximum or minimum **value** of the graph? ____

d. What is the axis of symmetry of the graph? _______

Student Activity

Graphical Meaning of Complex Zeros

The x-intercepts of a graph can also be called the *zeros* (where the y-values are zero). When we try to find the x-intercepts of a quadratic function, and the result is a pair of complex numbers, does that give us any useful information for graphing? It turns out that the answer is **yes!**

Example 1: $y = x^2 - 4x + 29$

a) Solve $x^2 - 4x + 29 = 0$ using the quadratic formula, as if you were finding the x-intercepts of the graph. The solution will be a pair of complex numbers.

b) If the zeros are given by $p \pm qi$, then for this problem, $p =$ ____ and $q =$ ____. We will also need the value of the leading coefficient: $a =$ ____.

c) Now we will graph the parabola, using only the values of p, q, and a.

- First we graph the vertex. The x-coordinate of the vertex is p and the y-coordinate of the vertex is aq^2. Write the vertex: (____, ____). Then graph the vertex.

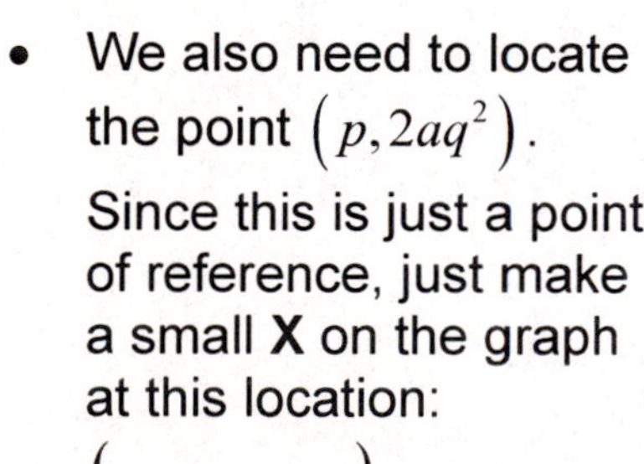

- We also need to locate the point $(p, 2aq^2)$. Since this is just a point of reference, just make a small **X** on the graph at this location: (____, ____).

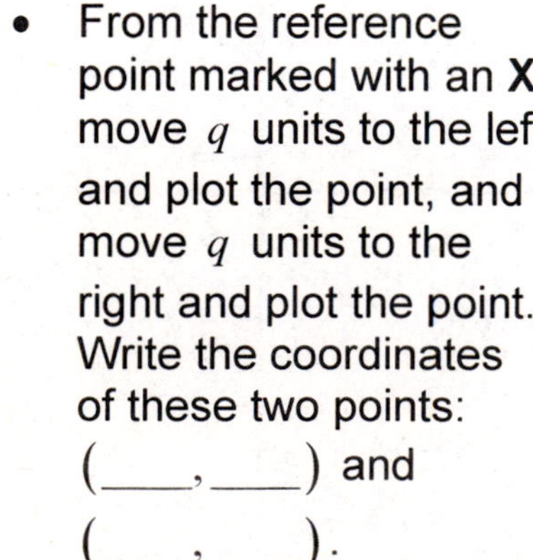

- From the reference point marked with an **X**, move q units to the left and plot the point, and move q units to the right and plot the point. Write the coordinates of these two points: (____, ____) and (____, ____).

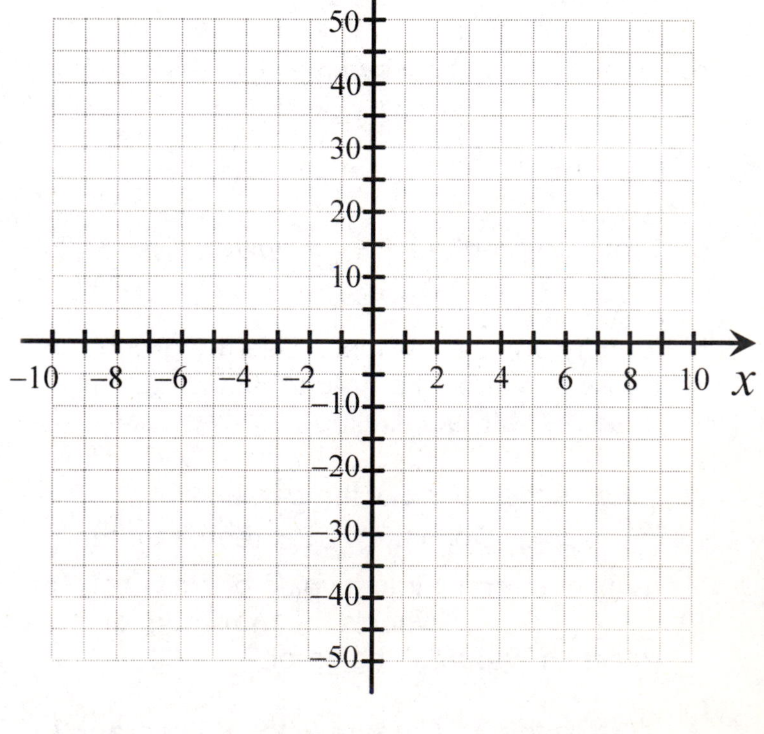

- You should now have three points to use to sketch the graph of the parabola.

Example 2: $y=-2x^2+4x-20$

a) Solve $-2x^2+4x-20=0$ using the quadratic formula, as if you were finding the *x*-intercepts of the graph. The solution will be a pair of complex numbers.

b) If the zeros are given by $p \pm qi$, then for this problem, $p=$ ____ and $q=$ ____. We will also need the value of the leading coefficient: $a=$ ____.

c) Now we will graph the parabola.

- Graph the vertex: $(p, aq^2)=($____, ____$)$.
- Locate $(p, 2aq^2)=($____, ____$)$ with a small **X**.

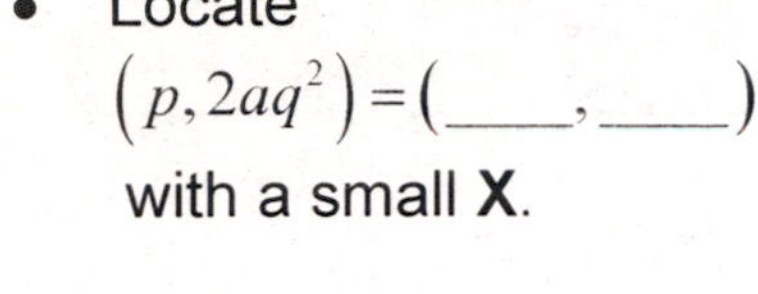

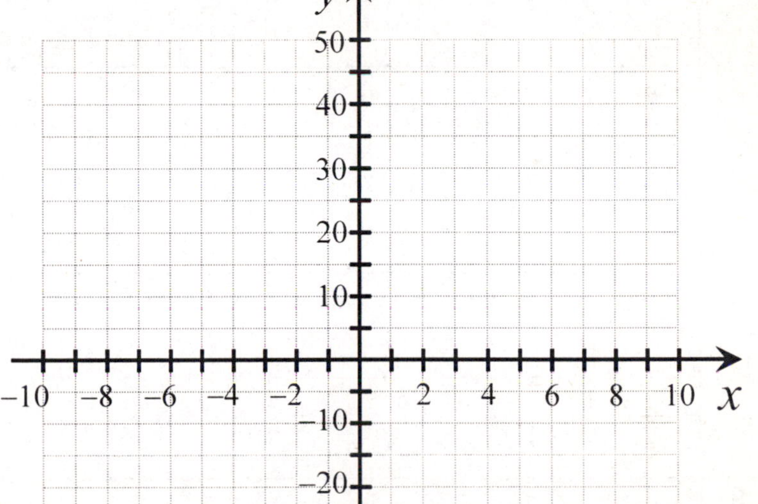

- From the **X**, plot points q units to the left and q units to the right. The points are (____, ____) and (____, ____).
- You should now have three points to use to sketch the graph of the parabola.

Connection to complex zeros:

a) Below each graph, write the complex zeros for the graph in the form $p \pm qi$. **Then calculate** $p \pm q$ and write those two values below the graphs as well.

b) Draw a horizontal dashed line on each graph that goes through the vertex of the parabola. Then draw a perfect reflection of the parabola over this dashed line. We'll call this new parabola the *imaginary parabola* for lack of a better name.

c) Where does the *imaginary parabola* cross the *x*-axis? In other words, what are its zeros? _____ and _____ Surprise!

Assess Your Understanding

Quadratics

For each of the following, describe the strategies or key steps that will help you **start** the problem. You do **not** have to complete the problems.

		What will help you to start this problem?
1.	Complete the perfect square trinomial: $x^2 - 14x +$ _____	
2.	Solve $x^2 + 6x + 5 = 0$ by completing the square.	
3.	Solve $x^2 - 10 = 0$ using the quadratic formula.	
4.	Solve: $16(x-3)^2 = 9$	
5.	Solve $15x^2 = 4 - 17x$ using the quadratic formula.	
6.	Solve: $x - 8\sqrt{x} + 12 = 0$	
7.	Solve $x^2 + 10 = 7x$ by factoring.	

		What will help you to start this problem?
8.	How many solutions does $2x^2 + x - 21 = 0$ have? Are the solutions rational, irrational, or complex numbers?	
9.	Solve: $x^2 - 8x = 20$	
10.	Does the graph of $y = 3 - x^2$ open up or down?	
11.	Solve: $x^4 - 2x^2 - 35 = 0$	
12.	Find the maximum value of $y = -x^2 + 2x + 3$.	
13.	Find the vertex of the graph of $y = 3x^2 - 3x$..	
14.	Write $y = x^2 - 6x + 8$ in standard form.	
15.	Find the x-intercept(s) of the graph of $y = 3x^2 - 3x - 6$.	

Metacognitive Skills

Quadratics

Metacognitive skills refer to the ability to judge how well you have learned something and to effectively direct your own learning and studying. This is a self-evaluation tool designed to help you focus your studying and to improve your metacognitive skills with regards to this math class.

Fill the 1st column out **before** you begin studying. Fill the 2nd column out after you study for your test.

Go back to this assessment after your test and circle any of the ratings that you would change – this identifies the "disconnects" between what you **thought** you knew well and what you **actually** knew well.

Use the scale below to assign a number to each topic.

5 *I am confident I can do any problems in this category correctly.*
4 *I am confident I can do most of the problems in this category correctly.*
3 *I understand how to do the problems in this category, but I still make a lot of mistakes.*
2 *I feel unsure about how to do these problems.*
1 *I know I don't understand how to do these problems.*

Topic or Skill	Before Studying	After Studying
Solving a quadratic equation using factoring.		
Solving a quadratic equation using the square root property.		
Using the square root property and understanding where the double-sign comes from.		
Simplifying an expression written with a double-sign ($\pm$).		
Completing a perfect square trinomial.		
Solving a quadratic equation by completing the square.		
Knowing the quadratic formula.		
Solving a quadratic equation by using the quadratic formula.		
Finding the discriminant for a quadratic equation.		
Using the discriminant to determine the type and number of solutions.		
Looking at a quadratic equation and choosing the most appropriate method to solve the equation (factoring, square root property, completing the square, or the quadratic formula).		
Recognizing an equation that is quadratic in form.		
Solving an equation that is quadratic in form by making an appropriate u-substitution.		
Knowing the general shape of the graph of a quadratic function, including whether the parabola opens up or down.		
Finding the vertex of a parabola by first finding the x-coordinate.		
Finding the x- and y-intercepts of a parabola.		
Graphing a quadratic function of the form $y = ax^2$.		
Graphing a quadratic function in standard form: $y = a(x-h)^2 + k$.		
Finding the vertex if the quadratic function is written in standard form.		
Writing a quadratic function in standard form by completing the square.		

RAT: Rational Expressions and Equations

Guided Learning Activity

Undefined Expressions

You just stand there with that blank expression. Sometimes I seriously wonder if you are even trying to be rational...

A **rational expression** is an expression of the form $\frac{A}{B}$ where A and B are polynomials and B does not equal 0. To evaluate a rational expression, you may find it helpful to first create the parentheses skeleton.

Example 1:

Evaluate $\frac{x^2+4x-5}{x^2-9}$ for $x=2$ and $x=-3$.

Parentheses skeleton: $\frac{(\ \ )^2+4(\ \ )-5}{(\ \ )^2-9}$

Evaluate for $x=2$: $\frac{(2)^2+4(2)-5}{(2)^2-9}=\frac{4+8-5}{4-9}=\frac{7}{-5}$

Evaluate for $x=-3$: $\frac{(-3)^2+4(-3)-5}{(-3)^2-9}=\frac{9-12-5}{9-9}=\frac{-8}{0}=$ undefined

Expression	Parentheses Skeleton	Evaluate the expression for ...				
		$x=1$	$x=4$	$x=0$	$x=-2$	$x=\frac{1}{3}$
$\frac{3x-12}{x^2+2x}$						
$\frac{3x^2+8x-3}{x^2-5x+4}$						
$\frac{x^2-16}{x^2+2x}$						

It is often easier to perform the evaluation if the rational expression is factored first. This way, you can quickly see the values that create a factor of zero in the numerator or denominator.

Example 2: Evaluate $\frac{x^2+4x-5}{x^2-9}$ for $x=0$, $x=1$, and $x=3$.

Factored form: $\frac{(x+5)(x-1)}{(x+3)(x-3)}$.

Evaluate for $x=0$: $\frac{(0+5)(0-1)}{(0+3)(0-3)}=\frac{(5)(-1)}{(3)(-3)}=\frac{-5}{-9}=\frac{5}{9}$

Evaluate for $x=1$: $\frac{(1+5)(1-1)}{(1+3)(1-3)}=\frac{(6)(0)}{(4)(-2)}=\frac{0}{-8}=0$

Evaluate for $x=3$: $\frac{(3+5)(3-1)}{(3+3)(3-3)}=\frac{(8)(2)}{(6)(0)}=\frac{16}{0}=$ undefined

As soon as you see a **factor** of zero in the numerator or denominator, you can quickly find the value of the answer.

Expression	Factored Form of the Expression	Evaluate the expression for ...				
		$x=1$	$x=4$	$x=0$	$x=-2$	$x=\frac{1}{3}$
$\dfrac{3x-12}{x^2+2x}$						
$\dfrac{3x^2+8x-3}{x^2-5x+4}$						
$\dfrac{x^2-16}{x^2+2x}$						

Recall that we can solve an equation like $(x+3)(x-5)=0$ using the zero property. Because of the Zero Factor Property, either $x+3=0$ or $x-5=0$. This leads us to solutions of -3 and 5.

Example 3: Where is the expression $\dfrac{x^2+4x-5}{x^2-9}$ undefined?

This expression is undefined when the denominator is equal to zero.
We can answer the question by solving the equation $x^2-9=0$.
Factor first: $(x+3)(x-3)=0$
Set each factor equal to zero and solve:

$x+3=0$ OR $x-3=0$

$x=-3$ $x=3$

The expression $\dfrac{x^2+4x-5}{x^2-9}$ is undefined for -3 and 3.

Expression	Factored Form of the Expression	Set the denominator = 0. Solve the resulting equation.	Where is the expression undefined?
$\dfrac{3x-12}{x^2+2x}$			
$\dfrac{3x^2+8x-3}{x^2-5x+4}$			
$\dfrac{x^2-16}{x^2+2x}$			

Student Activity

Match Up on Simplifying Rational Expressions

Directions: Match each of the expressions in the squares in the table below with an equivalent simplified expression from the top. If an equivalent expression is not found among the choices A through E, then choose F (none of these).

A 1 **B** -1

C $x+5$ **D** $\frac{x}{3}$

E $3x$ **F** None of these

I can't believe I'm saying this, but I really miss the good old days when fractions only involved numbers ...

$\frac{9x^3+15x}{3x^2+5}$	$\frac{x^2-25}{x-5}$		
$\frac{x-1}{1-x}$	$\frac{x^2+x}{3x+3}$		
$\frac{(3x+2)(x+1)}{3x^2+5x+2}$	$\frac{x-1}{1+x}$	$\frac{3x^3-6x^2}{x-2}$	$\frac{3x+1}{1+3x}$
$\frac{3x^3-27x}{(x+3)(x-3)}$	$\frac{x^3+2x^2+x}{3x^2+6x+3}$	$\frac{x^2+10x+25}{x+5}$	$\frac{3x-1}{1-3x}$
$\frac{x^2+6x+5}{x+1}$	$\frac{x^2+25}{x^2-25}$	$\frac{x-8}{-x+8}$	$\frac{18x^2-3x}{-1+6x}$

Student Activity

The Ones Recycling Center

Directions: In each pair of expressions, make a decision about whether the expression is equivalent to 1 or -1. Then sort each expression into the correct recycling bin below. If an expression does not belong in either recycling bin, just leave it out! The first one has been done for you.

The 1 Bin: The numerator and denominator are equivalent.
The -1 Bin: The numerator and denominator are opposites.

$\frac{x-1}{1-x}$	$\frac{x+2}{2+x}$	$\frac{x-3}{x+3}$	$\frac{x^2-1}{1-x^2}$
$\frac{(x+3)^2}{x^2+6x+9}$	$\frac{3(x+4)}{(x+4)(3)}$	$\frac{v^2-1}{v^2+1}$	$\frac{-x-3}{x+3}$
$\frac{-x+3}{3-x}$	$\frac{y^2+1}{1+y^2}$	$\frac{x^2+2x-1}{x^2+2x+1}$	$\frac{x^2-2x-3}{3+2x-x^2}$
$\frac{a-b}{-a+b}$	$\frac{x^2+3x+1}{3x+1+x^2}$	$\frac{5-x}{5-x}$	$\frac{(x+3)^2}{x^2+9}$

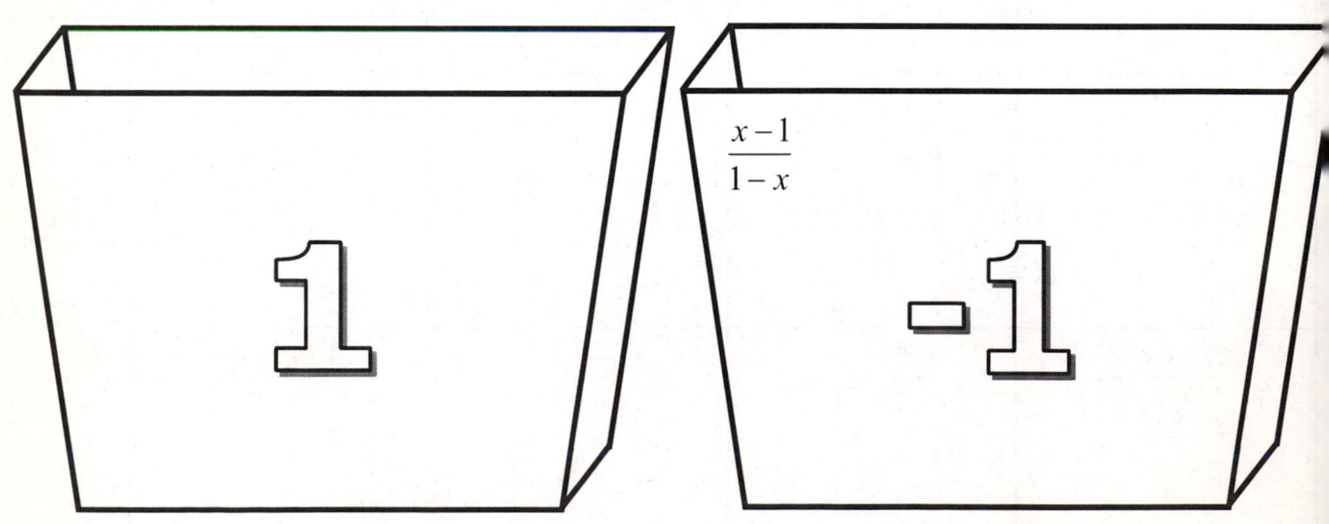

Student Activity

Which of These is Not Like the Others?

Directions: When you look at your answer to a problem and compare it to the answer in the back of the book or a friend's answer, you might find that they are not quite the same. This does not mean that one of them is *really* different though. In each row of the table, all the expressions are equivalent except for one of them. **Circle the expressions that are the same and place an X over the "oddball" expression in each row.** The first one has been done for you.

Here's two hints if you're really stuck:

- You could evaluate all the expressions for the same given value and see which expressions have the same result.
- You could try factoring out a -1 if the numerator and denominator look suspiciously like they might be opposites.

1.	$\frac{-4}{-x}$ (crossed out)	$\frac{-4}{x}$ (circled)	$\frac{4}{-x}$ (circled)	$-\frac{4}{x}$ (circled)
2.	$-\frac{x+3}{x-4}$	$\frac{-x-3}{x-4}$	$\frac{-x-3}{-x+4}$	$\frac{x+3}{4-x}$
3.	$\frac{x-5}{5-x}$	$\frac{x-5}{x-5}$	$-\frac{x-5}{x-5}$	$\frac{x-5}{-x+5}$
4.	$-\frac{-x-4}{-x-4}$	$\frac{x+4}{x+4}$	$\frac{x+4}{4+x}$	$-\frac{-x-4}{x+4}$
5.	$-\frac{x+1}{1-x}$	$\frac{1+x}{x-1}$	$\frac{x+1}{x-1}$	$\frac{1+x}{1-x}$
6.	$\frac{x-4}{(x+1)(x+3)}$	$\frac{x-4}{x^2+4x+3}$	$\frac{4-x}{x^2+4x+3}$	$-\frac{4-x}{x^2+4x+3}$

Student Activity

Does Position Matter?

Directions: Simplify each expression below. Circle the multiplication example that matches the result of the shaded division problem.

Expression	Simplified Fraction	Decimal Equivalent
$\frac{1}{5}\cdot\frac{3}{4}$	$\frac{3}{20}$	0.15
$\frac{1}{5}\cdot\frac{4}{3}$		
$\frac{5}{1}\cdot\frac{4}{3}$		
$\frac{5}{1}\cdot\frac{3}{4}$		
$\frac{1}{5}\div\frac{3}{4}$		

Expression	Simplify the Expression
$\frac{1}{x+3}\cdot\frac{x+3}{2}$	
$\frac{1}{x+3}\cdot\frac{2}{x+3}$	
$\frac{x+3}{1}\cdot\frac{x+3}{2}$	
$\frac{x+3}{1}\cdot\frac{2}{x+3}$	
$\frac{1}{x+3}\div\frac{2}{x+3}$	

Expression	Simplify the Expression
$\frac{2x+3}{x}\cdot\frac{1}{x}$	
$\frac{x}{2x+3}\cdot\frac{1}{x}$	
$\frac{2x+3}{x}\cdot\frac{x}{1}$	
$\frac{x}{2x+3}\cdot\frac{x}{1}$	
$\frac{2x+3}{x}\div\frac{1}{x}$	

Question: When dividing fractions, does it matter whether you use the reciprocal of the first or second fraction? In other words, do you get the same result?

Student Activity

Paint by Factors of 1

Directions: For each expression, perform the multiplication or division and simplify by factoring the numerator and denominator **completely** and removing factors equal to 1.

As you remove factors of 1, like $\frac{x+2}{x+2}$ or $\frac{x^2}{x^2}$, shade in the corresponding squares in the grid below. The first one has been done for you. There's a surprise when you're finished!

1. Multiply: $\frac{2x-4}{x+3}\cdot\frac{x+3}{2x+8}=\frac{2(x-2)(x+3)}{2(x+3)(x+4)}=\frac{\cancel{2}^1(x-2)\cancel{(x+3)}^1}{\cancel{2}_1\cancel{(x+3)}_1(x+4)}=\frac{x-2}{x+4}$ (shade $\frac{x+3}{x+3}$ & $\frac{2}{2}$)

2. Divide: $\frac{3x^2-6x}{x^3+2x^2}\div\frac{-x^2-2x+8}{x^4+6x^3+8x^2}$

3. Multiply: $\frac{5x^4+2x^3}{2x^4+3x^3}\cdot\frac{2x^3+x^2-3x}{5x^2-3x-2}$

4. Multiply: $\frac{5x^2-125}{x^3+5x^2}\cdot\frac{6x^2+2x}{5x-25}$

5. Divide: $\frac{21x^2-49x-42}{4x^2-28x+48}\div\frac{21x+14}{8x^2-32x}$

$\frac{12}{12}$	$\frac{4x-3}{4x-3}$	$\frac{x+3}{x+3}$	$\frac{x^2}{x^2}$	$\frac{2-5x}{2-5x}$	$\frac{x^2+16}{x^2+16}$
$\frac{7x+3}{7x+3}$	$\frac{x+2}{x+2}$	$\frac{x^3}{x^3}$	$\frac{x-2}{x-2}$	$\frac{-9}{-9}$	$\frac{10x+1}{10x+1}$
$\frac{3}{3}$	$\frac{x-7}{x-7}$	$\frac{5}{5}$	$\frac{2x+3}{2x+3}$	$\frac{4x+3}{4x+3}$	$\frac{8x+3}{8x+3}$
$\frac{4-5x}{4-5x}$	$\frac{5x-7}{5x-7}$	$\frac{x+5}{x+5}$	$\frac{7}{7}$	$\frac{13}{13}$	$\frac{25}{25}$
$\frac{7x+2}{7x+2}$	$\frac{x+25}{x+25}$	$\frac{3x+2}{3x+2}$	$\frac{x+4}{x+4}$	$\frac{3x-4}{3x-4}$	$\frac{x+8}{x+8}$
$\frac{2-9x}{2-9x}$	$\frac{8x-3}{8x-3}$	$\frac{4}{4}$	$\frac{2}{2}$	$\frac{x^2+36}{x^2+36}$	$\frac{4x-17}{4x-17}$
$\frac{4x-5}{4x-5}$	$\frac{x+10}{x+10}$	$\frac{x-3}{x-3}$	$\frac{x-4}{x-4}$	$\frac{x^2+4}{x^2+4}$	$\frac{7x-3}{7x-3}$
$\frac{3x}{3x}$	$\frac{5x+2}{5x+2}$	$\frac{x-5}{x-5}$	$\frac{x}{x}$	$\frac{x-1}{x-1}$	$\frac{2x-1}{2x-1}$

Student Activity

Testing Equality

Directions: In each table there is a rational expression and a simplified expression. Use a test value to check the equality (or equivalency) of the two expressions. The first one has been started for you!

Note: Because of some unique properties of 0, 1 and 2 ($0+0=0\cdot 0$, $1\cdot 1=1$, $2+2=2\cdot 2$), it is usually best **not** to use these for test values.

1.

Given expression:		**Possible simplification:**
$\dfrac{x^2+3}{x}\cdot\dfrac{1}{x+2}$	$\overset{?}{=}$	$\dfrac{x^2+3}{x^2+2}$

What values make the denominator of the **given** expression zero? $0, -2$

Choose a simple test value for x that does **not** result in a zero denominator: Let $x=3$.

Evaluate the rational expression:		**Evaluate the possible simplification:**
$\dfrac{(3)^2+3}{(3)}\cdot\dfrac{1}{(3)+2}=\dfrac{12}{3}\cdot\dfrac{1}{5}=\dfrac{12}{15}=\dfrac{4}{5}$	$\overset{?}{=}$	

Are the two expressions equal? _____

2.

Given expression:		**Possible simplification:**
$\dfrac{x^3+3x^2}{x}\div\dfrac{x}{x+7}$	$\overset{?}{=}$	$\dfrac{x^2}{(x^3+3x^2)(x+7)}$

What values make the denominator of the **given** expression zero? _______

Choose a simple test value for x that does **not** result in a zero denominator: _______

Evaluate the rational expression:		**Evaluate the possible simplification:**
	$\overset{?}{=}$	

Are the two expressions equal? _____

3. **Given expression:** $\dfrac{x^2-x-12}{x-2} \div \dfrac{x+3}{x-2}$ $\overset{?}{=}$ **Possible simplification:** $x-4$

What values make the denominator of the **given** expression zero? ________

Choose a simple test value for x that does **not** result in a zero denominator: _______

Evaluate the rational expression:	$\overset{?}{=}$	**Evaluate the possible simplification:**

Are the two expressions equal? _____

4. **Given expression:** $\dfrac{2x^2-5x-12}{x-4} \cdot \dfrac{2x+3}{x-4}$ $\overset{?}{=}$ **Possible simplification:** $\dfrac{4x^2+12x+9}{x-4}$

What values make the denominator of the **given** expression zero? ________

Choose a simple test value for x that does **not** result in a zero denominator: _______

Evaluate the rational expression:	$\overset{?}{=}$	**Evaluate the possible simplification:**

Are the two expressions equal? _____

Student Activity

Thread of Like Terms

Directions: Simplify each of the expressions that follow.

1.

7 miles + 5 miles	$7x+5x$	$7x^2+5x^2$	$7x^2y+5x^2y$
$\frac{7}{13}+\frac{5}{13}$	$\frac{7}{x}+\frac{5}{x}$	$\frac{7}{xy}+\frac{5}{xy}$	$\frac{7}{x+2}+\frac{5}{x+2}$

2.

$\frac{9}{11}-\frac{2}{11}$	$9w^2-2w^2$	$\frac{9}{ab^2}-\frac{2}{ab^2}$	$\frac{9}{y}-\frac{2}{y}$
$\frac{9}{x-4}-\frac{2}{x-4}$	$9ab-2ab$	$9y-2y$	$9\text{ ft}^2-2\text{ ft}^2$

3.

$\frac{8}{a^2b}-\frac{1}{a^2b}$	$8xy-xy$	$\frac{8}{z}-\frac{1}{z}$	$8a^2-a^2$
$\frac{8}{x+8}-\frac{1}{x+8}$	$8w-w$	8 cm − 1 cm	$\frac{8}{9}-\frac{1}{9}$

4.

$\frac{4}{a}-\frac{3}{a}$	$4\text{ in}^3-3\text{ in}^3$	$\frac{4}{5}-\frac{3}{5}$	$\frac{4}{5x^2}-\frac{3}{5x^2}$
$\frac{4}{x+1}-\frac{3}{x+1}$	$4x^2y^2-3x^2y^2$	$4a-3a$	$4b^3-3b^3$

Question: What is the lesson to be learned here about addition and subtraction in algebra and mathematics in general?

Student Activity

Match Up with Like Denominators

Match-up: Complete each addition or subtraction problem and simplify the result. Then choose the letter that corresponds to the **numerator** of this result. If the numerator is not among the choices, then choose E (None of these). The first one has been done for you.

A $x+1$ **B** -1 **C** 1 **D** $x-1$ **E** None of these

$\frac{x}{3x+3}+\frac{1}{3x+3}$ $=\frac{x+1}{3x+3}=\frac{(x+1)}{3(x+1)}$ $=\frac{1}{3}$ C	$\frac{1}{x^3}-\frac{x^2+1}{x^3}$	$\frac{x^2}{x^2+x}-\frac{1}{x^2+x}$	$\frac{-3x^2+1}{(2x+1)^2}+\frac{3x^2+2x}{(2x+1)^2}$
$\frac{2x-1}{x}-\frac{3x-2}{x}$	$\frac{1-2x}{x}-\frac{2-3x}{x}$	$\frac{x^2+x}{x^2+1}+\frac{1-x^2}{x^2+1}$	$\frac{x^2+x}{x^2-1}+\frac{1-x^2}{x^2-1}$
$\frac{x^2}{x^2-1}+\frac{2x+1}{x^2-1}$	$\frac{x}{2x+1}-\frac{x+1}{2x+1}$	$\frac{2x^2-2x}{(x-1)^2}-\frac{2x-2}{(x-1)^2}$	$\frac{2x^2-2x}{2(x-1)^2}-\frac{2x-2}{2(x-1)^2}$

Student Activity

Determining the LCD

Directions: For each of the expressions below, start by factoring both denominators. If a denominator **doesn't** factor, write that denominator **in parentheses**. Then find the LCD for the expression.

For example, the denominators in $\frac{x}{x+1}+\frac{3}{4x+4}$ would be factored like this:

$\frac{x}{(x+1)}+\frac{3}{4(x+1)}$ and the LCD would be $4(x+1)$.

	Expression	Expression with Factored Denominators	LCD
1.	$\frac{4}{x+3}+\frac{2}{x^2+3x}$	$\frac{4}{\_\_\_\_}+\frac{2}{\_\_\_\_}$	
2.	$\frac{5x}{x^2-25}-\frac{5}{x+5}$	$\frac{5x}{\_\_\_\_}-\frac{5}{\_\_\_\_}$	
3.	$\frac{3}{x^2+x-12}+\frac{2x}{x^2-x-6}$	$\frac{3}{\_\_\_\_}+\frac{2x}{\_\_\_\_}$	
4.	$\frac{x+3}{5x^3+25x^2}+\frac{2x}{3x+15}$	$\frac{x+3}{\_\_\_\_}+\frac{2x}{\_\_\_\_}$	
5.	$\frac{x}{x-2}-\frac{3}{2-x}$	$\frac{x}{\_\_\_\_}-\frac{3}{\_\_\_\_}$	
6.	$\frac{10}{8x^2-14x-15}-\frac{15}{8x^2-20x}$	$\frac{10}{\_\_\_\_}-\frac{15}{\_\_\_\_}$	
7.	$\frac{x^2+9}{3x^4-27x^2}+\frac{3x}{x^3-3x^2}$	$\frac{x^2+9}{\_\_\_\_}+\frac{3}{\_\_\_\_}$	
8.	$\frac{4}{15x^3+12x^2}-\frac{y}{3x^2}$	$\frac{4}{\_\_\_\_}-\frac{y}{\_\_\_\_}$	

Student Activity

The Missing Form of 1

Directions: In each equation below, there is a missing multiple with a form of 1 and a missing numerator. Fill in the empty spaces to complete each expression.

Example: $\frac{x-2}{x+3}\cdot\left[\frac{\quad}{\quad}\right]=\frac{\quad}{(x+3)(x-2)}$ becomes $\frac{x-2}{x+3}\cdot\left[\frac{x-2}{x-2}\right]=\frac{x^2-4x+4}{(x+3)(x-2)}$.

Hint for problems 6-10: Try factoring the denominator on the right hand side first.

1. $\frac{x+1}{x+2}\cdot\left[\frac{\quad}{\quad}\right]=\frac{\quad}{3x(x+2)}$

2. $\frac{x-5}{x+5}\cdot\left[\frac{\quad}{\quad}\right]=\frac{\quad}{(x+5)(x-5)}$

3. $\frac{x}{2-x}\cdot\left[\frac{\quad}{\quad}\right]=\frac{\quad}{x-2}$

4. $\frac{x^2+3x}{3x+2}\cdot\left[\frac{\quad}{\quad}\right]=\frac{\quad}{(3x+2)(x+3)}$

5. $\frac{4}{x+3}\cdot\left[\frac{\quad}{\quad}\right]=\frac{\quad}{(x+3)(x+4)(x+5)}$

6. $\frac{3x}{x+1}\cdot\left[\frac{\quad}{\quad}\right]=\frac{\quad}{x^2-1}$

7. $\frac{2x+3}{x+7}\cdot\left[\frac{\quad}{\quad}\right]=\frac{\quad}{x^2+14x+49}$

8. $\frac{16x}{4-x}\cdot\left[\frac{\quad}{\quad}\right]=\frac{\quad}{16-8x+x^2}$

9. $\frac{x^2}{5x+4}\cdot\left[\frac{\quad}{\quad}\right]=\frac{\quad}{30x^3+24x^2}$

10. $\frac{8+x}{4x+3}\cdot\left[\frac{\quad}{\quad}\right]=\frac{\quad}{20x^2+7x-6}$

Student Activity

The Reunion

Directions: Now that we've made it through addition, subtraction, multiplication, and division, it seems only fair to bring the whole gang back together for a reunion. Simplify each expression according to its operation!

$\frac{x-1}{x+2}+\frac{x+3}{x+2}$	
$\frac{x-1}{x+2}-\frac{x+3}{x+2}$	
$\frac{x-1}{x+2}\cdot\frac{x+3}{x+2}$	
$\frac{x-1}{x+2}\div\frac{x+3}{x+2}$	

$\frac{x}{x-3}+\frac{x-1}{3-x}$	
$\frac{x}{x-3}-\frac{x-1}{3-x}$	
$\frac{x}{x-3}\cdot\frac{x-1}{3-x}$	
$\frac{x}{x-3}\div\frac{x-1}{3-x}$	

$\frac{12}{x+2}+\frac{4x+8}{x+1}$	
$\frac{12}{x+2}-\frac{4x+8}{x+1}$	
$\frac{12}{x+2}\cdot\frac{4x+8}{x+1}$	
$\frac{12}{x+2}\div\frac{4x+8}{x+1}$	

Student Activity

Tempting Expressions

Directions: In all the expressions below, you will find yourself tempted to do incorrect mathematics by visually-pleasing expressions. Of course, just because it looks good, that doesn't mean it is the right thing to do. ☺ Think carefully about the steps involved in each expression.

1. Add: $\frac{5}{x}+\frac{x}{5}$

2. Subtract: $\frac{2}{x+2}-\frac{1}{x}$

3. Add: $\frac{x}{x+4}+\frac{x+4}{x}$

4. Divide: $\frac{x-1}{x+6}\div\frac{x+6}{x^2-36}$

5. Simplify: $\frac{2x+8}{2x+4}$

6. Multiply: $\frac{x^2+4x-12}{x^2+4x-32}\cdot\frac{x^2+10x+16}{x^2+10x+24}$

7. Subtract: $\frac{3x}{x+2}-\frac{x+2}{3x+3}$

8. Simplify: $\frac{5x^2+25x}{5x^2-125}$

9. Subtract: $\frac{x^2}{x^2-9}-9$

10. Add: $\frac{2}{x+1}+\frac{2}{1-x}$

Student Activity

Tic-Tac-Toe on Complex Fraction Pieces

Directions: In every box in the tic-tac-toe grid, there is a rational expression and a simplified form. If the two are equivalent, then circle the simplified form (thus placing an **O** on the square). If the expression has **not** been simplified correctly, then put an **X** over the incorrect simplification.

Tic-tac-toe Game#1:

$x^2\left(\frac{3}{x}\right)$ $3x$	$(2-x)\left(\frac{1}{x-2}\right)$ -1	$\left(\frac{2}{x}\right)\left(\frac{x^2}{4}\right)$ $2x$
$(24x^3)\left(\frac{3}{8x^2}\right)$ $9x$	$(15x)\left(\frac{4}{25x^2}\right)$ $\frac{4}{5x}$	$\left(\frac{x+3}{2(x-3)}\right)(x-3)$ $\frac{x+3}{2}$
$\left(\frac{3-2x}{x^3}\right)(x^3)$ $3x^3-2x^4$	$(81x^4)\left(\frac{9}{9x^4}\right)$ 81	$\left(\frac{1}{x-1}\right)(1-x)$ 1

Tic-tac-toe Game#2:

$(8x^2)\left(\frac{3}{4x}-2\right)$ $6x-16x^2$	$x^2\left(\frac{5}{x^2}-\frac{2}{x}\right)$ $5-\frac{2}{x}$	$x^3\left(\frac{5}{x^2}+4x\right)$ $5x+4x^2$
$x\left(4+\frac{x+1}{x}\right)$ $5x+1$	$\left(\frac{x}{5}\right)\left(\frac{25}{x}+5\right)$ $5+5x$	$\left(\frac{x^2}{3}\right)\left(6-\frac{9}{10x^2}\right)$ $2x^2-\frac{3}{10}$
$(32x^4)\left(\frac{1}{16x}-\frac{1}{32x^4}\right)$ $2x^3-1$	$\left(\frac{24x}{y}\right)\left(\frac{y}{8x}-y\right)$ $3-24x$	$(42x^2)\left(\frac{1}{6x}+\frac{1}{7x}\right)$ 5

Student Activity

Double the Fun on Complex Fractions

Directions: For each of the complex fractions below, simplify the expression using both methods. The result should be the same for both methods. If they are not, go back and look for a mistake. The first one has been **started** for you.

Method I: Using division. **Method II:** Multiplying by the LCD.

1.	**Method I**	$\dfrac{\frac{1}{x}+4}{\frac{5}{x}-1}=\dfrac{\frac{1}{x}+4\left(\frac{x}{x}\right)}{\frac{5}{x}-1\left(\frac{x}{x}\right)}=\dfrac{\frac{1+4x}{x}}{\frac{5-x}{x}}=$
	Method II	$\dfrac{\frac{1}{x}+4}{\frac{5}{x}-1}=\dfrac{\left(\frac{1}{x}+4\right)}{\left(\frac{5}{x}-1\right)}\cdot\dfrac{\frac{x}{1}}{\frac{x}{1}}=$
2.	**Method I**	$\dfrac{5-\frac{1}{x^2}}{\frac{2}{x}+x}$
	Method II	$\dfrac{5-\frac{1}{x^2}}{\frac{2}{x}+x}$
3.	**Method I**	$\dfrac{\frac{1}{x+1}+1}{1+\frac{2}{x}}$
	Method II	$\dfrac{\frac{1}{x+1}+1}{1+\frac{2}{x}}$

Student Activity

One of Us is Wrong!

Directions: Lou and Stu have both simplified the given complex fraction. Unfortunately, they have different answers. 1) Choose a test value for x and use it to evaluate the complex fraction and the students' answers. Give the student that is correct a ☺ for their work. 2) Then determine where the unfortunate student made the error.

1. Given expression	Lou's Work	Stu's Work
$\dfrac{\frac{2x+5}{x+2}}{\frac{x+2}{2x+5}}$ Test value: $x =$ ____	$\dfrac{\frac{2x+5}{x+2}}{\frac{x+2}{2x+5}} = \dfrac{2x+5}{x+2} \div \dfrac{x+2}{2x+5}$ $= \dfrac{2x+5}{x+2} \cdot \dfrac{2x+5}{x+2} = \dfrac{(2x+5)^2}{(x+2)^2}$	$\dfrac{\frac{2x+5}{x+2}}{\frac{x+2}{2x+5}} = \dfrac{\frac{2x+5}{\cancel{x+2}^{1}}}{\frac{\cancel{x+2}_{1}}{2x+5}}$ $= \dfrac{\cancel{2x+5}^{1}}{\cancel{2x+5}_{1}} = 1$
Expression value:	Expression value:	Expression value:

2. Given expression	Lou's Work	Stu's Work
$\dfrac{2x - \frac{1}{2x}}{\frac{1}{2x} + 2x}$ Test value: $x =$ ____	$\dfrac{2x - \frac{1}{2x}}{\frac{1}{2x} + 2x} = \dfrac{\cancel{2x}^{1} - \frac{1}{\cancel{2x}^{1}}}{\frac{1}{\cancel{2x}^{1}} + \cancel{2x}^{1}}$ $= \dfrac{1-1}{1+1}$ $= \dfrac{0}{2} = 0$	$\dfrac{2x - \frac{1}{2x}}{\frac{1}{2x} + 2x} = \dfrac{\left(2x - \frac{1}{2x}\right)}{\left(\frac{1}{2x} + 2x\right)} \cdot \dfrac{\frac{2x}{1}}{\frac{2x}{1}}$ $= \dfrac{2x\left(\frac{2x}{1}\right) - \frac{1}{\cancel{2x}_{1}}\left(\frac{\cancel{2x}^{1}}{1}\right)}{\frac{1}{\cancel{2x}_{1}}\left(\frac{\cancel{2x}^{1}}{1}\right) + 2x\left(\frac{2x}{1}\right)} = \dfrac{4x^2 - 1}{1 + 4x^2}$
Expression value:	Expression value:	Expression value:

3. Given expression	Lou's Work	Stu's Work
$\dfrac{\frac{x^2}{2}}{\frac{2}{x} + \frac{4}{x}}$ Test value: $x =$ ____	$\dfrac{\frac{x^2}{2}}{\frac{2}{x} + \frac{4}{x}} = \dfrac{x^2}{2} \div \left(\dfrac{2}{x} + \dfrac{4}{x}\right)$ $= \dfrac{x^2}{2} \cdot \left(\dfrac{x}{2} + \dfrac{x}{4}\right) = \dfrac{x^2}{2} \cdot \left(\dfrac{2x}{4} + \dfrac{x}{4}\right)$ $= \dfrac{x^2}{2} \cdot \left(\dfrac{3x}{4}\right) = \dfrac{3x^3}{8}$	$\dfrac{\frac{x^2}{2}}{\frac{2}{x} + \frac{4}{x}} = \dfrac{\frac{x^2}{2}}{\frac{6}{x}} = \dfrac{x^2}{2} \div \dfrac{6}{x}$ $= \dfrac{x^2}{2} \cdot \dfrac{x}{6} = \dfrac{x^3}{12}$
Expression value:	Expression value:	Expression value:

Student Activity

Checking Solutions to Rational Equations with a Calculator

You can use a calculator to check if the solution that you find to a rational equation is correct. For example, let's see if -10 is a solution of the equation $\frac{4}{x+2}=\frac{5}{x}$.

a) Construct the parentheses skeleton for the equation: $\frac{4}{(\;)+2}=\frac{5}{(\;)}$

b) Use extra parentheses or brackets to group the numerator or denominator if they contain more than one term: $\frac{4}{[(\;)+2]}=\frac{5}{(\;)}$

c) Insert the possible solution: $\frac{4}{[(-10)+2]}=\frac{5}{(-10)}$

d) Enter each side into your calculator, using extra parentheses to group the numerator or denominator if there is more than one term:

Left side: **4/((-10)+2)** Right side: **5/(-10)**

In this case, both sides result in -0.5, so -10 **is** a solution to the equation $\frac{4}{x+2}=\frac{5}{x}$.

Directions: Fill in the table below. If the one side of the equation is *undefined* for a value, say so (your calculator will likely give you an "error" message if this is the case). The first row has been done for you.

Equation	Parentheses skeleton and extra parentheses needed for calculator	Evaluate for...	Left side of equation	Right side of equation	Is this value a solution?
$\frac{5}{3+2x}=\frac{2}{x+1}$	$\frac{5}{[3+2(\;)]}=\frac{2}{[(\;)+1]}$	$x=1$	1	1	yes
		$x=0$			
		$x=-3$			
$\frac{9}{x}=\frac{12}{x-1}$		$x=1$			
		$x=0$			
		$x=-3$			
$\frac{1-x}{x}=\frac{x-1}{3}$		$x=1$			
		$x=0$			
		$x=-3$			
$\frac{x}{x+3}=3x+\frac{x}{x+1}$		$x=1$			
		$x=0$			
		$x=-3$			

Student Activity

Can't Use That!

Directions: Examine each of the equations below and decide what values cannot be allowed because of a "division by zero" problem. For the "disallowed" values, shade in the corresponding value in the grid below. The first one has been done for you.

1. $\dfrac{x+5}{x-2}=\dfrac{3}{x}$; $x\neq 0,2$
2. $\dfrac{x-4}{x+3}+\dfrac{2}{x-2}=5$
3. $\dfrac{x}{3}+\dfrac{2}{x-6}=1$
4. $\dfrac{3}{2x-1}=\dfrac{4}{x+5}$
5. $\dfrac{3}{4x}+\dfrac{1}{2}=\dfrac{2-x}{4-x}$
6. $\dfrac{5}{x+7}-\dfrac{3x}{x-10}=6$
7. $\dfrac{x+3}{x-3}=\dfrac{4}{3x+2}$
8. $1+\dfrac{x}{5}=\dfrac{5x}{x-8}$
9. $\dfrac{1}{4x-1}=\dfrac{4}{x}$
10. $\dfrac{9}{9+x}-2=\dfrac{1-x}{2x}$
11. $\dfrac{4}{x}=\dfrac{2x+1}{10}$
12. $\dfrac{x+6}{x-7}-\dfrac{9}{3x}=4$
13. $\dfrac{x+1}{x+5}+\dfrac{2}{5x}=3$
14. $\dfrac{x-3}{3x-4}+\dfrac{2x}{x+3}=8$
15. $\dfrac{x}{12}+5=\dfrac{x+2}{x-2}$

$x\neq 0,1,4$	$x\neq -3,0,2$	$x\neq 0,2,4$	$x\neq -5,0,2$	$x\neq 0,2$
$x\neq 0,\frac{1}{4},1$	$x\neq -3,2,4$	$x\neq 0,8$	$x\neq 0,4$	$x\neq 2,4$
$x\neq 0,\frac{1}{4}$	$x\neq -3,2$	$x\neq 8$	$x\neq 6$	$x\neq -5,0$
$x\neq -2,3$	$x\neq -3,-\frac{2}{3},3$	$x\neq -7,10$	$x\neq 0,6$	$x\neq 3,6$
$x\neq 0$	$x\neq -\frac{2}{3},3$	$x\neq 0,7$	$x\neq \frac{4}{3},3$	$x\neq 2$
$x\neq 0,1,9$	$x\neq -5,\frac{1}{2}$	$x\neq -6,0$	$x\neq -7,0,10$	$x\neq -2,0,2$
$x\neq 0,9$	$x\neq -5,1$	$x\neq 0,-1$	$x\neq -3,0,3,4$	$x\neq -1,0,5$

Student Activity

One Step at a Time for Rational Equations

When you are solving rational equations, the first steps can be the hardest ones. In this activity we practice taking those first few steps. Find the LCD, multiply by the LCD on both sides of the equation, distribute (where necessary), simplify each side of the equation, and then you finally have an equation that can be solved with prior methods.

Example:

	Equation: $\frac{3}{x}+\frac{4}{x+2}=3-\frac{1}{x}$ LCD: $x(x+2)$ $x(x+2)\left(\frac{3}{x}+\frac{4}{x+2}\right)=\left(3-\frac{1}{x}\right)\cdot x(x+2)$	
	Left side	**Right side**
Distribute	$x(x+2)\left(\frac{3}{x}\right)+x(x+2)\left(\frac{4}{x+2}\right)$	$(3)(x)(x+2)-\frac{1}{x}(x)(x+2)$
Simplify	$3(x+2)+4x$ $3x+6+4x$ $7x+6$	$3x(x+2)-(x+2)$ $3x^2+6x-x-2$ $3x^2+5x-2$
And reunite!	$7x+6=3x^2+5x-2$	

1.

	Equation: $\frac{3x+1}{x-3}=\frac{6x+2}{x^2-4x+3}$ LCD: ____________	
	Left side	**Right side**
Distribute		
Simplify		
And reunite!		

2.

Equation: $1-\frac{27}{x^2+x-12}=-\frac{1}{x-3}$ LCD: ____________	
	Left side	**Right side**
Distribute		
Simplify		
And reunite!		

Working to keep your equations clear of fractions

3.

	Equation: $\frac{2x+5}{4x+5}=\frac{-2}{2x+1}-\frac{5}{4x+5}$ LCD: ____________	
	Left side	**Right side**
Distribute		
Simplify		
And reunite!		

Student Activity

Match Up on Simple Rational Equations

Directions: Solve each rational equation and check the result. Then choose the letter that corresponds to this result. Some problems may have more than one answer. If there is no solution, then choose E (No Solution). If the result is not among the choices, then choose F (None of these). The first one has been done for you.

Nice try, but you can't just erase the hard parts.

A 2 | **B** −1

C 1 | **D** $\frac{1}{2}$

E No Solution | **F** None of These

$\frac{5x+5}{x+4}=2$ $5x+5=2(x+4)$ $5x+5=2x+8$ $3x=3$ $x=1$ **(C)**	$3=\frac{8-x}{x+2}$	$\frac{4x-2}{x+2}=4$
$\frac{x-3}{2x+3}=5$	$\frac{3}{x}+\frac{4}{x+2}=-\frac{1}{x}$	$\frac{5}{2}=\frac{11+x}{x+5}$
$\frac{x+2}{x-4}=\frac{2}{x-3}$	$\frac{x-4}{x+1}=-\frac{2x+7}{(x+2)(x+1)}$	$\frac{3x^2+5}{7x+5}=2$

Student Activity

Working with the Smaller Pieces

Practice with Uniform Motion Problems

We use the formula $d = rt$ to solve motion problems, where *d* is the distance, *r* is the rate, and *t* is the time.

1. A husband and wife go jogging together. The husband can run 2 miles an hour faster than the wife.

a. If the wife runs at a rate of x miles per hour, what is the rate of the husband? __________

b. If they run 10 miles, write an expression that describes the *time* it takes for the wife to run this distance. __________

c. If they run 10 miles, write an expression that describes the *time* it takes for the husband to run this distance. __________

2. Two trains (the Midnight Express and the Night Special) leave a station traveling in opposite directions. The Midnight Express is 10 mph faster than the Night Special, which travels at 50 mph. How far apart are the trains after two hours? __________ Every hour the trains are __________ more miles apart.

3. The same two trains as the previous problem leave a station traveling in the *same* direction. How far apart are the trains after two hours? __________ Every hour the trains are __________ more miles apart.

Practice with Shared-work Problems

Work completed, work rate, and time can be related using the formula: $w = rt$
Often the "work completed" is represented by 1 (for 1 job completed).

4. Suppose Joann can clean her house in 2 hours and Marty can clean the house in 5 hours. If JoAnn and Marty work together to clean the house, will it take
a. less than 2 hours **b.** between 2 and 5 hours **c.** more than 5 hours

5. If it takes you 2 hours to clean one pool, how long will it take you to clean 3 pools of the same size? __________

6. It normally takes you 2 hours to clean the pool. If you have help to clean it, will it take you
a. more than 2 hours **b.** less than 2 hours

7. Saundra can type 5 pages in 30 minutes. How long should it take her to type 8 pages? __________

8. Shell can paint 500 square feet of wall in 40 minutes. What is his work rate in square feet per minute? __________

Guided Learning Activity

Understanding Shared-Work Problems

Work completed = rate of work · time completed or $W = RT$ where

W = fraction of one job that is completed

R = Rate of work = $\dfrac{\text{amount of work}}{\text{time}}$

T = time (in the same units as the rate of work)

Start each problem below by first designating what a "job" is. Then fill in the blank boxes in the table, write an equation that can be used to solve the problem, and find the solution.

1. If the community pool is filled from the reserve water tower, it will take 6 hours to fill the pool. If the pool is filled from the city water pipes, it will take 4 hours to fill the pool. How long will it take to fill the pool if water from both the reserve water tower and the city water pipes are used?

Job: Fill one community pool	**Rate**	**Time**	**Work completed**
Reserve water tower	$\dfrac{1\text{ pool}}{6\text{ hours}} = \dfrac{1}{6}$	x hours	$\dfrac{x}{6}$
City water pipes			

Equation: **Solution:**

2. It takes Mr. Dupree 5 hours to audit 12 bank files. His associate, Ms. Carmichael can audit 16 files in 4 hours. If they work together, how long will it take them to audit 30 files?

Job:	**Rate**	**Time**	**Work completed**
Mr. Dupree			
Ms. Carmichael			

Equation: **Solution:**

3. Macon Construction can erect 100 square feet of rock facing on an exterior wall in 8 hours. Thomson Masonry can erect 200 square feet of rock facing in 12 hours. If you hire the two crews to work together, but Thomson Masonry starts working 2 hours after Macon Construction, how long will it take them to erect 250 square feet of rock facing?

Job:	**Rate**	**Time**	**Work completed**
Macon Construction			
Thomson Masonry			

Equation: **Solution:**

Student Activity

Working with the Language of Proportions

Directions: Fill in the table using the language of proportions. Find the units for each ratio and set up the proportion. You do not need to solve the problem. The first one has been done for you.

Problem	Fill in the Missing Values		Proportion
	Ratio 1	Ratio 2	$\frac{A}{B}=\frac{C}{D}$
1. If 12 doughnuts cost \$2.50, how many doughnuts can you buy for \$5.00?	$\frac{12 \text{ doughnuts}}{\$2.50}$	$\frac{x \text{ doughnuts}}{\$5.00}$	$\frac{12}{2.50}=\frac{x}{5.00}$
2. A chocolate chip cookie recipe uses 3 cups of flour to make 5 dozen cookies, how much flour is required to make 12 dozen cookies?	☐ cups flour / ☐ dozen cookies	☐ cups flour / ☐ dozen cookies	
3. If it takes Joel 4.5 hours to run a 26.2 mile marathon, how many minutes does it take him to run 1 mile?	☐ miles / ☐ minutes	☐ miles / ☐ minutes	
4. If jeans are on sale at three pairs for \$50, how much will seven pairs cost?	☐ pairs / \$☐	☐ pairs / \$☐	
5. If Jamie gets paid \$1000 for a 40-hour work week, how many hours will he have to work to earn \$800?	\$☐ / ☐ hours	\$☐ / ☐ hours	
6. There are 1160 calories in a 32 oz. chocolate shake from McDonalds. How many calories are in the 24 oz. shake? Source: www.mcdonalds.com	☐ calories / ☐ oz.	☐ calories / ☐ oz.	
7. According to American Red Cross guidelines, CPR should consist of cycles of 30 chest compressions followed by 2 breaths. If a rescuer has given 270 compressions, how many breaths should she have given? Source: www.redcross.org	☐ compressions / ☐ breaths	☐ compressions / ☐ breaths	

Student Activity

Proportion Heteronyms

Directions: In writing, there are words that are spelled the same but have different pronunciations and different definitions; these are called heteronyms. Many mathematical expressions and equations look similar but are really very different (almost like mathematical heteronyms). In each set of "proportion heteronyms" below, first **identify** whether you are being given an equation or an expression. If it is an expression, simplify it. If it is an equation, solve it. **Do your work (simplifying or solving) on another sheet of paper.** The first line has been done for you.

Heteronym Variation I			
	$\frac{5}{x}=\frac{2}{3}$	Equation.	$x=\frac{15}{2}$
	$\frac{5}{x}-\frac{2}{3}$		
	$\frac{5}{x}+\frac{2}{3}$		
	$\frac{5+2}{x+3}$		
	$\frac{5}{x}-\frac{2}{3}=0$		
	$\frac{5}{x}\div\frac{2}{3}$		

Heteronym Variation II			
	$\frac{x+1}{8}+\frac{x}{2}$		
	$\frac{x+1}{8}\div\frac{x}{2}$		
	$\frac{x+1}{8}=\frac{x}{2}$		
	$x+\frac{1}{8}=\frac{x}{2}$		
	$\frac{x}{8}+1=\frac{x}{2}$		
	$\frac{x}{8}+1+\frac{x}{2}$		

Student Activity

Equation Lineup

Directions: We have solved many different types of equations at this point in our learning. For each equation below, categorize the type of equation, write the strategies and key steps for remembering how to solve it, then solve the equation. The first one has been started for you.

Equation Types: Linear, quadratic, rational

1. $3x^2 - x = 4$	**2.** $-7(x+2) = 1 + 5(x-1)$
Type of Equation? Quadratic	**Type of Equation?**
Strategies and Key Steps Get =0 on one side of the equation, then factor to solve.	**Strategies and Key Steps**
Solve it!	**Solve it!**

3. $\frac{x+4}{x-4} = \frac{x-1}{x+3}$	**4.** $-\frac{3}{8} = \frac{1}{4} - \frac{5x}{16}$
Type of Equation?	**Type of Equation?**
Strategies and Key Steps	**Strategies and Key Steps**
Solve it!	**Solve it!**

5. $2x^2 = 3(x+3)$

Type of Equation?

Strategies and Key Steps

Solve it!

6. $\dfrac{x}{x+4} - \dfrac{1}{x+1} = 1$

Type of Equation?

Strategies and Key Steps

Solve it!

7. $5(2x+1) = 3(x+4) + 7x$

Type of Equation?

Strategies and Key Steps

Solve it!

8. $x^2 = 10x$

Type of Equation?

Strategies and Key Steps

Solve it!

Guided Learning Activity

Orienting Similar Triangles

Similar triangles are triangles with the same shape, but not necessarily the same size. These pairs of triangles can be easy to spot when the orientation of the triangles is the same.

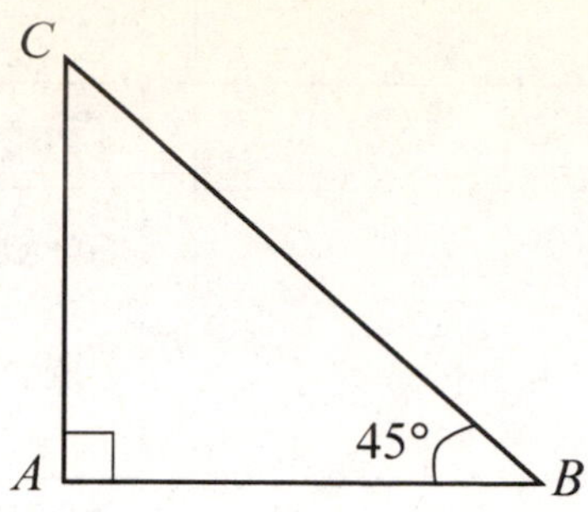

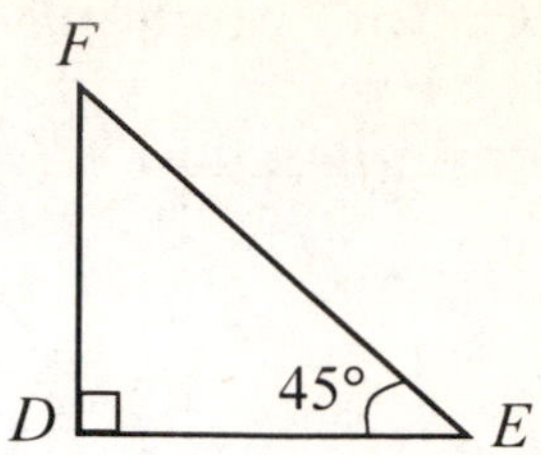

To the right, the two triangles are similar, and we would write that relationship like this: $\Delta ABC \sim \Delta DEF$.

For two triangles to be similar, two of the angle measures in the triangles must be the same (this forces all three angles to be the same).

Example:

The sum of the angles in a triangle is ______.

Find the missing angles in triangles ΔJKL and ΔMNP to the right.

P
21°
33°
L
J
126°
126°
K
M

Are the two triangles similar? _____

Which angles are equal? $\angle J =$ _____ $\angle K =$ _____ $\angle L =$ _____

Use a different color to highlight each pair of similar angles.

Use a different color to highlight each pair of similar sides.

Which sides are similar? $JK \sim$ ____ $KL \sim$ ____ $LJ \sim$ ____

Because of these side similarities, we can set up equal ratios of similar sides (fill in the missing denominators): $\frac{JK}{\_\_\_} = \frac{KL}{\_\_\_} = \frac{LJ}{\_\_\_}$

This can be written as three different proportions. What are they?

$\frac{\_\_\_}{\_\_\_} = \frac{\_\_\_}{\_\_\_}$ $\frac{\_\_\_}{\_\_\_} = \frac{\_\_\_}{\_\_\_}$ $\frac{\_\_\_}{\_\_\_} = \frac{\_\_\_}{\_\_\_}$

Triangles	Are the triangles similar? If so, state the side similarities.	Find the lengths of the side(s) indicated by variables.
1. 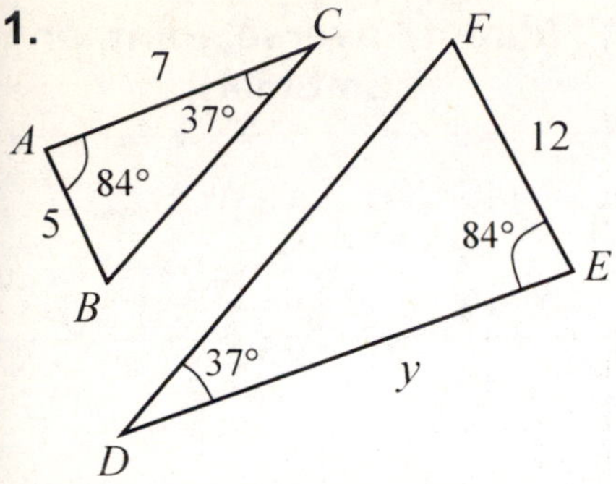 A, B, C: 7, 5, 37°, 84°; D, E, F: 12, y, 84°, 37°		
2. 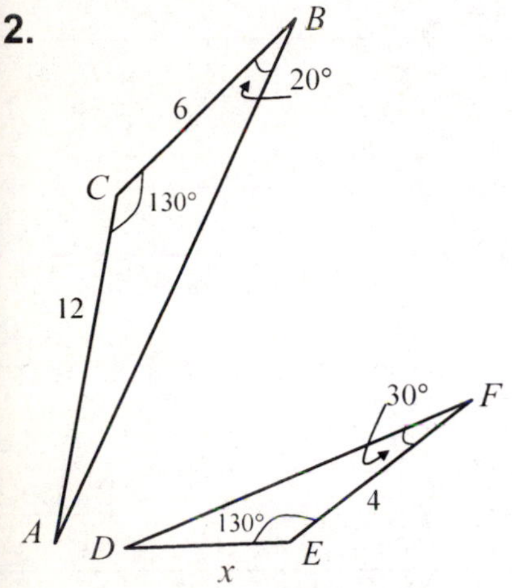 A, B, C: 6, 12, 20°, 130°; D, E, F: 30°, 4, 130°, x		
3. A, B, C: 20, 21, 46°; D, E, F: 62°, 15, 8 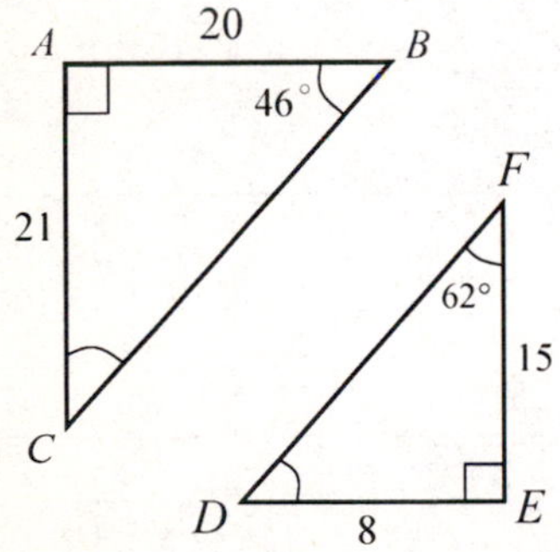		
4. E, F, D: 6, y, z, 55°; C, A, B: x, 2.5, 55°, 1.5 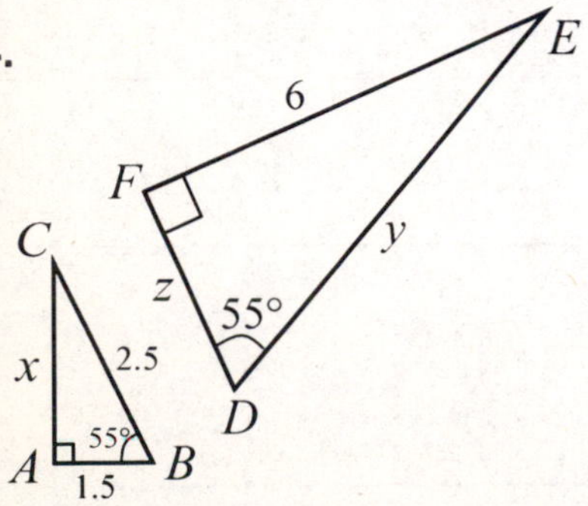		

Student Activity

Charting the Language of Variation

Directions: Fill in the blank spaces in the chart. The first one has been done for you.

In Words	In Equation Form	Type of Variation (direct, inverse, joint, or combined)
1. The amount of sales tax, s, on an item varies directly as the price, p, of the item.	$s = kp$	direct
2. The time, t, it takes to clean the house is __________ proportional to the number of people, n, who help clean it.	$t = \frac{k}{n}$	
3. The area of a circle, A, varies directly with as square of the radius, r.		
4. The time, t, it takes to arrive at a destination is __________ proportional to the rate of travel, r.	$t = \frac{k}{r}$	
5. The force of wind, F, on a billboard varies jointly as the area of the billboard, A, and the square of the wind velocity, v.		
6. The voltage, V, in an electrical circuit with a fixed resistance is __________ proportional to the current, I.	$V = kI$	
7. The mass, m, of an iron solid varies directly as the volume, V.		
8. The number of supports, n, needed to hold up a bridge varies __________ as its length, ℓ.	$n = k\ell$	
9. The price of gasoline, p, varies inversely as the number of gallons of crude oil, g, that are refined.		
10. The pressure, P, of a gas is directly proportional to the temperature, T, and inversely proportional to the volume, V.		

Guided Learning Activity

Constant of Variation... an Oxymoron

What does the word *constant* mean? ____________________

In mathematics, what is a *constant*? ____________________

What does the word *variation* mean? ____________________

In mathematics, what is a variable? ____________________

When we work with variation problems, we use the letter k to represent the constant of variation. But what does it mean when we say *constant of variation*? Let's investigate.

Hourly Pay:

1. Fill in the tables for Monte, Clay, and Lisa to tell them how much gross pay (before taxes and other fees are removed) they will take home for working the given number of hours. Then write an equation for each that relates their gross pay, P, to the hours worked, h.

Monte: $7.50/hour

Hours	Pay
10	
15	
20	
30	
40	

Clay: $22/hour

Hours	Pay
10	
15	
20	
30	
40	

Lisa: $18.25/hour

Hours	Pay
10	
15	
20	
30	
40	

Monte: $P =$ ______ Clay: $P =$ ______ Lisa: $P =$ ______

2. In this example, what does k represent? ____________________

3. Write an equation for: "gross pay is directly proportional to hours worked" using k as the constant of variation. ____________

a. Lisa makes $365 for working 20 hours. Use this data and the equation above to solve for the constant k.

$k =$ ________ Look familiar?

b. Now write the variation equation with the numeric value of k in place: ____________

c. Use this equation to find Lisa's gross pay if she works 25 hours. ____________

d. Could you use the equation from part b to find gross pay if Monte works for 25 hours? Why or why not?

Density: The density, ρ, of an object is its mass per unit volume, that is $\rho = \frac{m}{V}$. If the object is made of a uniform material, then the density is constant.

1. Calculate the density (in g/cm^3) for the given mass and volume of each material in the tables below.

Balsa Wood:

Mass	12 g	15 g	24 g	39 g	45 g
Volume	100 cm^3	125 cm^3	200 cm^3	325 cm^3	375 cm^3
Density					

Ice:

Mass	229.25 g	366.8 g	458.5 g	495.18 g	687.75 g
Volume	250 cm^3	400 cm^3	500 cm^3	540 cm^3	750 cm^3
Density					

2. What is constant for Balsa wood? ____________ What is constant for ice? ____________

3. Is the density constant for Balsa wood the same as the density constant for ice? _____

Now we'll turn this around.

4. Write an equation for: **"**mass is directly proportional to volume" using k as the constant of variation. ____________

a. The mass of 100 cm^3 of balsa wood is 12 g. Use this data and the equation you just wrote to solve for k.

$k =$ ________ Look familiar?

b. Now write the variation equation with the numeric value of k in place: ____________

c. Use this equation to find the volume of a 500 g piece of balsa wood.

d. Could you use the equation from part b to find the volume of 500 g of ice? Why or why not?

5. See if you can now explain what it means when we say *constant of variation.*

Assess Your Understanding

Rational Expressions and Equations

For each of the following, describe the strategies or key steps that will help you **start** the problem. You do **not** have to complete the problems.

		What will help you to start this problem?
1.	Subtract: $\frac{4x}{x+1} - \frac{2x+3}{x+1}$	
2.	Simplify: $\frac{x^2 - 2x - 35}{7x+35}$	
3.	Add: $\frac{3}{x+2} + \frac{4}{x+3}$	
4.	Multiply: $\frac{x^2+8x}{6x+18} \cdot \frac{12x-24}{x^2+6x-16}$	
5.	Where is $\frac{x+3}{x^2-16}$ undefined?	
6.	Add: $\frac{4x+5}{3x+5} + \frac{2x+5}{3x+5}$	
7.	Divide: $\frac{x^2+4x}{x+8} \div \frac{x^3+4x^2}{x^2+6x-16}$	
8.	Simplify: $\frac{x^2+9x+14}{4-x^2}$	

9.	Subtract: $\dfrac{x^2+3}{x^2-4}-\dfrac{x+5}{x-2}$	
10.	Multiply: $\dfrac{3x^2+12x}{x-6}\cdot\dfrac{x^2-4x-12}{x^2+2x}$	
11.	If you add the same number to the numerator and denominator of $\dfrac{7}{9}$, the result is $\dfrac{5}{6}$. What is the number?	
12.	Add: $\dfrac{4}{x+3}+\dfrac{1}{3x}$	
13.	Solve: $\dfrac{2}{x+6}+\dfrac{5}{x^2+6x}=1$	
14.	Simplify: $\dfrac{\frac{2}{3}+\frac{5}{x}}{\frac{1}{2x}-4}$	
15.	On a map, a distance of 1.5 inches represents 1.125 miles. How many miles does 2.2 inches on the map represent?	

Metacognitive Skills

Rational Expressions and Equations

Metacognitive skills refer to the ability to judge how well you have learned something and to effectively direct your own learning and studying. This is a self-evaluation tool designed to help you focus your studying and to improve your metacognitive skills with regards to this math class.

Fill the 1st column out **before** you begin studying. Fill the 2nd column out after you study for your test.

Go back to this assessment after your test and circle any of the ratings that you would change – this identifies the "disconnects" between what you **thought** you knew well and what you **actually** knew well.

Use the scale below to assign a number to each topic.

5 *I am confident I can do any problems in this category correctly.*
4 *I am confident I can do most of the problems in this category correctly.*
3 *I understand how to do the problems in this category, but I still make a lot of mistakes.*
2 *I feel unsure about how to do these problems.*
1 *I know I don't understand how to do these problems.*

Topic or Skill	Before Studying	After Studying
Factoring polynomials.		
Correctly identifying opposite factors.		
Simplifying a rational expression.		
Finding the values where a rational expression is undefined.		
Multiplying rational expressions.		
Dividing rational expressions.		
Simplifying expressions that involve units of measurement.		
Adding or subtracting rational expressions with like denominators.		
Finding the LCD of two (or more) rational expressions.		
Building an equivalent rational expression with a desired denominator.		
Understanding why you need a common denominator when you add or subtract rational expressions.		
Identifying denominators that are opposites.		
Adding or subtracting rational expressions with unlike denominators.		
Identifying a complex fraction.		
Simplifying a complex fraction by simplifying the numerator & denominator and then dividing.		
Simplifying a complex fraction by finding the LCD for all terms in the complex fraction and then multiplying by LCD/LCD.		
Finding the values where a rational *equation* is undefined and knowing why it is important to find these values.		
Clearing the fractions from a rational equation.		

Continued on the next page.

Use the scale below to assign a number to each topic.

5 *I am confident I can do any problems in this category correctly.*
4 *I am confident I can do most of the problems in this category correctly.*
3 *I understand how to do the problems in this category, but I still make a lot of mistakes.*
2 *I feel unsure about how to do these problems.*
1 *I know I don't understand how to do these problems.*

Topic or Skill	Before Studying	After Studying
Understanding why you cannot "clear the fractions" from a rational expression.		
Solving a linear equation that may result from a rational equation.		
Solving a quadratic equation that may result from a rational equation.		
Checking the solutions of a rational equation.		
Solving a formula containing rational expressions for one of the variables.		
Setting up a table of information for a uniform motion application problem.		
Setting up a table of information for a shared-work application problem.		
Setting up a table of information for an investment application problem.		
Converting the information in a table to an equation that can be solved.		
Writing ratios from sentences.		
Creating a proportion from an application problem.		
Solving a proportion.		
Understanding the specific properties that allow a proportion equation to be solved differently from other rational equations.		
Knowing the types of triangles that you can apply the Pythagorean Theorem to, and being able to correctly set up the Pythagorean Theorem for those triangles.		
Knowing what makes two triangles similar.		
Setting up a proportion of sides for two similar triangles.		
Understanding what the constant of variation is.		
Converting the language of variation problems to variation equations containing a constant of variation.		
Solving for k in a variation equation.		
Using the variation equation to solve for something else in a problem.		

RAD: Radical Expressions and Equations

Student Activity

Nightmare on Radical Street

Directions:
Each of the radical expressions listed below is either real or non-real.

If the expression is real and **is** a perfect square, then place it in the appropriate **rational** number box.

If the expression is real and **is not** a perfect square, then place it in the appropriate **irrational** number box.

If the expression is **non-real**, then cross it out to eliminate Scotty's nightmares!

Do this activity without a calculator!

Never again would Scotty make the mistake of eating spicy food before he finished his algebra homework.

$\sqrt{\frac{64}{49}}$ $\quad -\sqrt{-1}$ $\quad \sqrt{\frac{1}{2}}$ $\quad \sqrt{\frac{4}{9}}$ $\quad \sqrt{-9}$ $\quad \sqrt{\frac{16}{25}}$ $\quad \sqrt{36}$ $\quad \sqrt{\frac{39}{2}}$ $\quad \sqrt{1}$ $\quad \sqrt{9}$

$\sqrt{3}$ $\quad \sqrt{16}$ $\quad \sqrt{6}$ $\quad \sqrt{45}$ $\quad \sqrt{-4}$ $\quad \sqrt{14}$ $\quad \sqrt{0}$ $\quad \sqrt{\frac{60}{2}}$ $\quad \sqrt{-25}$ $\quad \sqrt{50}$

Rational Numbers		Irrational Numbers	
$\square = \frac{4}{5}$	$\square = 6$	$0 < \square < 1$	$4 < \square < 5$
$\square = 4$	$\square = 3$	$1 < \square < 2$	$5 < \square < 6$
$\square = \frac{2}{3}$	$\square = 0$	$2 < \square < 3$	$6 < \square < 7$
$\square = \frac{8}{7}$	$\square = 1$	$3 < \square < 4$	$7 < \square < 8$

Student Activity

Square Roots Using a Calculator

aOften the $\sqrt{\ }$ function is found ABOVE one of the keys, and students will need to use a [2nd] button to use the $\sqrt{\ }$ function.

Your calculator probably falls into one of two groups:

Group 1: Radicand first

On these calculators, you would find $\sqrt{9}$ by keying in [9] then [$\sqrt{\ }$]. It is likely that this will immediately give you the answer 3 without pressing any other keys.

Group 2: Radical Symbol first

On most graphing calculators and some of the newer scientific calculators, you would do the keystrokes like this: [$\sqrt{\ }$] and then [9]. Then you probably see $\sqrt{(9}$ in your display. You have to close the parentheses by pressing [ )] and then [ENTER] to evaluate.

Now fill out the table below, by finding the proper keystrokes to do the evaluation on your calculator. In the last column, square the calculator result to see if you get the radicand back. The first one has been done for you.

	Expression	Calculator Keystrokes and result	Check by squaring the result
a.	$\sqrt{8}$	Keystrokes will vary.	$(2.8284)^2 \approx 7.9998 \approx 8$
b.	$\sqrt{169}$		
c.	$\sqrt{\frac{16}{25}}$		
d.	$\sqrt{\frac{1}{2}}$		
e.	$\sqrt{\frac{2}{3}}$		
f.	$\sqrt{-4}$		
g.	$\sqrt{\frac{4}{9}}$		
h.	$\sqrt{1.5625}$		

Student Activity

Match Up on Simplifying Radical Expressions

Directions: Match each of the expressions in the squares in the table below with its simplified form from the top. Assume all variables represent positive numbers. If the simplified form is not found among the choices A through D, then choose E (none of these).

A $3x^2$

B $2x$

C $-5x$

D $4x^3$

E None of these

$\sqrt{9x^4}$	$\sqrt[3]{-125x^3}$	$\sqrt[4]{81x^8}$
$\sqrt[6]{64x^6}$	$\sqrt[3]{64x^9}$	$\sqrt{-25x^2}$
$2x^2 + x^2$	$\sqrt[3]{27x^6}$	$\sqrt{\dfrac{36x^2}{9}}$
$\sqrt{9x}$	$\sqrt{x^2} + x$	$\sqrt{16x^6}$
$-\sqrt[4]{625x^4}$	$\sqrt[3]{12x^9}$	$2 \cdot \sqrt[3]{8x^9}$

Guided Learning Activity

Radical Graphing

The graph of $f(x)=\sqrt{x}$, is a curve with a definite endpoint.

1. Draw the graph of $f(x)=\sqrt{x}$ by plotting points.

x	$f(x)$
9	
4	
1	
0	
−1	
−4	
−9	

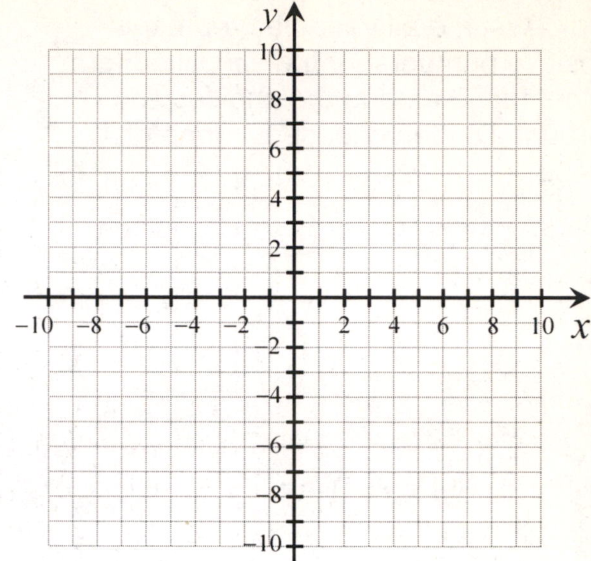

What is the domain of f? ________

What is the range of f? ________

Why are there no points for $x<0$?

The graph of $f(x)=\sqrt[3]{x}$, is a curve that has a similar shape to the curve $y=x^3$.

2. Draw the graph of $f(x)=\sqrt[3]{x}$ by plotting points.

x	$f(x)$
8	
1	
0	
−1	
−8	

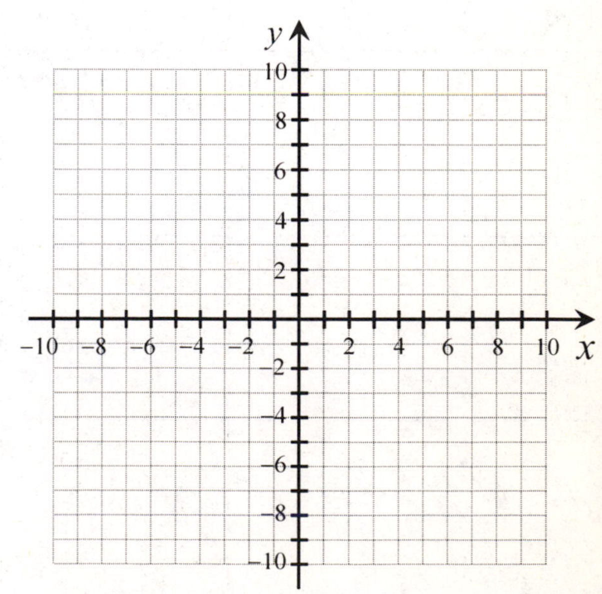

What is the domain of f? ________

What is the range of f? ________

These radical functions can be translated and reflected in the exact same way as $y=|x|$, $y=x^2$, and $y=x^3$.

3. Draw the graph of $g(x)=\sqrt{x}-3$ by shifting the graph of $f(x)=\sqrt{x}$ ______ 3 units.

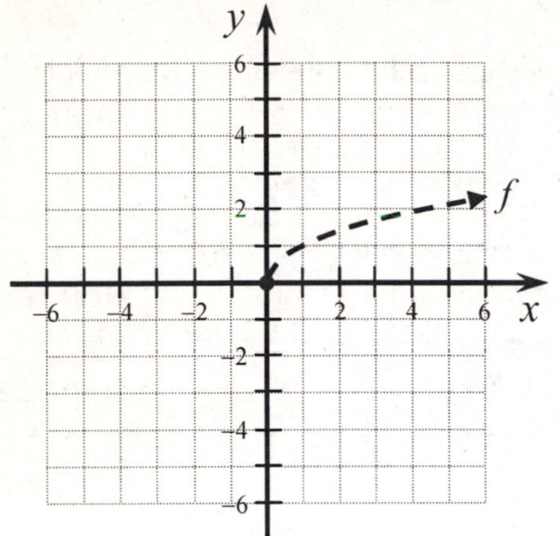

What is the domain of g? _______

What is the range of g? _______

5. Draw the graph of $g(x)=-\sqrt{x}$ by reflecting the graph of $f(x)=\sqrt{x}$ over the __-axis.

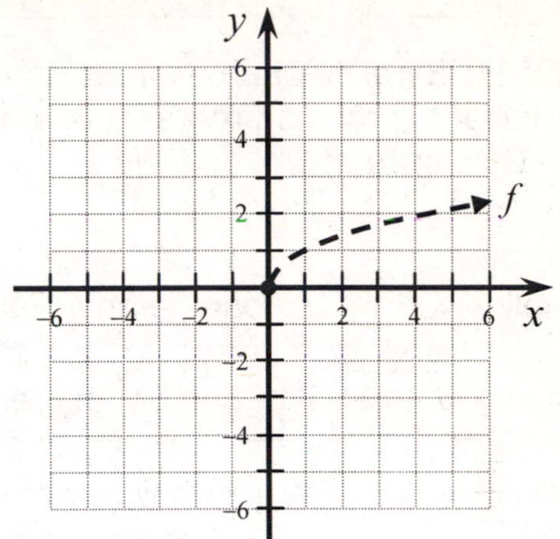

What is the domain of g? _______

What is the range of g? _______

4. Draw the graph of $g(x)=\sqrt[3]{x+2}$ by shifting the graph of $f(x)=\sqrt[3]{x}$ ______ 2 units.

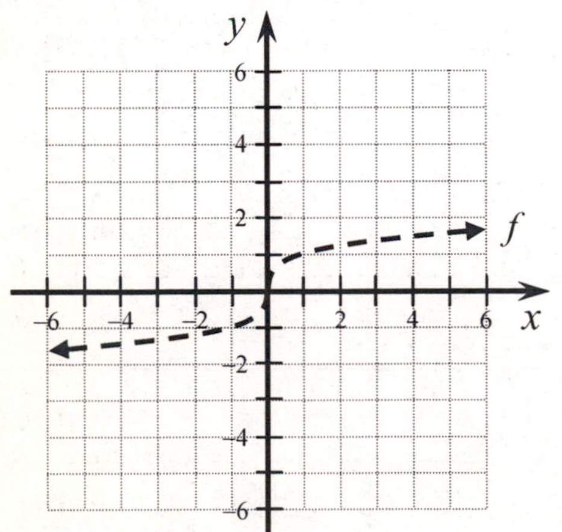

What is the domain of g? _______

What is the range of g? _______

6. Draw the graph of $g(x)=\sqrt[3]{x+1}-4$ by shifting the graph of $f(x)=\sqrt[3]{x}$ ______ 1 unit and ______ 4 units.

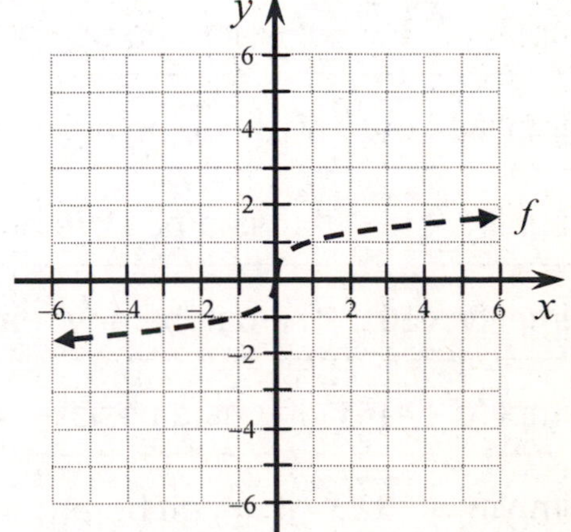

What is the domain of g? _______

What is the range of g? _______

Student Activity
Difficult Choices

If the variables can represent any real number and the radical expression represents a real number:

$\sqrt[n]{x^n} = |x|$ if n is even.

$\sqrt[n]{x^n} = x$ if n is odd.

If the variables only represent positive values, it is not necessary to use absolute values for the even roots.

Directions: Simplify each expression and choose the simplified expression from A or B. If the radical expression represents a non-real number (an imaginary number), then choose C (non-real).

	A	B	C
1. Simplify $\sqrt{36x^2}$ if x can be any real number.	$6x$	$6\|x\|$	Non-real
2. Simplify $\sqrt{-36x^2}$ if x can be any real number.	$6x$	$-6x$	Non-real
3. Simplify $\sqrt{36x^2}$ if x is a positive number.	$6x$	$6\|x\|$	Non-real
4. Simplify $\sqrt[3]{8x^6}$ if x can be any real number.	$2x^2$	$\|2x^2\|$	Non-real
5. Simplify $\sqrt[3]{-8x^6}$ if x can be any real number.	$-2x^2$	$2x^2$	Non-real
6. Simplify $\sqrt{49x^2}$ if x can be any real number.	$7x$	$7\|x\|$	Non-real
7. Simplify $\sqrt{49x^2}$ if x is a positive number.	$7x$	$7\|x\|$	Non-real
8. Simplify $\sqrt{16x^4}$ if x can be any real number.	$4x^2$	$4\|x^2\|$	Non-real
9. Simplify $\sqrt{-16x^4}$ if x can be any real number.	$-4x^2$	$\|-4x^2\|$	Non-real
10. Simplify $\sqrt[3]{64x^3}$ if x can be any real number.	$4x$	$4\|x\|$	Non-real
11. Simplify $\sqrt[4]{16x^4}$ if x is a positive number.	$2x$	$2\|x\|$	Non-real
12. Simplify $\sqrt[4]{16x^4}$ if x can be any real number.	$2x$	$2\|x\|$	Non-real
13. Simplify $\sqrt[4]{-16x^4}$ if x is a positive number.	$2x$	$2\|x\|$	Non-real
14. Simplify $\sqrt[5]{-32x^5}$ if x can be any real number.	$-2x$	$2\|x\|$	Non-real
15. Simplify $\sqrt[5]{32x^5}$ if x can be any real number.	$2x$	$2\|x\|$	Non-real

Student Activity
Paint by Rational Exponents

Directions: Convert each expression according to the directions. Then shade in the box that contains your answer. The first one has been done for you. There's a surprise when you're finished!

Convert to radical notation:

1. $(xy)^{1/2} = \sqrt{xy}$
2. $(25+x^2)^{1/2}$
3. $5x^{1/2}$
4. $(-x)^{1/5}$
5. $\dfrac{y}{x^{-1/2}}$

Convert to rational exponent notation:

6. $2\sqrt{xy}$
7. $5\sqrt[4]{x^3}$
8. $\sqrt{3x^5}$
9. $\sqrt[6]{(10x+7)^7}$
10. $\sqrt[5]{-ab^4}$

$3x^{2/5}$	$(3x^5)^{1/2}$	$(3x)^{5/2}$	$\sqrt{25+x^2}$	$5+x$
$\dfrac{1}{\sqrt[5]{x}}$	$\sqrt[5]{-x}$	$-\sqrt{x^5}$	$y\sqrt{x}$	$-\dfrac{y}{\sqrt{x}}$
$(xy)^2$	$10x+7^{6/7}$	$10x^{7/6}+7^{7/6}$	$x\sqrt{y}$	$a^{1/5}b^{-4/5}$
$5x^{4/3}$	$\sqrt{5x}$	$\sqrt{xy}$	$(2xy)^2$	$ab^{-4/5}$
$5x^{3/4}$	$\dfrac{r}{\sqrt{t}}$	$10x+7^{7/6}$	$2(xy)^2$	$(-ab^4)^{1/5}$
$(5x)^{3/4}$	$5\sqrt{x}$	$-t\sqrt{r}$	$2(xy)^{1/2}$	$(2xy)^{1/2}$
$(5x)^{\frac{4}{3}}$	$-r\sqrt{t}$	$(10x+7)^{7/6}$	$(10x+7)^{6/7}$	$2xy^{1/2}$

Student Activity

Language of the Other Roots

Directions: For each set, fill in the missing boxes in the first row with numbers that make the two expressions equal. Write out a statement that describes the relationship between the two expressions in the second row. The first two have been started for you.

Expression written with root or power	simplified expression
Write out a statement in words.	

$\sqrt[3]{125}$	5
The cube root of ____ is ____.	

$(\quad)^{3}$	216
The cube of ____ is ____.	

$(\quad)^{\frac{1}{4}}$	2

$\sqrt[4]{}$	9

	27
The cube of ____ is ____.	

$\sqrt[3]{}$	7

	$\frac{1}{16}$
_____ raised to the -2 power is _____.	

$(\quad)^{-2}$	
3 raised to the ____ power is _____.	

$\frac{1}{5^2}$	

$1^{-\frac{3}{4}}$	

$(-1000)^{\frac{1}{3}}$	

$(\quad)^{\frac{3}{2}}$	
_____ raised to the $\frac{3}{2}$ is 8.	

Student Activity

Rational Exponents Using a Calculator

By now you should be familiar with the keys on your calculator used to evaluate expressions like $\sqrt{196}$, $\sqrt[5]{3125}$. Your calculator can also evaluate expressions containing rational exponents.

Consider the following expression: $\left(\sqrt[4]{625}\right)^3$

To evaluate this expression with a calculator, you would first need to write an equivalent expression with a rational exponent:

$$\left(\sqrt[4]{625}\right)^3 = 625^{3/4}$$

Most calculators have an exponent key that looks like $\boxed{\wedge}$. When entering a rational exponent, you should always use parentheses around the exponent. Enter each of the following expressions exactly as they are shown:

$625\boxed{\wedge}3/4 =$ ________________

$625\boxed{\wedge}(3/4) =$ ________________

You should see two *very* different answers! Which one is correct? If you calculate $\sqrt[4]{625}$ then cube the result you get 125. This is the correct answer. Your calculator will always follow the order of operations, so without parentheses around the 3/4, your calculator will cube 625 *then* divide the result by 4. The result is a very large number!

Now try finding these with your calculator. The first row has been done for you.

Radical Notation	Rational Exponent Notation	Keystrokes to Enter into Calculator	Answer
1. $\left(\sqrt[3]{64}\right)^2$	$64^{2/3}$	$64\boxed{\wedge}(2/3)$	16
2.	$-243^{6/5}$		
3. $\dfrac{1}{\sqrt[3]{8^2}}$			
4.	$1024^{-2/5}$		
5. $\dfrac{1}{\left(\sqrt[4]{1296}\right)^{-3}}$			
6.	$625^{3/4}$		
7. $\left(\sqrt[5]{32}\right)^{-3}$			

Student Activity

Escape the Rational Exponent Matrix

Directions: Assume all variables represent positive numbers. Begin at the box marked START. By shading in adjacent pairs of squares that contain equivalent expressions, you will eventually find the path to "escape" this matrix of boxes. The first "step" in the path and a couple steps in the middle have been taken for you.

START $(36b^6)^{1/2}$	$18b^3$	$9b^3$	$x+\sqrt{x}$	$7^{2/3}$
$6b^3$	$36b^3$	$3y$	$\sqrt[4]{3y}$	$x^{9/5}$
$x^{3/4}\cdot x^2$	$\sqrt[4]{x^{11}}$	$(3y^{2/3})^3$	$9y^2$	$(\sqrt{2x})^3$
$\sqrt{x^3}$	$x^{4/11}$	$27y^2$	$\sqrt[3]{3y^2}$	$2x^{4/3}$
$x^{4/3}$	$x^{-4/3}$	$\frac{x^3}{x^{7/3}}$	$\sqrt[3]{x^2}$	$x^{5/6}(x^{1/6}+x^{1/2})$
$x^{1/3}$	$\frac{1}{x^7}$	x^7	$x^{1/9}$	$x+\sqrt[3]{x^4}$
7	$\frac{x^3}{8}$	$\left(\frac{x^5}{32}\right)^{3/5}$	$y^{1/20}$	$\sqrt[4]{\sqrt[5]{y}}$
$\frac{1}{y^2}$	$\sqrt[10]{y^5}$	y^2	$y^{5/4}$	y^{20}
$3\sqrt{y}$	$y^{1/2}$	49	ESCAPE $343^{2/3}$	$\sqrt{y^{400}}$

Student Activity

Radical Heteronyms

Directions: In writing, there are words that are spelled the same but have different pronunciations and different definitions; these are called heteronyms. Many mathematical expressions look similar but are really very different (almost like mathematical heteronyms). Simplify each of the expressions below, paying close attention to the use of parentheses, the mathematical operations, and notation. Assume all variables represent positive numbers.

1.

9^{-1}	$9^{1/2}$	9^{2}	9^{-2}	-9^{-1}	$9^{-1/2}$

2.

$\sqrt{25-9}$	$(25-9)^{1/2}$	$25^{1/2}-9^{1/2}$	$(25-9)^{-1}$	$25^{-1}-9^{-1}$

3.

$x^{1/2}x^{1/3}$	$\left(x^{1/2}\right)^{1/3}$	$x^{1/2}+x^{1/3}$	$\dfrac{x^{1/2}}{x^{1/3}}$	$\left(\dfrac{1}{2}x\right)\left(\dfrac{1}{3}x\right)$	$\dfrac{1}{3}\left(\dfrac{1}{2}x\right)$

4.

$\left(16a^{4}\right)^{-2}$	$\left(16a^{4}\right)^{1/2}$	$\left(16a^{4}\right)^{1/4}$	$\dfrac{1}{2}\left(16a^{4}\right)$	$16\left(a^{4}\right)^{1/2}$

5.

$\left(\dfrac{9}{4}\right)^{-2}$	$\left(\dfrac{9}{4}\right)^{1/2}$	$\left(\dfrac{9}{4}\right)^{-1/2}$	$\left(\dfrac{9}{4}\right)^{-1}$	$\dfrac{9^{1/2}}{4}$	$\dfrac{9^{-1/2}}{4}$

Student Activity

Factor Trees

Directions: Use a factor tree to find the prime factorization for each number. Build branches of the tree using factor pairs. When a prime number is reached in one of the branches, circle it in red to indicate that it is one of the "apples" (prime factors) on the tree. To write the prime-factored form, collect all the "apples" on the tree. The factor tree shown here tells us that the prime factorization of 80 is $2\cdot2\cdot2\cdot2\cdot5$. To find $\sqrt{80}$, we find pairs of factors in $\sqrt{2\cdot2\cdot2\cdot2\cdot5}$ and bring them outside the radical as a single factor to get $2\cdot2\sqrt{5}$ or $4\sqrt{5}$. To make a particularly colorful and "treelike" factor tree below, use brown for all the "branches," green for all the numbers, and red to circle and shade in the prime numbers. In the factor tree below, your "trees" will go in all directions to make the branches. The number 90 has been prime factored for you.

80
8 10
4 2 2 5
2 2

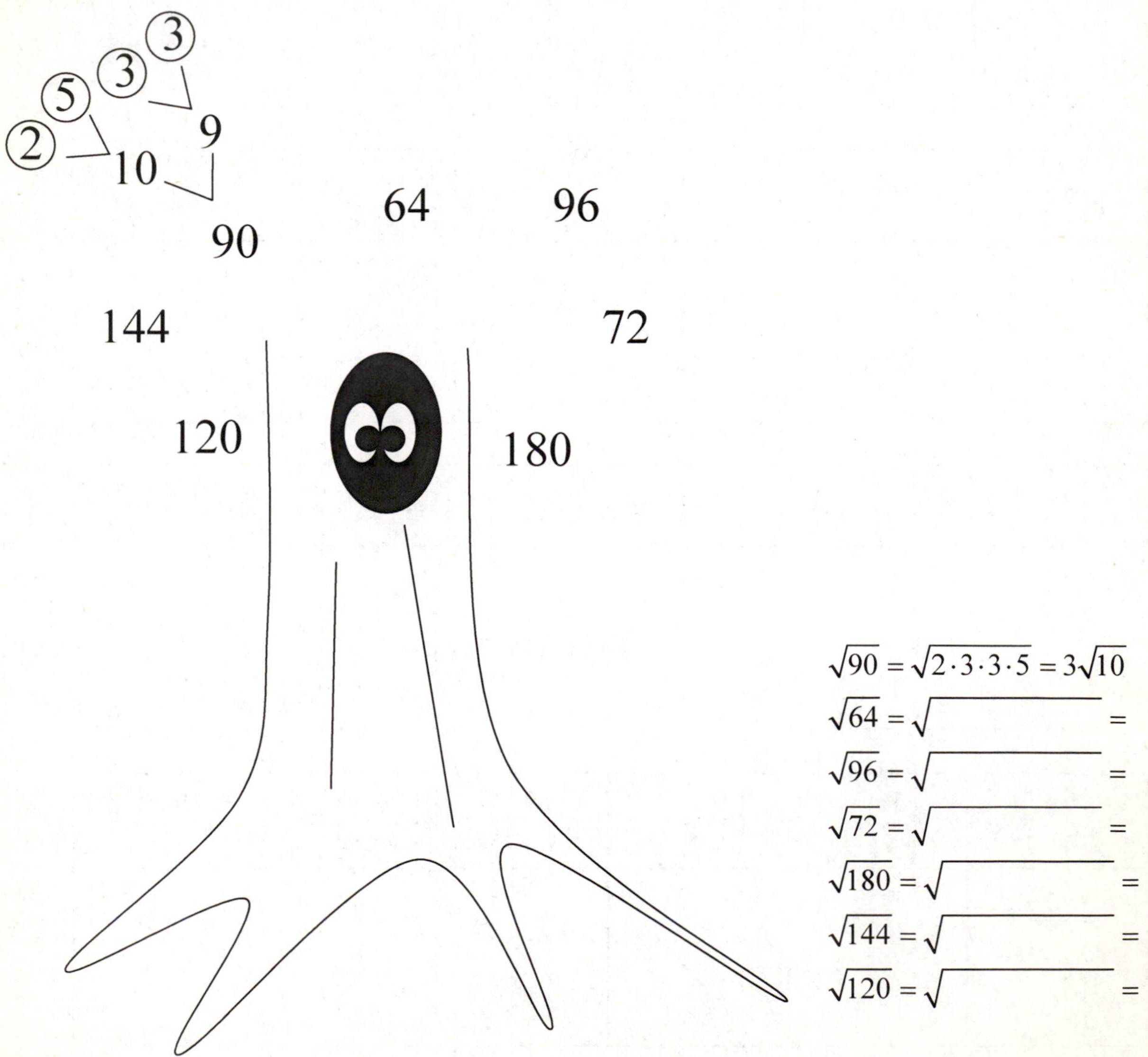

Guided Learning Activity

Higher on the Factor Tree

Recall that a **square root** is usually written without an index, so $\sqrt{a} = \sqrt[2]{a}$. In a square root we had to find **pairs** of a factor in order to simplify the radicand.

Similarly, in a **cube root**, like $\sqrt[3]{a}$, we would need to find **triples** of a factor in order to simplify the radicand.

And in a **fourth root**, $\sqrt[4]{a}$, we would need to find **quadruples** of a factor in order to simplify the radicand.

Notice that the index of the radical tells us how many matching factors we need to find in order to simplify.

$$\to\sqrt[2]{a} \qquad \to\sqrt[3]{a} \qquad \to\sqrt[4]{a}$$

One way to search for these pairs, triples, quadruples, etc. is to use factor trees to find the prime factorization for each radicand.

Example: Simplify: $\sqrt[3]{162}$.

We look for triples since the index is three.

Using a factor tree, we factor 162.

$$\sqrt[3]{162} = \sqrt[3]{3 \cdot 3 \cdot 3 \cdot 3 \cdot 2} = 3\sqrt[3]{3 \cdot 2} = 3\sqrt[3]{6}$$

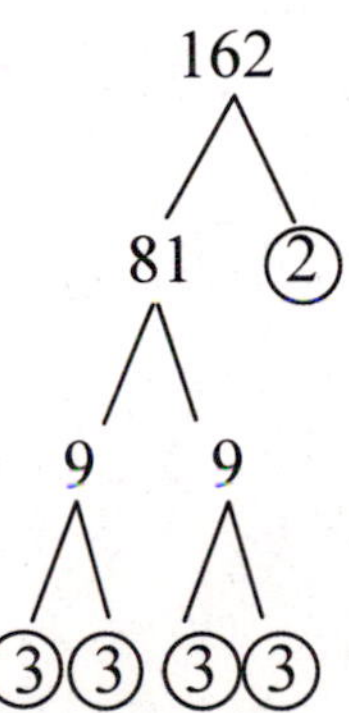

Example: Simplify: $\sqrt[4]{80}$

We look for quadruples since the index is four.

Using a factor tree, we factor 80.

$$\sqrt[4]{80} = \sqrt[4]{5 \cdot 2 \cdot 2 \cdot 2 \cdot 2} = 2\sqrt[4]{5}$$

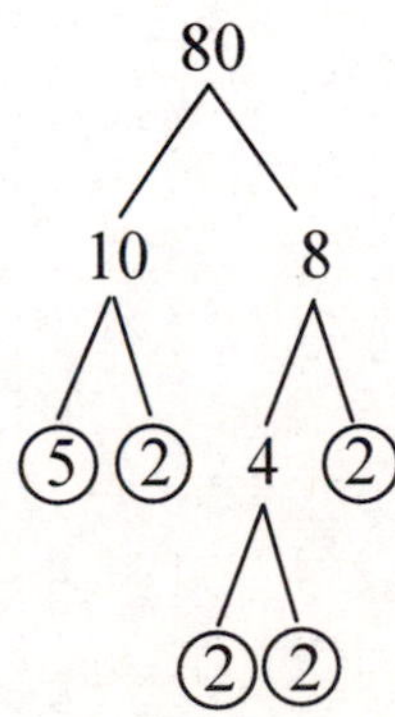

Directions: Use factor trees to help simplify the following higher order roots.

1. $\sqrt[3]{3{,}000}$

4. $\sqrt[3]{625}$

2. $\sqrt[4]{432}$

5. $\sqrt[4]{512a^8}$

3. $\sqrt[5]{243}$

6. $\sqrt[3]{216x^5}$

Student Activity

Match Up on Simplifying Radicals

Directions (READ them): Simplify each radical expression. Then look at the remaining radicand and choose the letter that corresponds to the remaining radicand.

For example, for $2\sqrt{3}$, $x\sqrt{3}$, $5\sqrt{3}$, and $x^2\sqrt{3}$, you would choose B. If there is no radical left after simplification, choose E.

If none of these choices are correct, choose F. Assume that all variables represent nonnegative numbers.

A 2 **B** 3

C 5 **D** x

E No radical left **F** None of these

Radical Expressionists

$\sqrt{72}$	$\sqrt{80}$	$\sqrt{81}$	$\sqrt[3]{375}$
$\sqrt{90}$	$\sqrt{x^7}$	$\sqrt{125}$	$\sqrt[3]{x^4}$
$\sqrt{200}$	$\sqrt{45}$	$\sqrt{30}$	$\sqrt[3]{3000x^3}$
$\sqrt{108}$	$\sqrt{121}$	$\sqrt{16x}$	$\sqrt[3]{1024}$

Student Activity

Thread of Like Terms II

Directions: Simplify each of the expressions that follow.

1.

$3x^2 + 5x^2$	$\frac{3}{11} + \frac{5}{11}$	$3\sqrt{x} + 5\sqrt{x}$	$3\sqrt[3]{3} + 5\sqrt[3]{3}$

2.

$x^3 - 5x^3$	$\frac{1}{3} - \frac{5}{3}$	$\sqrt{5} - 5\sqrt{5}$	$\sqrt[3]{3} - 5\sqrt[3]{3}$

3.

$\frac{8}{x+8} - \frac{1}{x+8}$	$8xy - xy$	$8\sqrt{y} - \sqrt{y}$	$8\sqrt[3]{2} - \sqrt[3]{2}$

4.

$4x^2y^2 - 6x^2y^2$	$\frac{4}{5x^2} - \frac{6}{5x^2}$	$4\sqrt{xy} - 6\sqrt{xy}$	$4\sqrt[3]{15} - 6\sqrt[3]{15}$

5. What is the lesson to be learned here about addition and subtraction in algebra and mathematics in general?

6. Now be a little more careful with this group of expressions. If the expression can't be simplified, just say so!

$8\sqrt{3} + 2\sqrt{3}$	$6\sqrt{2} + 6\sqrt{3}$	$\sqrt{5} + \sqrt{5}$	$\sqrt{5}\sqrt{5}$
$3\sqrt{2} \cdot \sqrt{2}$	$3\sqrt{2} + \sqrt{2}$	$\sqrt[3]{x} - \sqrt[3]{x}$	$\sqrt{3} + \sqrt{3} + \sqrt{3}$
$3\sqrt[3]{ab} - \sqrt[3]{ab}$	$-5\sqrt{2} + 5\sqrt{2}$	$5\sqrt{x} + 5\sqrt{y}$	$\sqrt{4} \cdot \sqrt{4}$

Student Activity

Paint by Radicals

Directions: Simplify each expression and shade in the corresponding square in the grid below (that contains the correctly simplified version). Assume all variables represent positive numbers. The first one has been done for you. There's a surprise when you're finished!

1. Simplify $\sqrt{72}-\sqrt{50}=6\sqrt{2}-5\sqrt{2}=\sqrt{2}$

2. Simplify: $\sqrt{72}+\sqrt{50}$

3. Simplify: $\sqrt{16+9}$

4. Simplify: $\sqrt{40}-\sqrt{90}$

5. Simplify: $10\sqrt{20}+\sqrt{5}$

6. Simplify: $\sqrt{300}-10\sqrt{3}$

7. Simplify: $\sqrt{75}+\sqrt{12}$

8. Simplify: $\sqrt{48y^2}-y\sqrt{27}$

9. Simplify: $\sqrt{75x^2+25x^2}$

10. Simplify: $\sqrt{81b^2}+\sqrt{500b}$

11. Simplify: $-13+\sqrt{4x^2}+5\sqrt{2}-5\sqrt{2}$

12. Simplify: $10+10\sqrt[3]{5}+2\sqrt[3]{5}+2$

13. Simplify: $\sqrt{52b^2-3b^2}$

14. Simplify: $\sqrt{900}-\sqrt{80}+\sqrt{363}+\sqrt{80}$

15. Simplify: $\sqrt[3]{54}-\sqrt[3]{64}-3\sqrt[3]{2}$

7	$21\sqrt{5}$	5	0	$11\sqrt{2}$	$10x$
$\sqrt{5}$	$y\sqrt{3}$	$\sqrt{3y}$	$2\sqrt{2xy}$	$\sqrt{122}$	$-13+\sqrt{2}+2x$
$11b\sqrt{5}$	$\sqrt{2}$	$\sqrt{22}$	$30+11\sqrt{3}$	-8	$2x-13$
$-\sqrt{10}$	$7b$	$24\sqrt[3]{5}$	$19b\sqrt{10b}$	-4	$4-6\sqrt[3]{2}$
$\sqrt{-10}$	$12+12\sqrt[3]{5}$	$80+11\sqrt{3}$	$9b+10\sqrt{5b}$	$\sqrt{3}$	$7\sqrt{3}$

Student Activity

Radical Addition and Multiplication Tables

Here is a simple addition table with radical inputs. Fill in the missing boxes, use X if the inputs cannot be combined with the given operation.

Addition Table I:

$+$	$\sqrt{2}$	$\sqrt{3}$	$\sqrt{5}$	$\sqrt{6}$
$\sqrt{2}$	$2\sqrt{2}$	X	X	X
$\sqrt{3}$				
$\sqrt{5}$				
$\sqrt{6}$				

Now try this addition table that also involves variables! Again, use X if you cannot combine the input terms. You may want to simplify some of the radicals before you add. Assume $x > 0$.

Addition Table II:

$+$	$\sqrt{3x}$	$\sqrt{12x}$	$\sqrt{75x}$	$\sqrt{3x^2}$	$\sqrt{12x^2}$	$\sqrt{75x^2}$
$\sqrt{3x}$						
$\sqrt{12x}$						
$\sqrt{75x}$						
$\sqrt{3x^2}$						
$\sqrt{12x^2}$						
$\sqrt{75x^2}$						

Here is a simple multiplication table with radical inputs. Perform the multiplications and simplify each result.

Multiplication Table I:

$\bullet$	$\sqrt{2}$	$\sqrt{3}$	$\sqrt{5}$	$\sqrt{6}$
$\sqrt{2}$	2	$\sqrt{6}$	$\sqrt{10}$	$2\sqrt{3}$
$\sqrt{3}$				
$\sqrt{5}$				
$\sqrt{6}$				

Now try this multiplication table that also involves variables! Make sure to simplify the result of each multiplication. Assume $x > 0$.

Multiplication Table II:

$\bullet$	$\sqrt{3x}$	$\sqrt{12x}$	$\sqrt{3x^2}$	$\sqrt{12x^2}$	$\sqrt{3x^3}$	$\sqrt{12x^3}$
$\sqrt{3x}$						
$\sqrt{12x}$						
$\sqrt{3x^2}$						
$\sqrt{12x^2}$						
$\sqrt{3x^3}$						
$\sqrt{12x^3}$						

Student Activity

Tempting Radical Expressions

Directions: The expressions on this page might try to tempt you away from the strict mathematical rules you have learned about radical expressions. So, work carefully to simplify each expression and think about every move you make! If an expression cannot be simplified, then say so.

1. $\sqrt{25-9}$	**2.** $7\sqrt{3}-\sqrt{3}$	**3.** $5-2\sqrt{2}$
4. $\sqrt{8}+\sqrt{1}$	**5.** $\sqrt{25}\cdot\sqrt{25}$	**6.** $\sqrt{5}+\sqrt{5}$
7. $\sqrt[3]{8}+1$	**8.** $\sqrt[4]{4+12}$	**9.** $\sqrt{169-25}$
10. $5\sqrt{x}-\sqrt{x}$	**11.** $\sqrt{x^2}\cdot\sqrt{x^2}$	**12.** $\sqrt[3]{3}+\sqrt[3]{3}+\sqrt[3]{3}$
13. $\frac{\sqrt{3}}{2}+\frac{\sqrt{3}}{2}$	**14.** $\frac{\sqrt{3}}{2}\cdot\frac{\sqrt{3}}{2}$	**15.** $\sqrt[3]{\frac{8-7}{8}}$

Scrambled Answers:

$\sqrt[3]{9}$	12	$6\sqrt{3}$	x^2	4
$2\sqrt{5}$	$\frac{1}{2}$	$\frac{3}{4}$	Cannot be simplified further	2
$2\sqrt{2}+1$	$4\sqrt{x}$	$\sqrt{3}$	$3\sqrt[3]{3}$	25

Student Activity

Multiplication Madness Match Up

Directions: There are so many different ways to combine multiplication and radicals it can be easy to mix up the rules. Carefully work to simplify each of these expressions applying the appropriate rules as you go. Assume all roots are real numbers.

$\sqrt[n]{a} \cdot \sqrt[n]{b} = \sqrt[n]{a \cdot b}$ $\quad$ $a\sqrt[n]{b} \cdot c\sqrt[n]{d} = ac\sqrt[n]{bd}$ $\quad$ $\left(\sqrt[n]{a}\right)^n = a$

A $4x$ **B** $6+4\sqrt{2}$ **C** $36-x$ **D** $18\sqrt{10}$ **E** None of these

$\sqrt{2}\left(3\sqrt{2}+4\right)$	$3\sqrt{2} \cdot 6\sqrt{5}$	$\left(\sqrt[3]{4x}\right)^3$	$3\left(2+4\sqrt{2}\right)$	$\left(\sqrt{36}\right)^2 - \left(\sqrt{x}\right)^2$
$\left(\sqrt[3]{6}\right)^3 - \left(\sqrt[3]{x}\right)^3$	$\sqrt{2x} \cdot \sqrt{8x}$	$2\sqrt{5} \cdot 9\sqrt{2}$	$2\left(3+2\sqrt{2}\right)$	$4\sqrt{x} \cdot \sqrt{x}$
$\sqrt{2x} \cdot 2\sqrt{2x}$	$\left(6-\sqrt{x}\right)\left(6+\sqrt{x}\right)$	$\left(\sqrt{2}+2\right)^2$	$\sqrt{x}\left(36-\sqrt{x}\right)$	$\sqrt{5} \cdot 18\sqrt{2}$
$(6+x)(6-x)$	$\left(6-\sqrt{x}\right)^2$	$\sqrt[3]{64 \cdot x^3}$	$\left(9\sqrt{10}\right)^2$	$\left(\sqrt{6}\right)^2 + \sqrt{32}$

Student Activity

Double the Fun on Radical Expressions

Directions: When you simplify expressions that have multiplication or division of radicals, you sometimes have a choice. You could first simplify each radicand and then apply the operation, or you could first combine the radicals with the operation, and then simplify. For each of the expressions below, simplify the radical expression both ways (let $x > 0$). Your simplified expression should be the same for both methods. If they are not, you'll have to go back and look for a mistake. The first one has been done for you.

	Radical Expression	Simplify the radicands first	Combine the radicals first
1.	$\frac{\sqrt{18x}}{\sqrt{2x^3}}$	$\frac{\sqrt{18x}}{\sqrt{2x^3}} = \frac{3\sqrt{2x}}{x\sqrt{2x}} = \frac{3\cancel{\sqrt{2x}}^{1}}{x\cancel{\sqrt{2x}}_{1}} = \frac{3}{x}$	$\frac{\sqrt{18x}}{\sqrt{2x^3}} = \sqrt{\frac{18x}{2x^3}} = \sqrt{\frac{9}{x^2}} = \frac{3}{x}$
2.	$\sqrt{75x^2} \cdot \sqrt{48x^4}$		
3.	$\frac{\sqrt{108x^6}}{\sqrt{12x^4}}$		
4.	$\frac{\sqrt{625x^4}}{\sqrt{25x^8}}$		
5.	$\sqrt{12x} \cdot \sqrt{24x^2}$		
6.	$\frac{\sqrt[3]{192x^6}}{\sqrt[3]{3x^3}}$		

Student Activity

Rationalize with the Missing Form of 1

Directions: In each radical expression below, we need to rationalize the denominator. We do this rationalization by multiplying by a **form of one** that changes the denominator from a radical expression to a non-radical expression. Fill in the "ones" to rationalize the denominator of each expression and simplify the result.

Example: $\frac{6}{\sqrt{3}} \cdot \left[\frac{\quad}{\quad}\right]$ becomes $\frac{6}{\sqrt{3}} \cdot \left[\frac{\sqrt{3}}{\sqrt{3}}\right] = \frac{6\sqrt{3}}{3} = 2\sqrt{3}$.

Hint for problems 6-10: Remember to use a conjugate.

1. $\frac{8}{\sqrt{2}} \cdot \left[\frac{\quad}{\quad}\right]$

2. $\frac{5\sqrt{3}}{\sqrt{15}} \cdot \left[\frac{\quad}{\quad}\right]$

3. $\frac{\sqrt{5}}{\sqrt{10}} \cdot \left[\frac{\quad}{\quad}\right]$

4. $\sqrt{\frac{7}{2}} = \frac{\sqrt{\quad}}{\sqrt{\quad}} \cdot \left[\frac{\quad}{\quad}\right]$

5. $\frac{4}{\sqrt{18x}} = \frac{4}{\sqrt{\quad}} \cdot \left[\frac{\quad}{\quad}\right]$

6. $\frac{4}{\sqrt{2}+1} \cdot \left[\frac{\quad}{\quad}\right]$

7. $\frac{12}{3-\sqrt{5}} \cdot \left[\frac{\quad}{\quad}\right]$

8. $\frac{\sqrt{3}}{-2-\sqrt{3}} \cdot \left[\frac{\quad}{\quad}\right]$

9. $\frac{\sqrt{3}-2}{\sqrt{3}+4} \cdot \left[\frac{\quad}{\quad}\right]$

10. $\frac{\sqrt{6}}{x+\sqrt{2}} \cdot \left[\frac{\quad}{\quad}\right]$

Guided Learning Activity

Rationalizing Higher-order Roots

Believe it or not, rationalizing denominators with square roots is a pretty intuitive procedure. The procedure for rationalizing denominators with higher-order roots is not so obvious. Let's look at a comparison and see what's different.

Example with square roots:

To rationalize $\frac{6}{\sqrt{3}}$ we multiply by $\frac{\sqrt{3}}{\sqrt{3}}$, to get: $\frac{6}{\sqrt{3}} \cdot \frac{\sqrt{3}}{\sqrt{3}} = \frac{6\sqrt{3}}{\sqrt{3^2}} = \frac{6\sqrt{3}}{3} = 2\sqrt{3}$.

Example with cube roots:

If we followed exactly the same procedure as for square roots, here's what happens:

We multiply $\frac{6}{\sqrt[3]{2}}$ by $\frac{\sqrt[3]{2}}{\sqrt[3]{2}}$ to get: $\frac{6}{\sqrt[3]{2}} \cdot \frac{\sqrt[3]{2}}{\sqrt[3]{2}} = \frac{6\sqrt[3]{2}}{\sqrt[3]{2^2}} = ?$

The problem is that $\sqrt[3]{2^2}$ doesn't simplify. However, $\sqrt[3]{2^3}$ would simplify, equaling 2.

Let's try again:

1. Rationalize: $\frac{6}{\sqrt[3]{2}}$

It helps to think first of the radical denominator that you're going to aim for. Then you can figure out what to multiply by to get it. Fill in the blanks in the next few problems.

2. Rationalize $\frac{15}{\sqrt[3]{5}}$ by aiming for $\sqrt[3]{125}$ or $\sqrt[3]{5^3}$ in the denominator.

$$\frac{15}{\sqrt[3]{5}} \cdot \frac{\sqrt[3]{\quad}}{\sqrt[3]{\quad}} = \frac{\qquad}{\sqrt[3]{125}} = \frac{\qquad}{5} =$$

3. Rationalize $\frac{6}{\sqrt[4]{27}}$ by aiming for $\sqrt[4]{81}$ or $\sqrt[4]{3^4}$ in the denominator.

$$\frac{6}{\sqrt[4]{27}} \cdot \frac{\sqrt[4]{\quad}}{\sqrt[4]{\quad}} = \frac{\qquad}{\sqrt[4]{81}} = \frac{\qquad}{3} =$$

4. Rationalize $\frac{5}{2\sqrt[3]{25}}$ by aiming for $2\sqrt[3]{125}$ or $2\sqrt[3]{5^3}$ in the denominator.

$$\frac{5}{2\sqrt[3]{25}} \cdot \frac{\sqrt[3]{\quad}}{\sqrt[3]{\quad}} = \frac{\qquad}{2\sqrt[3]{125}} = \frac{\qquad}{2 \cdot 5} =$$

We always want to aim for a radical in the denominator that has a perfect power in the radicand that corresponds with the index of the root.

Let's practice deciding on a good denominator to aim for:

5. Rationalize $\frac{12}{\sqrt[5]{4}}$ by aiming for ________ in the denominator.

6. Rationalize $\frac{\sqrt[3]{3}}{\sqrt[3]{4}}$ by aiming for ________ in the denominator.

7. Rationalize $\frac{6x^2}{\sqrt[3]{3x}}$ by aiming for ________ in the denominator.

8. Rationalize $\frac{\sqrt[4]{2}}{\sqrt[4]{9m}}$ by aiming for ________ in the denominator.

Finally, we'll put it all together. Rationalize the denominators in the following problems.

9. Rationalize: $\frac{40}{\sqrt[3]{10}}$

10. Rationalize: $\frac{x}{\sqrt[4]{8x^2}}$

11. Rationalize: $\sqrt[3]{\frac{25}{3}}$ (Hint: first write this with two separate radicals.)

12. Rationalize: $\frac{\sqrt[4]{6}}{\sqrt[4]{9x^2}}$

Student Activity

The Radical Reunion

Directions: Now that we've made it through addition, subtraction, multiplication, and division, it seems only fair to bring the whole gang back together for a reunion. Assume all variables represent positive numbers. Simplify each of the following radical expressions. If an expression cannot be simplified, say so.

$\sqrt{50}+\sqrt{18}$	$\sqrt{50-18}$	$\sqrt{50+18}$	$\sqrt{50}\cdot\sqrt{18}$	$\dfrac{\sqrt{50}}{\sqrt{18}}$

$\sqrt{16a^2b}+\sqrt{9a^2b}$	$\sqrt{16a^2b-9a^2b}$	$\sqrt{16a^2b}\cdot\sqrt{9a^2b}$	$\dfrac{\sqrt{16a^2b}}{\sqrt{9a^2b}}$	$\left(\sqrt{16a^2b}\right)^2$

$\sqrt[3]{135xy^4}-\sqrt[3]{40x^4y}$	$\sqrt[3]{135xy^4-40x^4y}$	$\sqrt[3]{135xy^4}\cdot\sqrt[3]{40x^4y}$	$\dfrac{\sqrt[3]{135xy^4}}{\sqrt[3]{40x^4y}}$	$\left(\sqrt[3]{40x^4y}\right)^3$

Student Activity

Is it a Solution?

Directions: If the number in the square **IS** a solution of the equation, then circle the solution (placing an **O** on the square). If the given number **IS NOT** a solution, then put an **X** on the square.

$\sqrt[3]{7-x}=1-x$ -1	$(3-x)^{1/2}=x-3$ 3	$\sqrt{3-x}=x-3$ 0
$-3x=3\sqrt{-2x-1}$ 3	$-3x=3\sqrt{-2x-1}$ 0	$\sqrt{x}+\sqrt{x+3}=x-4$ 1
$x=\sqrt[4]{x^3+16}$ -1	$\sqrt{3x}+10=10$ 0	$-2=(3x-2)^{1/3}$ -2

Guided Learning Activity

Radical Isolation

When we solve radical equations, it is important to **isolate the radical** before using the power rule of equality. *But why?* Consider the radical equation $\sqrt{x+3}+4=6$.

Squaring first: Attempting to square both sides without isolating the radical leaves another radical term. This is messier looking than what we started with and we're not even done!

$$\sqrt{x+3}+4=6$$
$$\left(\sqrt{x+3}+4\right)^2=(6)^2$$
$$\left(\sqrt{x+3}+4\right)\left(\sqrt{x+3}+4\right)=36$$
$$x+3+8\sqrt{x+3}+16=36$$

This is worse than the original problem!

Isolating first: If we isolate the radical term first, and then square both sides of the equation, we can quickly solve for x. Checking the solution shows that $x=1$ is a valid solution to this radical equation.

$$\sqrt{x+3}+4=6$$
$$\sqrt{x+3}=2$$
$$\left(\sqrt{x+3}\right)^2=(2)^2$$
$$x+3=4$$
$$x=1$$

Now let's try some together!

1.

Isolate the radical:	Solve by squaring both sides:	Check:
$\sqrt{2x+3}+5=8$		

2.

Isolate the radical:	Solve by squaring both sides:	Check:
$8\sqrt{x+5}-12=28$		

3.

Isolate the radical:	Solve by squaring both sides:	Check:
$5+\sqrt{x+2}=0$		

4.

Isolate the radical:	Solve by cubing both sides:	Check:
$5+2x^{1/3}=13$		

5.

Isolate a radical:	Solve by cubing both sides:	Check:
$\sqrt[3]{3a+4}=\sqrt[3]{2a+7}$		

6.

Isolate the radical:	Solve by squaring both sides:	Check:
$\sqrt{x+7}-5=x$		

7.

Isolate a radical:	Squaring both sides:	
$\sqrt{x+4}+\sqrt{x-1}=5$		
Isolate the radical:	**Solve by Squaring both sides:**	**Check:**

Student Activity

Match Up on Solving Radical Equations

Directions: Match each of the equations in the squares in the table below with its solution from the top. If the solution is not found among the choices A through D, then choose E (none of these).

A 5 **B** 3 **C** −2 **D** No real number solution **E** None of these

$\sqrt{7-x}=2$	$3+\sqrt{x+6}=5$	$3\sqrt{x+9}=6$
$(2y-1)^{1/2}=3$	$8+\sqrt{x-1}=5$	$6\sqrt[3]{a-4}=6$
$x=2+\sqrt{14-x}$	$\sqrt[4]{3w}=\sqrt[4]{5+2w}$	$x+\sqrt[3]{3-12x}=x+3$
$-2\sqrt{3x-1}=10$	$x-1=(7-x)^{1/2}$	$x=\sqrt{4x-11}+2$

Student Activity
Working with the Language of Radicals

Fill in the table below. The first row has been done for you.

In words	Equation or Expression?	In Math Notation
1. the square root of the quantity $x+5$	Expression	$\sqrt{x+5}$
2. The quantity $3x+2$ squared is zero.		
3. The sum of 5 and the square root of x is 30.		
4.		$5(x+6)^2+8$
5.		$5\sqrt{x+6}+8$
6.		$7\sqrt{x}-7=42$
7. twice the square of the quantity $5x+9$		
8. 40 less the square root of x is 35.		
9.		$5\sqrt[3]{50-x}$
10. 5 times the cube root of the quantity $50-x$ is 25.		

Student Activity

Sail Into the Pythagorean Sunset

Directions: Almost all of the triangles in the figure below are right triangles. Using the sparse information you are given and the Pythagorean Theorem, work out the lengths of all the missing sides in the sailboat below. You should be able to work out the sides of the non-right triangles by piecing together the information you find from the right triangles. It might be helpful to start by finding the measure of a.

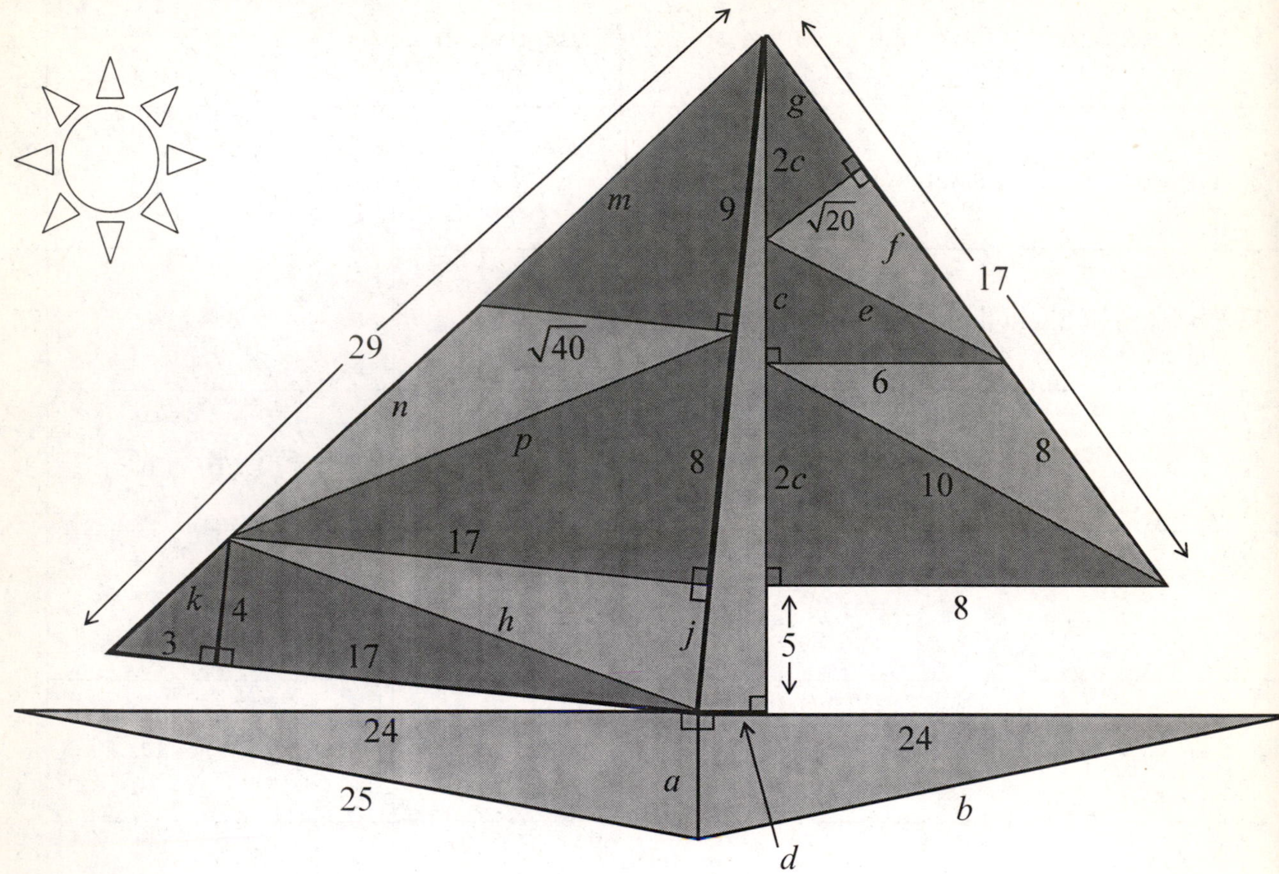

$a =$

$b =$ $h =$

$c =$ $j =$

$d =$ $k =$

$e =$ $m =$

$f =$ $n =$

$g =$ $p =$

Student Activity

These Triangles are Just Special

Several triangles have special names. In every diagram, the equal sides are represented with thick lines.

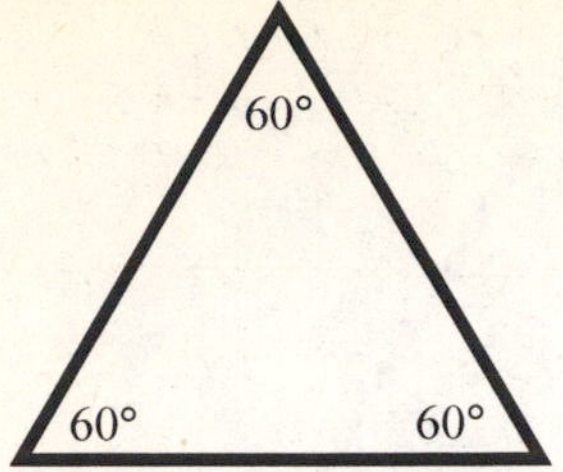

Equilateral Triangle
(all sides are equal,
all angles are equal)

Isosceles Triangle
(two equal sides,
two equal angles)

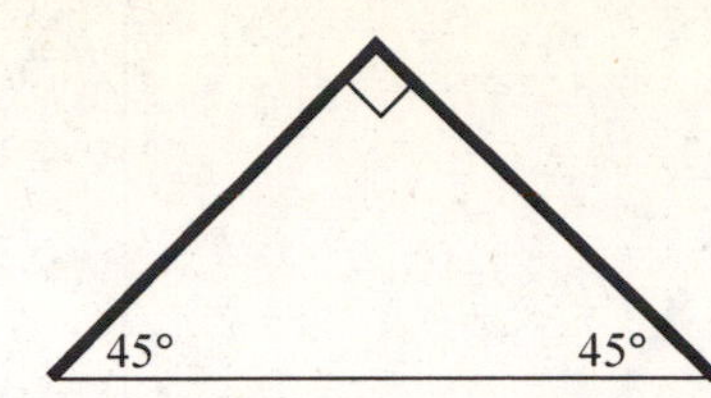

Isosceles Right Triangle
(one right angle and two equal angles)
45-45-90 Triangle

1. Use some geometric reasoning and the Pythagorean Theorem to work out the measure of each missing side for the triangles below.

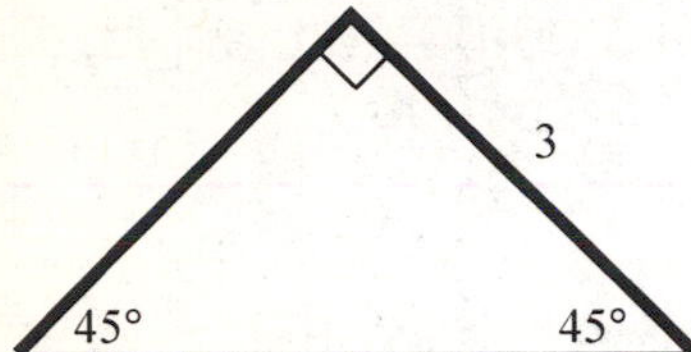

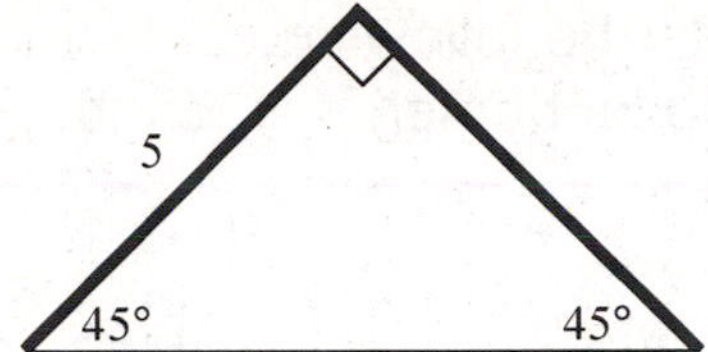

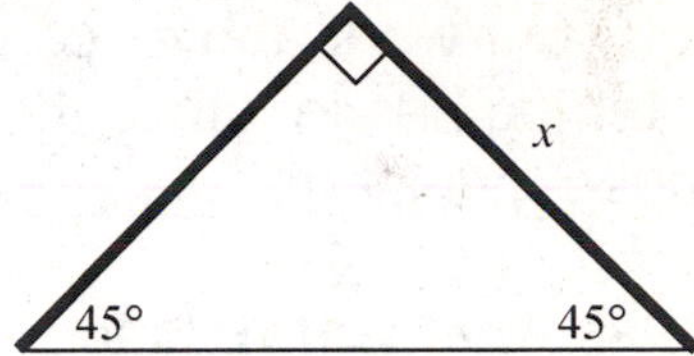

Based on your work above, complete the following conjecture: In an isosceles right triangle, if the legs have length l, then the hypotenuse has length ________.

If we take an equilateral triangle and cut it exactly down the middle, we get two congruent 30-60-90 triangles, demonstrated here:

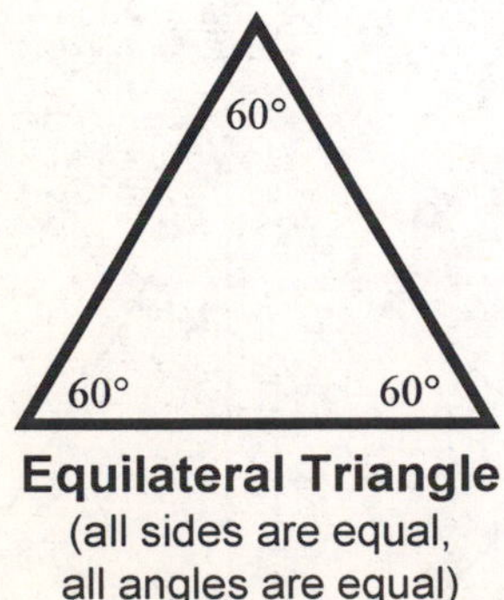

Equilateral Triangle
(all sides are equal,
all angles are equal)

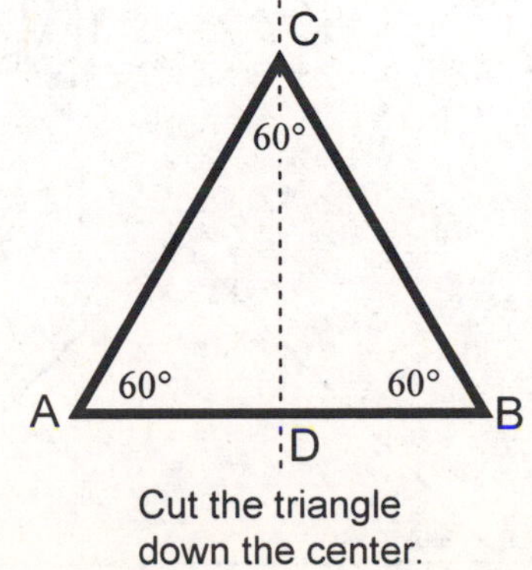

Cut the triangle
down the center.
$AD \cong DB = \frac{1}{2} AB$

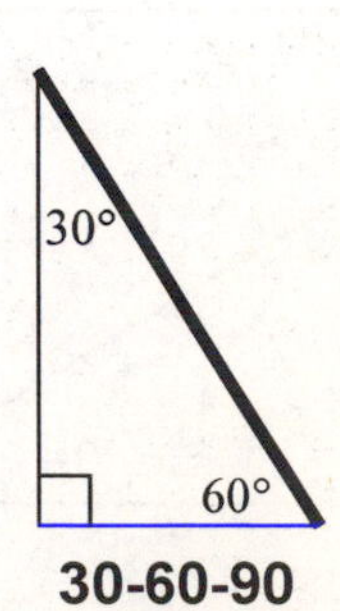

30-60-90 Triangle

2. Use some geometric reasoning and the Pythagorean Theorem to work out the measure of each missing side for the triangles below.

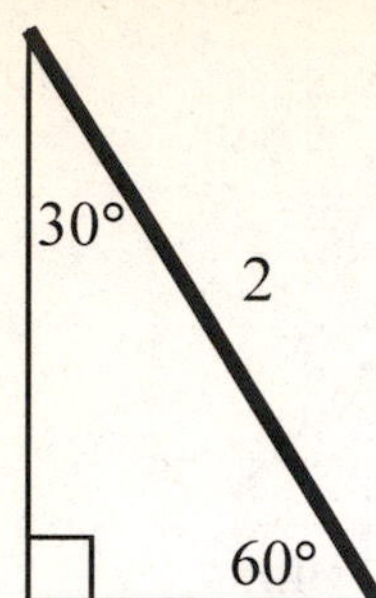

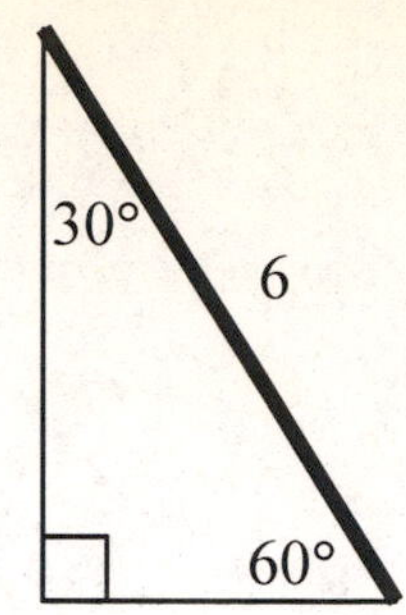

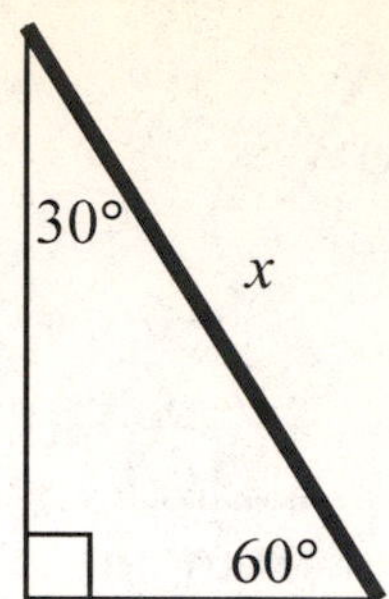

Based on your work above, complete the following conjecture: In a 30-60-90 triangle, if the shorter leg has length a, then the longer leg has length _______, and the hypotenuse has length _______.

3. Now use what you've learned about 30-60-90 and 45-45-90 triangles to work out the measure of the missing sides for the triangles below.

a.

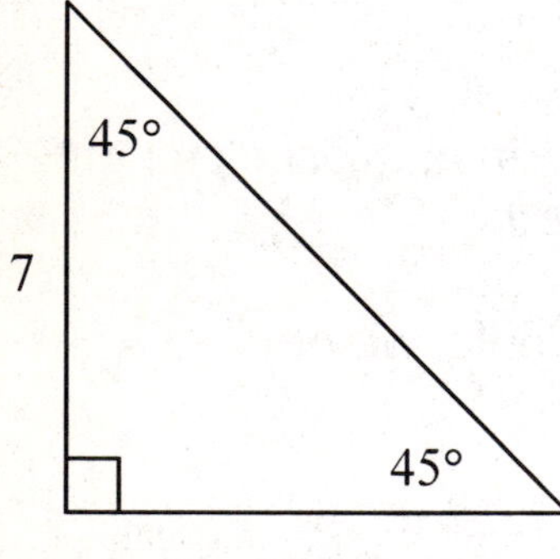

Fairy Tales that didn't make it:
Goldilocks and the Special Triangles

b.

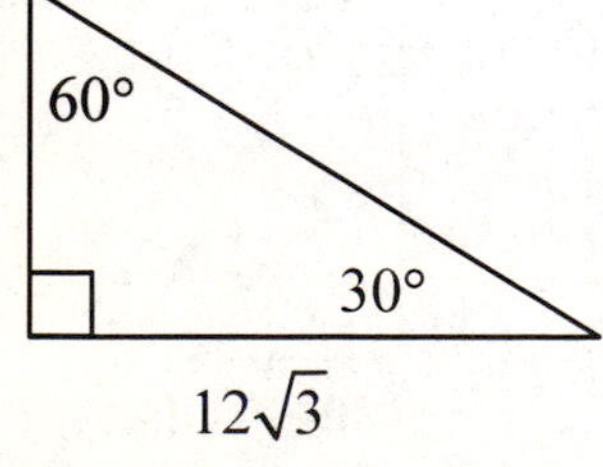

c.

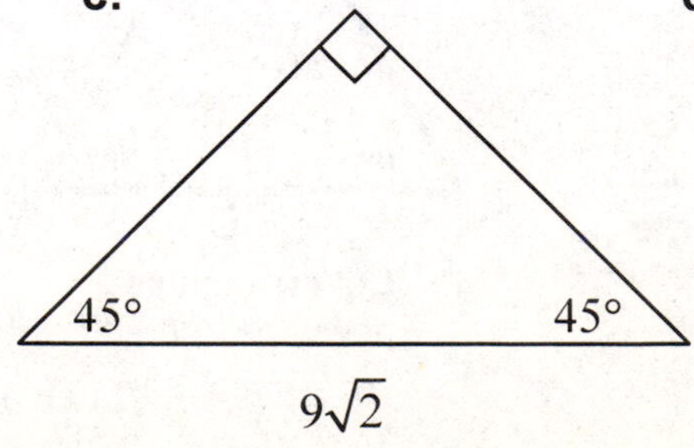

d.

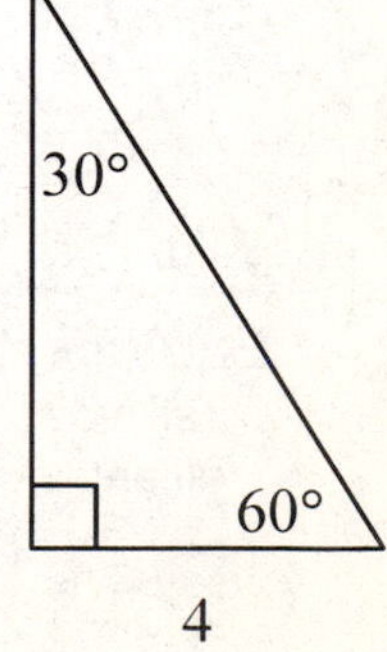

Guided Learning Activity

The Distance Formula

1. Find the length of side AB and side BC by measuring the distances on the graphing grid. Then use the Pythagorean Theorem to find the length of side AC.

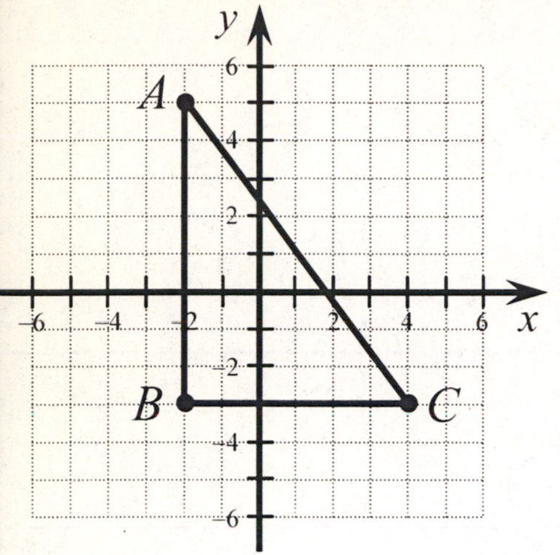

2. Find the length of line segment DE using a method similar to problem 1.

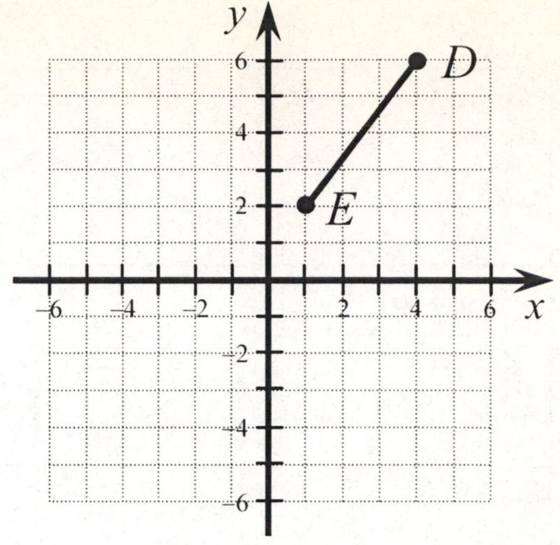

3. Find the length of the side d in the triangle below in terms of $x_1, x_2, y_1,$ and y_2.

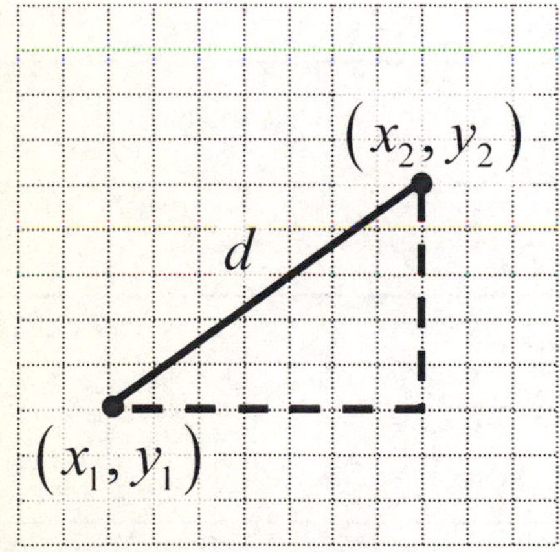

4. Now find the distance between $(4,-1)$ and $(7,-5)$ using the formula from Problem 3.

Assess Your Understanding

Radical Expressions and Equations

For each of the following, describe the strategies or key steps that will help you **start** the problem. You do **not** have to complete the problems.

		What will help you to start this problem?
1.	Multiply: $(3+\sqrt{5})(3-\sqrt{5})$	
2.	Divide: $\dfrac{\sqrt{120x^4y}}{\sqrt{12y^5}}$	
3.	Add: $\sqrt{8}+\sqrt{50}$	
4.	Simplify: $\sqrt{50}+\sqrt{18}$	
5.	Solve: $\sqrt{3x-1}+6=0$	
6.	Rationalize the denominator: $\dfrac{12}{6-\sqrt{3}}$	
7.	Is $\sqrt{8}$ rational or irrational?	
8.	Simplify: $\sqrt{120x^5}$	

		What will help you to start this problem?
9.	Evaluate: $4^{3/2}$	
10.	Find the distance between $(3,-6)$ and $(2,-4)$.	
11.	Solve: $\sqrt{3x-1}+6=0$	
12.	The smallest side of a 30-60-90 triangle is 6 cm. Find the lengths of the other two sides.	
13.	Simplify: $\sqrt[4]{80x^4y^{10}}$	
14.	Solve: $3\sqrt{2x-1}=6$	
15.	Simplify: $x^{1/3}x^{2/5}$	

Metacognitive Skills

Radical Expressions and Equations

Metacognitive skills refer to the ability to judge how well you have learned something and to effectively direct your own learning and studying. This is a self-evaluation tool designed to help you focus your studying and to improve your metacognitive skills with regards to this math class.

Fill the 1st column out **before** you begin studying. Fill the 2nd column out after you study for your test.

Go back to this assessment after your test and circle any of the ratings that you would change – this identifies the "disconnects" between what you **thought** you knew well and what you **actually** knew well.

Use the scale below to assign a number to each topic.

5 *I am confident I can do any problems in this category correctly.*
4 *I am confident I can do most of the problems in this category correctly.*
3 *I understand how to do the problems in this category, but I still make a lot of mistakes.*
2 *I feel unsure about how to do these problems.*
1 *I know I don't understand how to do these problems.*

Topic or Skill	Before Studying	After Studying
Recognizing perfect squares (like 1, 4, 9, 16, 25…).		
Recognizing perfect cubes (like 1, 8, 27, 64, 125…), perfect fourths (like 1, 16, 81, …), and perfect fifths (like 1, 32, 243, …).		
Ability to use the terms radical, radicand, radical symbol, square, and square root, appropriately.		
Understanding why $\sqrt{a}$ is not a real number if $a < 0$.		
Understanding why $\sqrt[n]{a}$ is non-real if n is even and $a < 0$.		
Understanding why $\sqrt[n]{a}$ is a real number as long as n is odd.		
Categorizing a radical as rational or irrational, real or imaginary.		
Using a calculator to approximate an irrational radical.		
Simplifying radical expression involving numbers and variables that are perfect powers (like $\sqrt{16x^2}$).		
Rewriting a radical expression with rational (fractional) exponents or vice versa.		
Knowing which part of a rational (fractional) exponent represents the root and which part is the power.		
Evaluating numbers with rational exponent powers, like $8^{3/2}$ or $9^{-1/2}$.		
Using a calculator to evaluate radicals with higher-order roots or numbers written with rational exponents..		
Applying exponent rules correctly to simplify expressions involving rational exponents.		

Continued on next page.

Topic or Skill	Before Studying	After Studying
Applying the product and quotient rules for radicals correctly; using these rules to simplify radical expressions like $\sqrt{\frac{9x^2}{16}}$ or $\sqrt{4 \cdot 25}$.		
Simplifying radical expressions by finding perfect squares or perfect powers inside the radicand (for example: $\sqrt{8}$, $\sqrt[3]{x^4}$, or $\sqrt{50x^3y^2}$).		
Adding or subtracting radical expressions by simplifying each radical and using like terms.		
Simplifying radical expressions involving multiplication and/or distribution like $3\sqrt{6} \cdot 2\sqrt{2}$, $2\sqrt{3}\left(4\sqrt{2} - 5\sqrt{3}\right)$, or $\left(4 + \sqrt{3}\right)\left(5 - \sqrt{2}\right)$.		
Simplifying a radical expression that is raised to a power, like $\left(2 + \sqrt{3}\right)^2$ or $\left(\sqrt{2x}\right)^2$.		
Rationalizing the denominator when there is a single term involving a square root in the denominator.		
Rationalizing the denominator when there are two terms involving at least one square root.		
Rationalizing the numerator of an expression involving square roots.		
Rationalizing the denominator when there is a single term involving a higher-order root (like a cube or fourth root).		
Solving an equation containing a single radical.		
Solving an equation with a rational exponent power.		
Solving an equation with more than one radical expression.		
Checking the solution to a radical equation.		
Recognizing when a radical equation cannot possibly have a real number solution.		
Solving application problems involving radical equations.		
Knowing when the Pythagorean Theorem can be applied and correctly using it.		
Applying the distance formula correctly.		
Knowing the special properties of 45-45-90 and 30-60-90 triangles.		

CPSIA information can be obtained
at www.ICGtesting.com
Printed in the USA
FFOW04n0014220813
1624FF

9 781111